Advances in seed science and technology for more sustainable crop production

It is widely recognised that agriculture is a significant contributor to global warming and climate change. Agriculture needs to reduce its environmental impact and adapt to current climate change whilst still feeding a growing population, i.e. become more 'climate-smart'. Burleigh Dodds Science Publishing is playing its part in achieving this by bringing together key research on making the production of the world's most important crops and livestock products more sustainable.

Based on extensive research, our publications specifically target the challenge of climate-smart agriculture. In this way we are using 'smart publishing' to help achieve climate-smart agriculture.

Burleigh Dodds Science Publishing is an independent and innovative publisher delivering high quality customer-focused agricultural science content in both print and online formats for the academic and research communities. Our aim is to build a foundation of knowledge on which researchers can build to meet the challenge of climate-smart agriculture.

For more information about Burleigh Dodds Science Publishing simply call us on +44 (0) 1223 839365, email info@bdspublishing.com or alternatively please visit our website at www.bdspublishing.com.

Related titles:

Biostimulants for sustainable crop production
Print (ISBN 978-1-78676-336-5); Online (ISBN 978-1-78676-338-9, 978-1-78676-339-6)

Improving organic crop cultivation
Print (ISBN 978-1-78676-184-2); Online (ISBN 978-1-78676-186-6, 978-1-78676-187-3)

Plant genetic resources: A review of current research and future needs
Print (ISBN 978-1-78676-451-5); Online (ISBN 978-1-78676-453-9, 978-1-78676-454-6)

Chapters are available individually from our online bookshop:
https://shop.bdspublishing.com

climate•SMART•publishing
IN AGRICULTURAL SCIENCE

BURLEIGH DODDS SERIES IN AGRICULTURAL SCIENCE
NUMBER 120

Advances in seed science and technology for more sustainable crop production

Edited by Dr Julia Buitink, INRAE, France and Professor Olivier Leprince, L'Institut Agro Rennes Angers, France

burleigh dodds
SCIENCE PUBLISHING

Published by Burleigh Dodds Science Publishing Limited
82 High Street, Sawston, Cambridge CB22 3HJ, UK
www.bdspublishing.com

Burleigh Dodds Science Publishing, 1518 Walnut Street, Suite 900, Philadelphia, PA 19102-3406, USA

First published 2022 by Burleigh Dodds Science Publishing Limited

Library of Congress Control Number: 2022934573

British Library Cataloguing in Publication Data
A catalogue record for this book is available from the British Library

ISBN 978-1-78676-917-6 (print)
ISBN 978-1-78676-920-6 (PDF)
ISBN 978-1-78676-919-0 (ePub)
ISSN 2059-6936 (print)
ISSN 2059-6944 (online)

DOI 10.19103/AS.2022.0105

Typeset by Deanta Global Publishing Services, Dublin, Ireland

Contents

Series list

Title | Series number

Achieving sustainable cultivation of maize - Vol 1 001
From improved varieties to local applications
Edited by: Dr Dave Watson, CGIAR Maize Research Program Manager, CIMMYT, Mexico

Achieving sustainable cultivation of maize - Vol 2 002
Cultivation techniques, pest and disease control
Edited by: Dr Dave Watson, CGIAR Maize Research Program Manager, CIMMYT, Mexico

Achieving sustainable cultivation of rice - Vol 1 003
Breeding for higher yield and quality
Edited by: Prof. Takuji Sasaki, Tokyo University of Agriculture, Japan

Achieving sustainable cultivation of rice - Vol 2 004
Cultivation, pest and disease management
Edited by: Prof. Takuji Sasaki, Tokyo University of Agriculture, Japan

Achieving sustainable cultivation of wheat - Vol 1 005
Breeding, quality traits, pests and diseases
Edited by: Prof. Peter Langridge, The University of Adelaide, Australia

Achieving sustainable cultivation of wheat - Vol 2 006
Cultivation techniques
Edited by: Prof. Peter Langridge, The University of Adelaide, Australia

Achieving sustainable cultivation of tomatoes 007
Edited by: Dr Autar Mattoo, USDA-ARS, USA and Prof. Avtar Handa, Purdue University, USA

Achieving sustainable production of milk - Vol 1 008
Milk composition, genetics and breeding
Edited by: Dr Nico van Belzen, International Dairy Federation (IDF), Belgium

Achieving sustainable production of milk - Vol 2 009
Safety, quality and sustainability
Edited by: Dr Nico van Belzen, International Dairy Federation (IDF), Belgium

Achieving sustainable production of milk - Vol 3 010
Dairy herd management and welfare
Edited by: Prof. John Webster, University of Bristol, UK

Ensuring safety and quality in the production of beef - Vol 1 011
Safety
Edited by: Prof. Gary Acuff, Texas A&M University, USA and Prof. James Dickson, Iowa State University, USA

Ensuring safety and quality in the production of beef - Vol 2 012
Quality
Edited by: Prof. Michael Dikeman, Kansas State University, USA

Achieving sustainable production of poultry meat - Vol 1 013
Safety, quality and sustainability
Edited by: Prof. Steven C. Ricke, University of Arkansas, USA

Achieving sustainable production of poultry meat - Vol 2 014
Breeding and nutrition
Edited by: Prof. Todd Applegate, University of Georgia, USA

Introduction

With the continued effects of climate change threatening the security of the global food system, there is a greater emphasis on ensuring successful crop establishment as a means of optimising agricultural production.

This volume considers how an improved understanding of seed quality, germination and seedling emergence can address this challenge. The book is split into two parts: Part 1 covers understanding seeds, from biology to the field, and includes chapters that examine ways to improve seed vigour. Part 2 chapters review how seed health and vigour can be controlled and tested, drawing attention specifically to seed phenotyping, testing and monitoring, preservation as well as seed conditioning and priming techniques.

Part 1 Understanding seeds: from biology to the field

The book opens with a chapter that provides an overview of seed dormancy and germination. Chapter 1 begins by introducing the five most frequently occurring types of dormancy: morphophysiological, morphological, physical, physiological and nondormancy before going on to examine the acquisition of dormancy, with special emphasis on the transcriptional network and the importance of compartmentation. The chapter then goes on to discuss the concept of ABA-GA balance which influences the seed's potential to germinate. Sections on breaking primary dormancy and secondary dormancy and dormancy cycling are also provided. The chapter next reviews the regulation of seed-to-seedling phase transition and biological variability, before concluding with an analysis of potential future research trends in seed biology.

Expanding on topics previously touched upon in Chapter 1, the next chapter examines the effects of the maternal environment in controlling seed dormancy. Chapter 2 first reviews dormancy response patterns to the effect of the maternal environment in seeds displaying both physiological and physical dormancy. It then goes on to analyse the physiological and molecular bases underlying the modulation of dormancy by the environment experienced by the mother plant during reproductive development. Sections on the ecological and agricultural implications of dormancy modulation by the maternal environment are also included.

Chapter 3 focuses on applying population-based threshold models to quantify and predict seed vigour attributes. The chapter begins by emphasising the importance of seed quality and behaviour in crop management and yield, weed control and transplant production. After a review of the different methods to analyse seed behaviour in population terms, the next section teaches the basic concepts of population-based threshold models and how they can describe seed behaviour in response to environmental factors that influence germination. The chapter also highlights the use of population-based models for physiological factors affecting seed germination and behaviour, such as dormancy, aging, seed vigour, respiration and hormones. A section on applying population-based threshold models is also provided.

The next chapter of Part 1 examines biotic sources of seed losses influencing germination and emergence success in crop plants and agricultural weeds. Chapter 4 first reviews the contribution of pathogens and predators to seed losses, distinguishing events that occur before and after seed dispersal. The chapter includes both weeds and crop plants because the greatest yield improvements can often result from targeting weed seed survival after crops are harvested. The chapter also highlights the strong potential for management practices to enhance predation losses of weed seeds relative to seed pathogen losses, particularly for species with long-lived dormant seeds.

The final chapter of Part 1 focuses on advances in understanding the genetic and environmental factors determining seed longevity. Chapter 5 begins by examining the distribution of seed life spans amongst species, then moves on to review how genetic diversity can be used to understand longevity. A section on the environmental effects of longevity is also provided, focusing specifically on plasticity and adaptation. The final section reviews aging protocols and the limits of extrapolating accelerated ageing conditions.

Part 2 Seed quality control and treatment

The first chapter of Part 2 examines advances in seed phenotyping using X-ray imaging. Chapter 6 begins by reviewing seed sample preparation which is then followed by an analysing of projection acquisition, drawing specific attention to radiography and tomography. With these two acquisitions in mind, the chapter moves on to review image reconstruction and how radiography and tomography can be used to do this. A section on image processing is also included, highlighting how visual assessment is crucial for successful seed analysis in phenotyping. The chapter concludes by emphasising the importance of developing current and new X-ray imaging techniques for the purpose of seed phenotyping.

The subject of Chapter 7 deals with testing seed health. The chapter discusses several methods for seed health testing which are broken down into four sections: traditional detection methods, serological detection techniques, nucleic acid-based detection methods and spectroscopy-based methods. The chapter also evaluates the use of next-generation sequencing for the detection of the most important seed transmitted pathogens. These methods are compared with TaqMan polymerase chain reaction methods.

Chapter 8 focuses on advances in preservation of seed vigour during storage. The chapter begins by posing the question how conditions during seed production can affect seed quality. The next sections examine post-harvest treatments to improve seed lot quality, maintaining seed viability and vigour during storage as well as the ways to predict viability and vigour during storage, drawing attention to monitoring and different testing methods. The chapter concludes with an overview of the challenges in seed longevity research and potential areas of development for future research.

The next chapter explores advances in enhancing seed defence mechanisms against pathogens. Chapter 9 reviews defence priming against plant pathogens when the priming agent is associated with seeds, concentrating on microbial endophytes and volatile organic compounds, and, where known, the mechanisms involved in defence responses. The chapter discusses the microbial and chemical priming agents that can be used to enhance seed vigour then goes on to consider transgenerational defence priming. The last section examines the use of beneficial microbe delivery for defence priming.

The final chapter of the book considers advances in seed priming techniques. Chapter 10 focuses on recent developments in priming from the point of view of the seed industry, based on patents and applications. It assesses the strengths and weaknesses of new techniques such as oxygen controlled priming, vapour heat priming, pulsed radio frequency, ultrasonics, cold plasma compared to established priming methods: drum, osmotic and solid matrix priming.

Acknowledgement

We wish to acknowledge the following for their help in reviewing particular chapters:

- Chapter 5: Dr Julia Zinsmeister, Institute of Biochemistry and Biophysics, Poland
- Chapter 7: Dr Valerie Grimault, GEVES, France

Part 1

Understanding seeds: from biology to the field

Chapter 1

Seed dormancy and germination: a critical update

Henk W. M. Hilhorst, Wageningen University & Research, The Netherlands and University of Cape Town, South Africa

1 Introduction

Seed dormancy and germination are adaptive traits that play a key role, together with flowering, in the timing of emergence of plants and their subsequent fitness. These developmental transitions can be seen as the basis for the establishment and maintenance of plant communities (Auge et al., 2018). Dormancy is defined as a physiological state resulting in the absence of germination of a viable seed under favourable conditions. Dormancy allows seeds to avoid germination during periods that are only transiently favourable.

Seed dormancy and (delayed) flowering time may have co-evolved to allow the colonization of habitats with seasonal fluctuations in the environment (Ritland, 1983). The most ancient signs of dormancy have been found in mature seeds of a Palaeozoic fossil conifer; a significant delay between fertilization and seed germination was interpreted as post-zygotic quiescence, a first step in the evolution of seed dormancy (Mapes et al., 1989). The evolutionary history of dormancy is complex as there exists a rich diversity in types and intensities of dormancy. This has resulted in various classifications of dormancy (Nikolaeva, 1999; Baskin and Baskin, 2004).

http://dx.doi.org/10.19103/AS.2022.0105.02

The five most frequently occurring types of dormancy are:

1 **Morphophysiological dormancy:** a combination of morphological and physiological dormancy. This is considered the ancestral state of dormancy.
2 **Morphological dormancy:** dormancy imposed by underdevelopment of the embryo which needs to complete development and grow inside the seed before radicle emergence.
3 **Physical dormancy:** dormancy imposed by dedicated physical structures of the embryo-surrounding tissues which require very specific environmental cues to allow water uptake and radicle emergence.
4 **Physiological dormancy:** dormancy imposed by one or more metabolic blocks in the seed that are removed by specific environmental cues. This type of dormancy is reversible.
5 **Nondormancy:** absence of dormancy in species in (near) constant environments.

It has been argued that the latter four dormancy types have evolved from morphophysiological dormancy, each adapting to specific habitats. Physiological dormancy is by far the most frequently occurring type of dormancy and often shows a transition to nondormancy, probably when a niche is occupied that does not demand seasonal dormancy (Willis et al., 2014). However, the intensity of dormancy is strongly dependent on the maternal environment. The observed transitions between physiological- and nondormancy may therefore not be permanent but rather dependent on the year of observation.

This chapter will mainly focus on the most dominant dormancy class, namely physiological dormancy, as well as germination, and review new insights into their regulation and mechanisms. For the other types of dormancy, the reader is referred to excellent reviews by, for example, Baskin et al. (2000) and Hudson et al. (2015) on physical dormancy and Baskin and Baskin (2014) on the other types of dormancy. Excellent reviews on seed dormancy in combination with germination have been published by Finch-Savage and Leubner-Metzger (2006), Graeber et al. (2012), Chahtane et al. (2017), Penfield (2017), Nonogaki (2018), Tuan et al. (2018) and Carrera-Castaño et al. (2020).

Despite numerous studies on physiological dormancy, predominantly in *Arabidopsis thaliana*, a comprehensive understanding of the phenomenon and its regulation is largely lacking. Current gene by gene discoveries, using mutants, and initial evidence from gene/protein regulatory networks, using whole-genome approaches, have not yet led to a comprehensive dormancy model. An important obstacle to progress in this field is a lack of non-destructive methodology to assess dormancy (or viability) of single seeds in a population.

Experiments in seed biology necessarily resort to seed lots (i.e. populations) for the assessment of quantitative biological properties such as dormancy and longevity. Germination is a qualitative, all-or-nothing phenomenon. However, dormancy is a quantitative trait described by such terms as shallow or deep, referring to the reluctance of a seed population to respond to a dormancy-breaking treatment. It therefore remains problematic to assign experimental output variables to inherent seed traits, unless comparing seed lots that germinate at 0% or 100%.

Nevertheless, studies on dormancy and germination continue, new technologies and approaches are developed, and a critical update seems timely. This review aims to (subjectively) pinpoint important and promising recent developments in seed science. In addition, it will draw attention to the huge complexity of seeds as biological systems. It will also briefly explore more holistic approaches to 'understand' the system (Nurse, 2008).

2 The acquisition of dormancy

Like various other functional seed attributes, such as germinability, desiccation tolerance and deposition of food reserves, primary dormancy is acquired during seed development, predominantly during the maturation and drying phases. A set of highly conserved transcription factors (TFs) with a B3 DNA-binding domain includes the master regulators of seed development and maturation (Carbonero et al., 2017):

- ABA-INSENSITIVE 3/ Viviparous-1 (ABI3/VP1);
- FUSCA3 (FUS3); and
- LEC2.

In addition, LEC1, encoding an atypical subunit of the nuclear transcription factor Y (NF-Y) CCAAT-binding transcription factor, is also considered a central regulator of many aspects of seed development (Pelletier et al., 2017). These 'LAFL' transcription factors not only control the thousands of genes involved in the seed development and maturation programs but may also have more specific functions. LEC1, for example, initiates the resetting or re-activation of *FLOWERING LOCUS C* (*FLC*) during early embryogenesis in *A. thaliana* so as to reset the parental 'memory of winter cold' (Tao et al., 2017).

2.1 *Targets of transcription factors in the acquisition of dormancy*

Experimental systems, such as yeast one-hybrid and two-hybrid screening techniques used to study interactions between TFs and their targets, have made it possible to map a multitude of (potential) physical DNA–protein interactions. This has been done for some of the master regulators. For example, LEC1 potentially interacts with some 1500 targets, as identified by chromatin immunoprecipitation (ChIP) and differential gene-expression analyses (Pelletier et al., 2017). Here we focus on ABI3 and its targets as this TF contributes to the abscisic acid (ABA) response which is strongly associated with dormancy as well as responses to abiotic stress (Hilhorst, 1995).

The first ABI3-regulon was identified by a combination of genome-wide ChIP-chip, transcriptome analysis, real-time quantitative reverse transcription-polymerase chain reaction and a transient promoter activation assay (Mönke et al., 2012). This regulon contained 98 ABI3 target genes, including FUS3 and LEC2, confirming that the LAFL genes also regulate each other (To et al., 2006). Two over-represented promoter binding motifs for bZIP and bHLH TFs were identified:

- the well-known G-box-derived ABRE (ABA-responsive element) motif ACGTG(T/G)C); and
- the RY/Sph motif CATGCA.

Not surprisingly, members of this ABI3 regulon appeared to be involved in such biological processes as embryonic development ending in seed dormancy, response to ABA and response to the stimulus. An important finding was that many of the ABI3 targets require ABA for their activation (Box 1).

Box 1

It has long been known that mutations in these TFs may lead to severe seed phenotypes, including loss of desiccation tolerance, loss of dormancy, no degradation of chlorophyll and impaired reserve food synthesis, reflecting a change of embryo identity to a vegetative developmental mode (Meinke et al., 1994; Lepiniec et al., 2018). Studies with these mutants underscore the status of these TFs as 'master regulators' of seed development. However, the question arises as to how this higher tier of regulation is translated into more specific downstream responses, by multitudes of genes, to regulate these distinct developmental processes, including dormancy. These various regulatory tiers include hormone signalling, (post)-transcriptional regulation and (post)-translational regulation, as well as epigenetic control (Fig. 1). In addition, intracellular compartmentation, tissue specificity and biophysical status of the cell add more layers of complexity to understanding the ultimate responses of the seed. With this complexity in mind, the common practice of gene by gene discovery, mainly by using mutants of a (limited) number of model species, seems to lack sufficient context to understand seed function. It is, however, encouraging that more and more studies recognize this complexity and make attempts to study genes in the context of cell identity, compartmentation and regulatory networks. New technologies create the possibilities to generate the data of this complexity, by *inter alia* (comparative) genomics, (single-cell) transcriptomics, ChIP-sequencing, translatomics and sophisticated imaging at high resolution. The role of bioinformatics in this is evident.

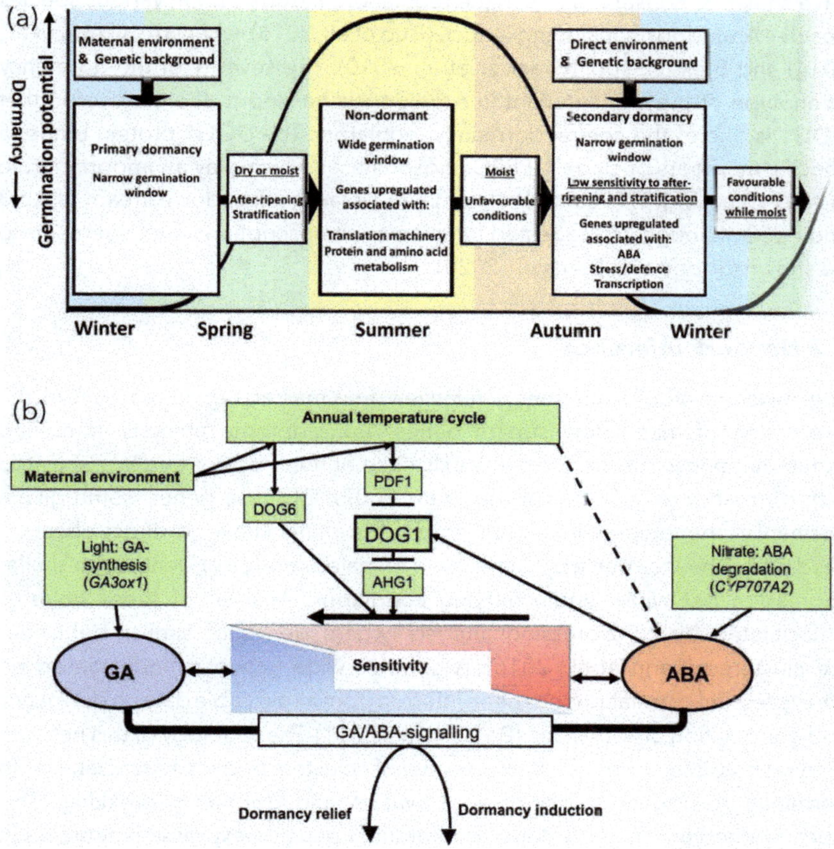

Figure 1 In the soil, seeds go through multiple dormancy cycles until they either germinate or lose viability. The driving force of the cycling are (seasonal) temperature shifts. (a) Genetic and environmental factors determine dormancy status which is reflected in transcriptomic changes dominated by transcripts associated with the translation machinery and protein synthesis. (b) Molecular model with key genes like *DOG1* controlling the cycling. Figure based on Buijs et al. (2020) (a) and Footitt et al. (2019) (b). With permission.

A more extensive attempt was made to identify ABI3 targets by ChIP-tiling array experiments to map binding sites for ABI3 genome wide (Tian et al., 2020). This approach yielded 317 directly activated genes and 87 directly repressed genes, also with overrepresented ACGTG(T/G)C) and RY/Sph motifs binding sites for B3 domain proteins. Among the directly activated genes, both by ABI3 and FUS3, were *DELAY OF GERMINATION1* (*DOG1*) and *REDUCED DORMANCY5* (*RDO5*), both positive regulators of dormancy. These genes had been identified by quantitative trait loci (QTL) analysis and, thus, are subject to natural variation (Bentsink et al., 2006, 2010). Extensive studies on DOG1 have shown its conserved role in the regulation of dormancy in various species besides Arabidopsis, including lettuce (Huo et al., 2016), wheat (Ashikawa et al., 2014) and Brassica spp. (Graeber et al., 2010). The severity of the dormancy phenotype of the *dog1* mutant in Arabidopsis has led to the conclusion that DOG1 is one of the central dormancy regulators. The DOG1 protein binds to specific members of clade A PP2C phosphatases which play an important role in ABA-signalling. This central role of DOG1 places it just downstream of ABI3 and FUS3 although its suggested function as a transcriptional factor remains to be shown (Soppe and Bentsink, 2020).

2.2 Network inference

The next question is whether a relatively few master regulators control the expression of many downstream genes. Transcriptome analysis combined with mutagenesis has revealed the existence of hundreds of genes associated with dormancy (e.g. Finch-Savage et al., 2007). Of these genes, some give a dormancy phenotype when knocked out but many show no direct effect on seed performance. Transcriptomes have revealed groups of genes with similar expression behaviour under varying conditions. Analysis of these patterns is facilitated by co-expression analysis, where genes of similar behaviour are clustered (Serin et al., 2016). A genome-wide network model based on co-expression correlations has been inferred from all available data on dormancy and germination of *A. thaliana* (Bassel et al., 2011). The topology of this network made it possible, for the first time, to identify clusters of genes associated with germination, dormancy, hormone activation and dormancy breaking. This study opened the way to further exploration of gene co-expression networks in relation to seed maturation processes, including dormancy.

A developmental study in seeds of *Medicago truncatula*, employing RNAseq and regulatory network inference, has resulted in the identification of several distinct clusters of genes closely associated with developmental intervals and, hence, with the related acquisition of seed traits such as longevity, desiccation tolerance and dormancy (Verdier et al., 2013). The dormancy and desiccation tolerance modules of the network could not be distinguished as these traits were acquired in the same interval. The module has been enriched with genes related to stress responses to desiccation, heat and temperature. These included genes encoding late embryogenesis abundant (LEA) proteins, heat shock and small heat shock proteins, as well as ABA-induced genes.

ABA is clearly the common denominator in the acquisition of dormancy and responses to stress (Hilhorst, 1995) and also explains the approximately 50% proportion of stress-associated genes in the dormancy gene set, identified by Finch-Savage et al. (2007). One of the genes identified in *M. truncatula* seeds by Verdier et al. (2013), initially predicted to play a role in seed longevity, encodes the seed-specific heat shock factor A9 (MtHSFA9) but appears to act as a negative regulator of the depth of dormancy during seed development via the modulation of ABA biosynthesis and sensitivity (Zinsmeister et al., 2020). This outcome underscores the intertwining of dormancy- and stress-associated regulation by ABA, presumably, but not exclusively, through ABI3. It is clear that network inference is a good approach to clarify a whole tier of regulation of distinct developmental processes, including dormancy (see Box 1).

2.3 Compartmentation

Seeds (and seed lots) are often considered incorrectly as a homogenous entity. Seed extracts of compounds like RNA, proteins and metabolites consist of a mixture derived from seed compartments like testa, endosperm, cotyledons and embryonic axis. When studying regulatory principles in seeds, this may lead to the masking of the essential processes and possibly erroneous conclusions (Box 2).

Box 2:

While more and more components of the regulation of dormancy and germination are being discovered, still largely through the use of (Arabidopsis) mutants, the question arises as to how the accumulated knowledge may provide us with insight in the dynamics and plasticity of seed performance and in interaction with the environment. Interpretation of results from studies with mutants has largely aimed at acquiring mechanistic insights in molecular interactions. However, if we adopt the idea that genes or proteins do not act in an isolated way but rather as entities within networks, a mutation, or a change in the environment, must be seen as a perturbation of that network. Hence, the question now is how such a perturbation may result in an altered phenotype or not at all. In other words, how is the flow of information changed by this perturbation?

For example, an early study on Arabidopsis seeds, dissected into seed coat and embryo, showed that ABA content of the seed coat (rather than the transient zygotic hormone content) correlated with the induction of primary dormancy during seed development (Kanno et al., 2010). The availability of laser-capture microdissection technology has accelerated the study of seed development in a tissue-specific context. Using this method, Belmonte et al. (2013) dissected Arabidopsis seeds into some ten regions and subregions which were all subjected to transcriptomic analysis and network inference. Apart from temporal clustering of gene networks along seed development, this approach added a spatial component which allowed the detection of subregion-specific mRNAs and prediction of the role of these various compartments in the acquisition of seed traits during development and maturation.

A network approach makes it possible to predict the key transcription factors controlling clusters of multiple genes per seed compartment (Belmonte et al., 2013). In trying to delineate the role of compartments (including dormancy) in the acquisition of seed traits during tomato seed development, Bizouerne et al. (2021) employed transcriptomic analysis of the dissected seed coat, endosperm and embryo. The study identified tissue-specific regulatory networks and hubs associated with the acquisition of desiccation tolerance, longevity and dormancy. Some gene clusters were enriched by the GO term 'negative regulation of germination', indicating the suppression of vivipary at early stages of seed maturation. An endosperm-specific cluster of genes correlated strongly with the release of dormancy during late seed maturation when ABA levels decrease (Hilhorst, 1995). GO terms associated with dormancy included mRNA processing terms, also associated with the chloroplast and mitochondria. Among these were genes related to light, circadian rhythm and flowering control, including PHYTOCHROME AND FLOWERING TIME 1 (PFT1), EMBRYO DEFECTIVE 1507, FLOWERING LOCUS C (FLC) and EARLY FLOWERING 4 (ELF4) (Bizouerne et al., 2021). Dormancy-related genes were not only located in the endosperm-specific clusters but also in those associated with the embryo. Interesting examples are SlDOG1-2 and PROCERA which are associated with dormancy and germination timing. How DOG1 is related to dormancy in tomato remains to be shown. Seeds of the *procera* mutant do not exhibit any DELLA activity and their germination is much faster than wild types, indicating the suppression of gibberellin (GA) signalling by PROCERA (Livne et al., 2015).

Such findings corroborate previous work on dormancy and germination of tomato and other species in which the seed coat (endosperm and testa) and embryo are considered to exert opposing forces which either result in the maintenance of dormancy or initiation of germination, controlled by ABA and GA, respectively. (Finch-Savage et al., 2007; Hilhorst and Downie, 1996).

3 To germinate or not to germinate: the abscisic acid-gibberellin balance

By the end of embryogenesis, embryos have generally acquired the capacity to grow (germinate) while seed fill and further completion of seed morphogenesis are still developing (Bewley et al., 2013). It is therefore important to permanently repress the induction of embryonic elongation associated with germination. Based on early work by Luckwill (1953), the concept of the ABA-GA (gibberellins) balance was developed, embodying the 'growth-inhibiting and growth-promoting substances' identified by Luckwill (1953). ABA- and GA-signalling act antagonistically, employing both hormone synthesis and hormone sensitivity. The outcome of this relationship determines whether the seed enters the seed-to-seedling stage of plant development or remains dormant. Key dormancy- and germination-associated loci identified in hormone mutants in *A. thaliana* include genes involved in GA and ABA biosynthesis, degradation, perception and signalling (Carrera-Castaño et al., 2020).

The ABA–GA balance has remained the core regulatory unit studied in various species other than Arabidopsis, such as lettuce (Argyris et al., 2008) and cereals like wheat (Tuan et al., 2018), but it is now becoming clear that other hormones, or rather 'plant growth regulators', also play important regulatory roles in the regulation of dormancy and germination. As an example, auxin plays a role in early embryogenesis, presumably by influencing dormancy by activating ABI3, mediated by AUXIN RESPONSE FACTOR 10/16 during maturation in Arabidopsis (Liu et al., 2013). Ethylene and brassinosteroids are also known to exert control of dormancy- and germination-related molecular signalling (reviewed in Graeber et al. (2012) and Corbineau et al. (2014)).

3.1 Stored mRNAs in dormancy and germination

Mature dry seeds may accumulate more than 10 000 mRNAs during seed maturation. These may be used for protein synthesis upon germination. Seed germination may occur in the presence of transcriptional inhibitors, such as cordycepin or α-amanitin. This suggests that the first steps of the germination process at least are independent of transcription and may rely on protein synthesis or (re)activation of stored proteins (reviewed in Bewley et al. (2013), Sano et al. (2020)). The nature of the mRNA population that is deposited depends on the seed's maturation history during which both internal and external factors contribute to this population. These stored mRNAs may be long-lived to remain translatable after decades of dry storage to ensure successful germination (Dure and Waters, 1965). A dry state appears to prevent mRNA decay, whereas in a hydrated environment, transcript half-lives range from minutes to days (Sajeev et al., 2019).

It is unlikely that all of the thousands of stored mRNAs are required for the germination process. They will also be involved in housekeeping activities in cells or are remnants of seed developmental processes. In rice, using RNA-seq, Sano et al. (2012) detected 529 mRNA candidates unique to germination, which accumulated in embryos between 10 and 20 days after flowering. Similarly in cotton, germination-specific mRNAs were detected between 30 and 50 days after flowering. In both species, this is the developmental phase starting just after the completion of embryo morphogenesis. This is when transcriptional inhibitors no longer inhibit precocious germination of embryos, suggesting the acquisition of transcriptional independence by these embryos (Sano et al., 2020).

It has been shown that stored mRNAs are also translated at later stages of germination (Law et al., 2012). For example, proteins involved in energy and cell structure are early products of seed-stored mRNAs that are required in the germination process. Fleming et al. (2018) convincingly demonstrated substantially greater mRNA transcript fragmentation in soybean seeds stored dry for 23 years as compared to 2 years of storage. Fragmentation occurred predominantly in the longer transcripts which were correlated with seed death whereas short transcripts kept their integrity. These results suggest that the storability of seeds may depend on how well the mRNAs required for translation upon germination are protected against damage during storage.

Several hundred stored mRNAs with high abundance in Arabidopsis seeds were found to be enriched in ABREs, approximately 1 kb upstream of their translational start site (Nakabayashi et al., 2005). ABA promotes dormancy and this suggests that these mRNAs contribute to the maintenance of the dormant state. In imbibing non-dormant Arabidopsis seeds, ABA levels start to decrease as early as 6 h which results in a lower expression of these ABRE-containing genes, resulting in a diminishing of dormancy maintenance. However, dormant seeds display *de novo* synthesis of ABA after an initial decrease, resulting in the maintenance of dormancy (Ali-Rachedi et al., 2004).

Dry after-ripening is one of the most common ways to break seed dormancy although, in some cases, this may require a long time to be effective (Bewley et al., 2013). For example, the deeply dormant Cvi ecotype of *A. thaliana* may take more than a year for complete loss of dormancy in the seed lot (Finch-Savage et al., 2007). The mechanism of after-ripening was unclear until recently when evidence was first obtained of the involvement of 'solid state' chemistry, *i.e.* occurrence of only passive and non-enzymatic reactions including oxidation and Amadori–Maillard reactions. The fact that dry seeds contain less than 10% of water implies that the cytoplasm of cells is turned into a near-solid glass state which poses severe restrictions to mobility and precludes active transcription/translation of mRNAs (Ballesteros and Walters, 2019).

In after-ripening of sunflower and wheat seeds, the oxidation of stored mRNAs has been associated with the alleviation of dormancy (Bazin et al.,

2011; Gao et al., 2013). Surprisingly, the mRNA oxidation was highly selective, targeting mRNAs involved in the maintenance of dormancy. This resulted in the formation of truncated non-functional proteins and thus seed dormancy release. Interestingly, the identified mRNAs differed between these two species. However, the targeting mechanism of this selective mRNA oxidation remains to be identified.

There is increasing interest in the occurrence of 'mobility islands' in the absence of free or weakly bound water, enabling metabolic activity in the dry state. This is based on the observation that metabolic activity in terms of transcription, translation and metabolism can occur at water content below 30% (~0.7 g H_2O g-1 DW). For example, the xanthophyll cycle driven by enzymatic conversion was still active in thalli of the desiccation-tolerant lichen *Flavoparmelia caperata* below 0.2 g H_2O g^{-1} DW (Carniel et al., 2021). While these authors suggested that enzyme activity at this low water content was enabled by the existence of a 'rubbery' state of the cytoplasm, an alternative explanation may be given by local pockets of mobility. Similarly, during dehydration of the desiccation-tolerant dicot *Craterostigma plantagineum*, respiration was maintained down to 20% relative water content (approx. 0.5 g H_2O g^{-1} DW) before a decline occurred (Xu et al., 2021).

Examples of such 'mobility islands' are oil droplets which are fluid in the cytoplasm of dry cells, thus potentially enabling long-range diffusion of molecules, including reactive oxygen species (ROS) and thus allowing chemical reactions (Ballesteros et al., 2020). Drying of a heterogeneous cell may also induce differences in concentrations of cellular compounds, thus providing a localized liquid environment, for example, by pockets of natural deep eutectic solvents. Another possibility is the occurrence of liquid–liquid phase transitions generating localized mobility in dry cells as demonstrated in the desiccation-tolerant brine shrimp *Artemia franciscana* enabled by the presence of a group 6 LEA proteins, potentially allowing molecular interactions such as protein binding (Belott et al., 2020). Such observations need further research to understand dry seed metabolism and its consequences for seed traits such as dormancy and longevity.

4 The breaking of primary dormancy

In general, dormancy is broken as a result of a stream of information from the environment, sometimes over long periods of time, targeting the right timing for subsequent growth of progeny. While a multitude of environmental factors involved in dormancy break has been described (Bewley et al., 2013), the (molecular) mechanisms by which this occurs are not yet clear. Dormancy break often requires more than one environmental factor. This is not only in cases of combinational dormancy such as morphophysiological dormancy where it might be expected that each component requires a different environmental factor.

4.1 The breaking of physiological dormancy

Seeds in the soil use seasonal temperature shifts as temporal information and prevailing environmental factors as spatial information (Finch-Savage and Footitt, 2012). In general, long-term (seasonal) temperature effects result in the sensitization of the seeds to the environment or a widening of the germination 'window' (Bewley et al., 2013; Buijs et al., 2020). For example, dry-after-ripening of Arabidopsis seeds increases the seed's sensitivity to nitrate and light (Finch-Savage et al., 2007). This phenomenon of sensitization to the environment has been shown for a variety of annual species (Baskin and Baskin, 2014). Dormancy break is characterized by the opening of the temperature window for germination. When this window overlaps with the field temperature, germination may occur but only if the prevailing spatial information is adequate to support seedling emergence and subsequent plant growth. For example, light and nitrate provide canopy gap-detecting mechanisms in order to avoid competition of the young seedling with neighbouring vegetation.

The various methods to break seed dormancy that are utilized in practice are often based on these environmental factors. These include dry storage or cold stratification, often in combination with natural chemical factors such as nitrate, nitrite, smoke components like karrikins, strigolactones, as well as artificial dormancy relievers like norflurazon and fluridone, and ROS-generating substances like methylviologen (Chahtane et al., 2017). These dormancy-releasing chemicals all affect the ABA–GA balance by inhibiting ABA biosynthesis (norflurazon and fluridone), promoting ABA degradation (nitrate, nitrite) or enhancing GA biosynthesis and signalling (karrikins and strigolactones). Gibberellins are frequently used directly to break seed dormancy but many dormant seeds are relatively insensitive to the hormone. However, how these changes in the hormones are accomplished at the molecular level remains to be demonstrated.

While it was suggested for a long time that (soil) micro-organisms may digest seed coats and thus alleviate dormancy, only recently has evidence been provided for such a mechanism. Sperber et al. (2017) demonstrated that the pericarp of *Lepidium didymum* poses a mechanical restraint which is the limiting factor for its germination. Dormancy is released by the activity of common fungi that weaken the pericarp along predetermined breaking zones by degrading non-lignified pericarp cells.

Such treatments are not always effective in alleviating dormancy. This is particularly true for bigger seeds with thick seed coats. Seeds with a hard and/or impermeable seed coat possess physical dormancy. In such cases mechanical scarification is common practice. Such treatment will damage the seed coat to such an extent that it enables seeds to take up water and/or weaken the resistance of the seed coat to embryo expansion. Methods include (dry) controlled crushing or sanding or a short treatment using boiling water or sulphuric acid.

4.2 The breaking of physical dormancy

Physical dormancy is caused by a water-impermeable palisade cell layer in seed (or fruit) coats. Dormancy break therefore depends on the permeabilization of the seed coat (Bewley et al., 2013). Physical dormancy is very common in seeds of the Leguminosae, but it is also found in seeds or fruits of 18 angiosperm families, including the Cannaceae, Convolvulaceae, Chenopodiaceae, Cistaceae and Malvaceae. Many of these species produce seeds with extremely hard coats in which water and oxygen entry is the limiting factor for germination (Baskin et al., 2000). In addition, seed hardness may protect against microbial attack and extend seed longevity and persistence in the soil seed bank (Dalling et al., 2011).

Upon imbibition, entry of water (and oxygen) is facilitated by the formation of a small opening (water gap) in a morpho-anatomically specialized zone in seeds (or fruits) known as the water-gap complex. Many types of water gap have been characterized, including chalazal, hilar, carpellar and lens gaps and slits, closed by a lid, cap or plug. Under laboratory conditions, application of wet heat of 50–100°C for short duration (minutes) is usually sufficient to open the gap and allow influx of water, initiating growth of the embryo (Gama-Arachchige et al., 2013). Such specific structures have also been observed in seeds with coat-enhanced dormancy. A good example is an NMR-based study of water entry into seeds of *Brassica napus*. It was shown that an endospermal lipid gap channels water to the radicle tip which is subsequently distributed to embryonic vasculature and cotyledon tissues (Munz et al., 2017).

However, the environmental factors causing dormancy break under field conditions are less well known and are highly variable among species. Physical dormancy may be broken in response to fire, diurnally fluctuating temperatures, changes in moisture, ingestion of seeds or fruits by animals, soil microorganisms or a combination of environmental conditions acting on seeds in the soil (Baskin and Baskin, 2014; Hudson et al., 2015). This extreme variation in dormancy breaking conditions is likely a reflection of a vast array of niche adaptations to ensure correct timing of emergence and to spread mortality risk (Willis et al., 2014).

Although the phenomenology of physical dormancy has been described for a substantial number of species, little is known about molecular mechanisms involved in dormancy break of these hard-coated seeds. Mutagenesis of the hard-seeded legume *Medicago truncatula* resulted in the identification of a mutant which had lost its physical dormancy (Chai et al., 2016). It appeared that the phenotype was caused by genes acting downstream from a class II KNOTTED-like homeobox (KNOX) gene, *KNOX4*, which regulates seed-coat cuticle development, presumably mediated by the CYP86A family of cytochrome P450-dependent fatty acid omega hydroxylases (Chai et al., 2016).

In many hard-coated species, cracks are also formed by environmental cues to allow water uptake. For example, cultivated soybeans show various degrees of seed coat permeability that are controlled by QTL. A QTL encoding calcineurin-like protein, GmHs1-1, was identified by map-based cloning and appeared to contribute to hardseededness in soybean by a mechanism yet to be determined (Sun et al., 2015). In addition, an endo-1,4-β-glucanase gene was identified by fine-mapping of soybean lines and reported to be responsible for the formation of microscopic cracks by changing cell wall structure, thus enabling influx of water (Jang et al., 2015). Interestingly, the expression levels of the *GmHs1-1* and *qHS1* homologues in *M. truncatula* did not show significant changes in the *knox4* mutant. This corroborates the high variability in mechanisms associated with the variation in phenotypes across species and thus the different pathways that control physical dormancy.

5 Secondary dormancy and dormancy cycling

Secondary dormancy may occur in seeds with physiological dormancy. When germination of non-dormant seeds is inhibited for prolonged periods of time, they may enter secondary dormancy. Inhibitory conditions include darkness or unfavourable temperatures. However, all conditions that inhibit germination may lead to the induction of secondary dormancy (Hilhorst, 1998). In the field, seeds may enter or leave secondary dormancy repeatedly in synchrony with seasonal cues to ensure correct timing of seedling emergence and establishment. Seeds in the soil seed bank are thus able to sense the continuous stream of environmental information about the suitability of the environment for successful progeny while proceeding through so-called dormancy cycles. The ecophysiology and modelling of seasonal dormancy cycling have been extensively described for various arable weeds (Baskin and Baskin, 2014).

Over the past two decades, attempts have been made to elucidate the underlying molecular and regulatory mechanisms of dormancy cycling. Of particular interest is the ability of seeds in the soil to sense the long-term fluctuations related to slow seasonal change, principally temperature (Probert, 2000), to determine the time of year and depth of dormancy. Such seasonal fluctuations coincide with fluctuations in the sensitivity of the seeds to prevailing environmental factors such as light, nitrate and soil water potential, which are indicators of the spatial environment and play a critical role in germination (Hilhorst, 2007). Each of these environmental factors may be required to remove a 'layer' of dormancy until the final layer, usually light, triggers germination (Finch-Savage et al., 2007). An increasing number of studies employing field sampling have made it clear that dormancy relief and induction under laboratory conditions do not readily mimic those under field conditions (Finch-Savage and Footitt, 2017; Buijs et al., 2020).

To better understand dormancy cycling, Finch-Savage and Footitt (2017) used contrasting Arabidopsis ecotypes, namely a winter vs. a summer annual, and a mapping population derived from these parents. They identified genome-wide expression patterns over a complete annual dormancy cycle in the field of both ecotypes. This highlighted several QTL associated with depth of dormancy and seedling emergence timing. These QTLs, named SET1 through to 5, collocated with such candidate genes as *PROTEIN PHOSPHATASE 2A SUBUNIT A2* (*PP2AA/PDF1*) (which acts upstream of *DOG1*), *PDF1* (which codes for one of the three scaffolding subunits of the PP2A family) and *DOG6* (likely *ANAC060*). The RNA-seq analysis in this study confirmed that DOG1 and ABA content do not determine timing of seedling emergence but are both required for emergence (Finch-Savage and Footitt, 2017). It was suggested that a changing sensitivity to ABA, in association with the amount of DOG1 protein present, controls timing of emergence. In this model, deep dormancy is maintained by high amounts of DOG1 and high ABA content, whereas reduced DOG1, ABA and sensitivity to ABA are associated with the removal of the successive layers of dormancy, resulting in the removal of the final layer of dormancy by light to allow completion of germination (Fig. 1).

In a burial experiment with Arabidopsis seeds, Buijs et al. (2020) monitored a complete annual cycle by RNA-seq of various near-isogenic lines containing various DOG introgressions (NILDOG1, 2, 3, 6 and 22) resulting in differing dynamics of dormancy cycling. The study showed that the effect of the maternal environment on depth of dormancy was completely abolished during the annual cycle, suggesting that the maternal environment determines the level of dormancy only to prevent germination in the wrong season upon seed shed. This study confirmed that dormancy cycling is driven by soil temperature, very likely through DOG1 as a mediator of the temperature signal. This had also been suggested for perception by developing seeds of maternal temperature (Penfield, 2017).

Large transcriptional changes were observed during dormancy cycling, particularly just prior to shifts in dormancy status, indicating that dormancy changes are highly dynamic. Globally, these changes correlated with previously identified changes in laboratory experiments, whereas the expression of genes related to GA signalling is reduced upon induction of secondary dormancy both in the lab and in the field; ABA-related gene expression, that is, *FUS3, ABI3 and ABI5* differed noticeably between the field and the laboratory experiments.

FUS3 was not expressed in the field data set and *ABI3* and *ABI5* expression did not change during dormancy cycling, whereas the abundance of these three transcripts was reduced during induction of secondary dormancy in the laboratory (Buijs et al., 2020). Interestingly, in this study, a large number of overrepresented genes were endosperm-specific according to the listing of these genes by Dekkers et al. (2013), suggesting that transcriptional changes in the endosperm are important for dormancy cycling in the field. In particular, genes in the GO categories related to translation are significantly upregulated in non-dormant seeds. This suggests that non-dormant seeds (only requiring light to germinate) are preparing for rapid seed germination (Buijs et al., 2020). These results also confirm that yet another layer must be added to the regulation of dormancy and germination, namely translational control.

6 The seed-to-seedling phase transition: germination

Upon the final environmental trigger (dormancy break) to initiate germination, the embryo switches from a developmental mode to a vegetative phase, characterized by growth and greening (Bewley et al., 2013). In order to do this, the regulation of seed development needs to be suppressed. Activity of the *LAFL* genes *LEC2* and *FUS3* is suppressed during early embryogenesis and upon germination by negative regulation through the transcription factors ARABIDOPSIS 6B-INTERACTING PROTEIN1-LIKE1/2 (ASIL1 and ASIL2) (Lepiniec et al., 2018). In addition, HIGH-LEVEL EXPRESSION OF SUGAR-INDUCIBLE GENE 2 (HSI2) and HSI2-like1 (HSL1), more commonly referred to as VP1/ABI3-Like1/2 (VAL1 and VAL2), may inhibit the LAFL network during seedling development and play a role in the termination of the developmental program of the embryo. VAL proteins also interact with chromatin or chromatin-modifying proteins, adding an epigenetic layer of regulation to LAFL expression, both in Arabidopsis and soybean (Lin et al., 2017). While the LAFL/VAL network components and their interactions have been extensively described (Lepiniec et al., 2018), it remains unclear how these higher-order transcriptional networks are informed by the environment, whether the prevailing environment or accumulated long-term (e.g. seasonal) environmental information.

6.1 Post-transcriptional regulation of germination

The study of post-transcriptional regulation of germination was sparked by the phenomenon of pre-formed mRNA in dry seeds which, upon imbibition, is readily translated to proteins, essential to early germination in sunflower and Arabidopsis (Layat et al., 2014; Basbouss-Serhal et al., 2015). Differential recruitment of mRNAs to monosomes and polysomes was observed during imbibition of dormant seeds and non-dormant seeds (those where only light is required for germination) (Basbouss-Serhal et al., 2015). Further studies have shown that regulation of dormancy in Arabidopsis is mainly under transcriptional control while germination of this species is under translational control (Sajeev et al., 2019). Translational regulation of germination allows a rapid response to developmental and environmental signals, either at a global or gene-specific level.

Polysome profiling studies of germinating sunflower seeds found translation activity of stored mRNAs to be increased compared with dormant seeds (Layat et al., 2014). In addition, extensive transcript-specific translational control was demonstrated during Arabidopsis seed germination, in which several thousands of mRNAs displayed changes in ribosome association (Bai et al., 2017). These studies have mostly focussed on the recruitment of long-lived mRNA populations of imbibing dry seeds. However, in field conditions, imbibition of dry seed only occurs upon seed shedding, after which the seeds in the soil go through successive cycles of re- and dehydration. This probably eliminates long-lived mRNAs which accumulated at the end of maturation. However, as described above, translational control was also observed during phases of low dormancy in the seed bank population. indicating readiness for (quick) germination (Buijs et al., 2020).

To add to the complexity of the regulation of dormancy and germination, a plethora of post-translational modifications of proteins have been described (Arc et al., 2011). A single gene can produce many different protein variants as a result of mRNA processing, protein proteolysis and more than 300 potential post-translational modifications. This leads to modification of protein function depending on localization, complex formation, stability and activity (Ytterberg and Jensen, 2010). Seed proteome analyses of a variety of wild and crop species have demonstrated a high proteomic complexity with a broad diversity of functions (Arc et al., 2011). Some of these post-translational modifications have been studied more closely in relation to seed functions such as dormancy, germination and longevity. For example, ROS have been linked to the selective oxidation of dormancy-associated proteins in Arabidopsis, thereby alleviating primary dormancy (Oracz et al., 2007).

It has long been known that the diatomic free radical nitric oxide (NO) can break seed dormancy (Bethke et al., 2006). A transient burst of NO upon germination has been observed in several species, including Arabidopsis in which it induced ABA catabolism (Liu et al., 2009). The direct targets of NO await identification but there is some evidence that NO binds to transition metals of metalloproteins (nitrosylation) or triggers modifications such as cysteine S-nitrosylation or tyrosine nitration (Arc et al., 2011).

To add to the list, protein modifications may alter redox properties of proteins, for example, by cleavage of disulphide bonds. Well-known protein modifications such as protein phosphorylation and dephosphorylation are of pivotal importance in hormone signalling in seeds. For example, sucrose nonfermenting-1 (Snf1)-related protein kinases (SnRKs) and protein phosphatases 2C (PP2Cs) are key regulators of ABA signalling in seeds (Hubbard et al., 2010). More post-translational protein modifications demonstrated for seeds are protein L-isoaspartyl formation, farnesylation and glycosylation (Arc et al., 2011).

6.2 Germination: temporal phases of activity

Several studies have shown that reactivation of metabolism upon imbibition occurs in two specific temporal phases during seed germination, coinciding with:

- hydration and mobilization of pre-existing cell constituents, including mRNAs, proteins and metabolites (Sano et al., 2020; Silva et al., 2017); and
- breaking of the seed coat allowing entry of abundant oxygen and removal of mechanical restraint, both promoting growth of the embryo.

Transcriptomic analysis, combined with co-expression analysis, of Arabidopsis seed germination revealed a first phase with large transcriptome changes in the progression from the dry state to a hydrated, active state (Dekkers et al., 2013). This first transcriptional phase dies out but increases again at testa rupture, the initiation of a second transcriptional phase. The switch between development and growth was marked by a repression of seed maturation genes (see above) and induction of growth-related genes. Interestingly, during the first transcriptional phase, gene expression of embryo and endosperm was surprisingly similar, but into the second phase, gene expression was associated with growth and senescence, respectively (Dekkers et al., 2013). Surprisingly, there were no clear indications of the involvement of either oxygen or light in the transcriptional phase transition. However, this study provided evidence that, upon testa rupture, mechano-sensing may be operational, engaging the *TOUCH3* and *TOUCH4* genes, which were highly upregulated in the endosperm (Dekkers et al., 2013). A similar conclusion was reached for the thick-walled endosperm of tomato during germination. The endosperm uses mechano-sensing to detect the increase in embryonic growth potential by promotive action of GA, mediated by the tomato ethylene response factor ERF1 (TERF1). In addition, one of the 'Touch' (*TCH*) genes xyloglucan endotransglycosylase/hydrolase (*XTH*) (a cell wall-associated gene) was also significantly enriched in the endosperm cap (Martínez-Andújar et al., 2012).

These findings suggest that the two opposing forces embryo growth potential increase and endosperm weakening, which collectively determine germination, are not independent forces. It also makes clear that the initial signal

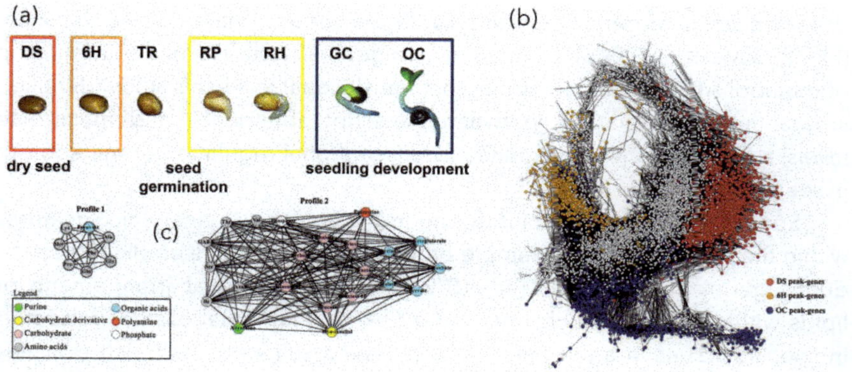

Figure 2 (a) Subdivision of the seed-to-seedling developmental stages: DS, dry seeds; 6H, 6-h imbibed; TR, testa rupture; RP, radicle protrusion; RH, root hair; GC, greening cotyledons; and OC, cotyledons fully opened. Colours refer to sub-networks of (b) Unweighted gene co-expression network for the seed-to-seedling transition. Circles (nodes) represent transcripts, and lines (edges) represent significant transcriptional interactions between the transcripts. Temporal analysis of network clusters was performed by colouring each gene by its maximum expression across seed-to-seedling transition developmental stages. (c) Metabolite profile networks, in which metabolites display higher connectivity to each other. Each edge in the networks indicates a source-target correlation. Profile 1: Network comprising stages DS to RP and Profile 2: stages RP to OC. Figure based on Silva et al. (2016, 2017). With permission.

towards germination, i.e. tipping of the ABA/GA balance, is likely located in the embryo. Finally, mechano-sensing, related to endosperm morphology, may represent yet another layer of regulation in seed dormancy and germination (Fig. 1).

Metabolite profiling across the seed-to-seedling transition in Arabidopsis revealed two clear transitional metabolic profiles (Silva et al., 2017). The inference of metabolite-metabolite correlation networks for each profile (belonging to a metabolic shift) revealed dominance of certain pathways based on the number of incoming edges in the network (Fig. 2):

- The first profile could be associated with the pathways of aminoacyl-tRNA biosynthesis, nitrogen metabolism, alanine, aspartate and glutamate metabolism, and carbon fixation in photosynthetic organisms.
- The second profile could be related to *alanine,* aspartate and glutamate metabolism, TCA cycle, aminoacyl-tRNA biosynthesis, galactose metabolism, and valine, leucine and isoleucine biosynthesis.

The two profiles are related, respectively, to metabolic preparation for germination and successful seedling establishment (Silva et al., 2017). Parallel co-expression analysis of the transcriptomes of this seed-to-seedling transition resulted in the inference of a transcriptional network, clearly showing the distinct clusters of genes associated with the various phases of the transition (Fig. 2; Silva et al., 2016). Contrary to the general assumption that regulation of metabolism is controlled at the post-translational level, this study identified several linear correlations between transcript abundance and metabolite levels, suggesting the occurrence of transcriptional regulation of metabolism in seeds (Silva et al., 2017).

Spatial metabolic dynamics during imbibition due to anatomical features within the seed is another example of the complexity of metabolic regulation during germination. Oilseed rape (*B. napus*) exploits spatial arrangements of lipids within the testa to channel water flow so that metabolism is rebooted in specific tissues in a temporal fashion. As a result, imbibition first activates endosperm respiration and lipid degradation. After this, embryo respiration starts once water reaches the embryonic axis, with sugars of the raffinose family being used as respiratory substrates to sustain elongation of the embryonic axis (Munz et al., 2017). This and comparable studies of seed traits also provide a glimpse of the complexity of the underlying control mechanisms of the various seed traits.

6.3 Germination: initiation of growth

The ultimate initiating signal of embryo expansion has long been enigmatic. What is this 'trigger' of growth after the seed has processed all the required information for germination? Since germination is irreversible, the process must be exceptionally well informed to minimize the probability of premature death, leading to relaxation of embryonic cell walls to allow water uptake and expansion of the cells. Cell wall loosening of embryonic cells is mediated by the moving apart of cell wall polymers, enabling expansion growth of the cells as a result of water influx into the vacuole. Cell wall constituents involved in embryonic cell wall loosening include expansins, xyloglucan endotransglycolases/hydrolases (see above), endo-(1,4)-β-d-glucanases and apoplastic ROS (Steinbrecher and Leubner-Metzger, 2017). Similar biochemical mechanisms occur in the cell walls of micropylar endosperms which highlight the requirement of endosperm weakening and rupture during germination. Endosperm weakening has been measured in species across various plant clades and families, including the Cucurbitaceae, Brassicaceae, Oleaceae, Solanaceae, Rubiaceae and Asteraceae, as well as in some monocots, indicating evolutionary conservation of the mechanism.

The regulation of endosperm weakening has been studied extensively in such species as Arabidopsis, tomato, lettuce and coffee (Nonogaki et al., 2007; Bewley et al., 2013). When comparing the model tomato and Arabidopsis systems, the mechanistic properties of endosperm weakening appear to be quite distinct. Tomato endosperm cap weakening is characterized by cell wall degradation through activity of hydrolytic enzymes, such as endo-β-mannanase, cellulase and β-1,3-glucanase. This is complemented by further cell wall modifications leading to cell separation, mediated by polygalacturonase and pectin methylesterase (Nonogaki et al., 2007). Like tomato, Arabidopsis endosperm weakening is mediated by a plethora of cell wall remodelling enzymes but primarily aiming at cell expansion to accommodate embryo growth prior to radicle protrusion, mediated by the spatiotemporal control of *EXPANSIN 2* (*EXPA2*). *EXPA2* expression is promoted by the NAC transcription factors NAC25 and NAC1L but repressed by the DELLA protein RGL2 (Sánchez-Montesino et al., 2019). DELLA protein inactivation is mediated by GAs and

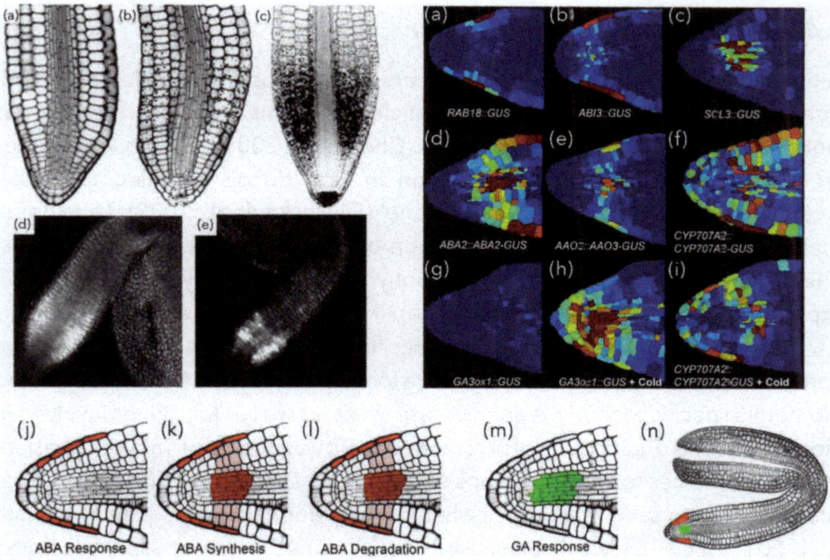

Figure 3 TOP LEFT: The elongation zone of Arabidopsis embryos during germination is marked by periodic acid-propidium iodide positive intracellular bodies (likely starch) at 7 (a) and 18 (b) h after imbibition just prior to the beginning of testa rupture at 24 h (c) and by enhanced fluorescence in the elongation zone well before testa rupture at 18 HAI in 2 GAL4-GFP enhancer trap lines (d, e). TOP RIGHT: Cellular basis for developmental fate switching in the elongation zone of dormant Arabidopsis seeds. Image shows the single-cell distribution of the ABA-responsive RAB18 promoter (a), the promoter of the ABA response-promoting transcription factor ABI3 (b), the GA-response proxy SCL3 promoter (c), the ABA biosynthesis proteins ABA2 (d) and AAO3 (e), the ABA degrading protein CYP707A2 (f), and the promoter activity of the GA biosynthesis enzyme GA3ox1 (g) in dormant embryos. (h) GA3ox1 promoter activity following dormancy break by 1 d at 4°C and (i) same as H for CYP707A2 protein abundance. BOTTOM: Schematics indicating ABA response (j), ABA biosynthesis (k), ABA degradation (l), and GA response (m) in the Arabidopsis radicle. (n) Whole-embryo highlighting of the subset of cells enriched for ABA response (red) and GA response (green). Figure based on Sliwinska et al. (2009) and Topham et al. (2017). With permission.

assumed to be a prerequisite for successful germination (Bewley et al., 2013). However, there is also evidence that homeostatic GA responses are regulated by both DELLA-dependent and -independent pathways (Livne et al., 2015).

Further studies on the GA-controlled expansin-mediated cell expansion of Arabidopsis endosperm identified 42 transcription factors (next to NAC25 and NAC1L) capable of binding promoter sequences on 7 EXPA genes, 13 of which also interacted with DELLA proteins. Binding of the transcription factors to the germination-associated EXPA promoters turned out to be highly redundant, possibly providing robustness to the germination process. All of these transcription factors are potential regulators of EXPA gene expression and, consequently, control cell wall biomechanics (Xu et al., 2020).

6.4 Regulation of growth initiation

Initially, elongation of embryo cells occurs (as a prelude to radicle protrusion) within a discrete region close to the radicle. While this specialized region was hinted at in expanding tomato embryos (Chen et al., 2001), subsequent studies on the initiation of embryo elongation in Arabidopsis revealed the exact location of this so-called 'elongation zone' (Sliwinska et al., 2009). This region is also characterized by the accumulation of discrete intracellular bodies, just prior to testa rupture, containing carbohydrates, very likely starch, and GFP expression of GAL4–GFP in enhancer trap lines (Sliwinska et al., 2009) (Fig. 3).

Based on these earlier studies, cell-specific responses to the environment have been studied in the elongation zone, with a focus on the ABA/GA balance. By using promoters specific to the ABA and GA responses, as well as key genes involved in biosynthesis and degradation of both hormones, it was possible to assign 'decision-making' properties to specific groups of cells with respect to the ABA/GA balance occurring during seed maturation, when primary dormancy is acquired (Topham et al., 2017). It was shown that responses to GA and ABA were spatially separated (in the vascular and epidermal tissues, respectively) but overlapped spatially with the elongation zone (Fig. 3). External admission of these hormones did not alter the expression of the ABA- and GA-response reporter systems RAB18::GUS and SCL3::GUS, respectively. This indicates that only hormone response defines the location of these signalling centres. Such a separation implies a role for transporters of both hormones. Indeed, applying dormancy-breaking cold to dormant seeds increased transport rates of both hormones (Topham et al., 2017).

Another important result from this study is that the dormancy/germination 'switch' in the radicle is informed by the surrounding endosperm, which contains all the decision-making components described for the embryo. Confirming the importance of transport and relocalization, Xia et al. (2019) demonstrated in dormant vs. non-dormant sunflower embryos that ABI5 and DELLA protein RGL2 (see above) were localized mainly in the nucleus in dormant seeds while they localized in the cytosol in germinating seeds. 1-Aminocyclopropane-1-carboxylic acid oxidase, a key enzyme in ethylene biosynthesis, was also only detected in the meristem of germinating seeds. These results underline the importance of relocalization in allowing changes in hormone sensitivity in the case of GA and ABA and on-site ethylene production.

What could be the evolutionary advantage of such a complex response mechanism in a multi-cellular system like the embryo? What is the benefit of spatial separation of hormonal response? Could it be that interactions between communicating cells (and tissues) provide the plant with more robust ways of integrating environmental variables to flipping the switch between dormancy and germination? In other words, the 'decision making' of irreversibly switching to the growth mode is optimized in this way, thereby minimizing the risk of premature death. Topham et al. (2017) developed this model from the effects of temperature

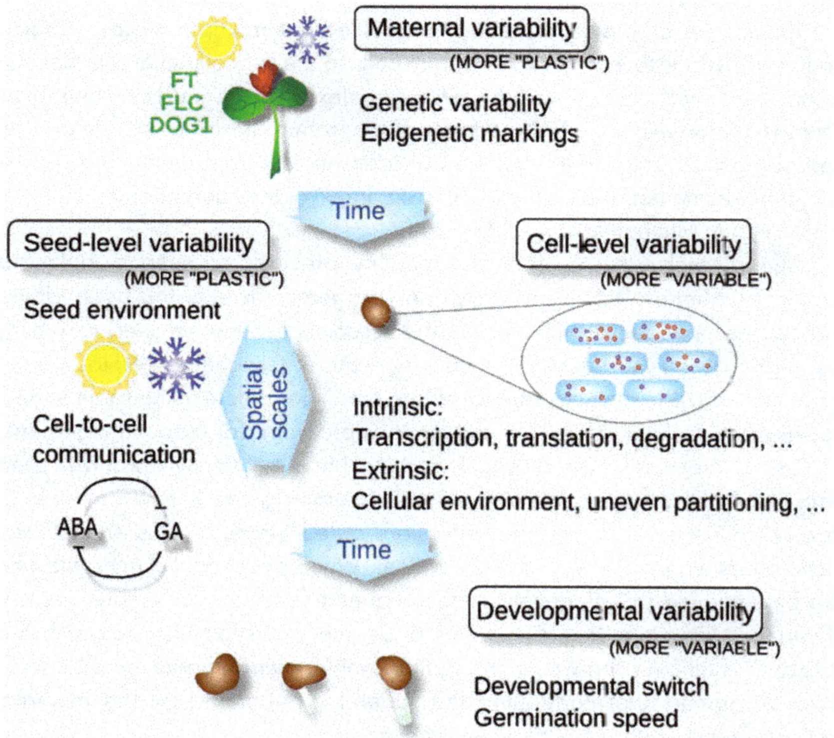

Figure 4 The sources and scales of variability in seeds ranging from the maternal environment (plastic), seed level variability (plastic), cell level variability (variable), to developmental variability (variable), in which the term 'plastic' refers to phenotypes that are consistently altered by the environment and 'variable' to a range of phenotypes observed across constant environmental conditions. From Mitchell et al. (2017). With permission.

variation on this switch and argued that this mechanism would benefit seeds in the soil when going through temperature-driven dormancy cycling.

7 Biological variability

A serious drawback in studying seed biology is the dependence on using seed lots, i.e. populations with inherent substantial biological variation. For example, non-destructive extraction of the enzyme endo-β-mannanase from endosperm fragments of single tomato seeds showed a remarkable variation in enzyme activities among individual seeds, exhibiting a range of up to four orders of magnitude. Variation of this magnitude would never be detected using pooled assays. It is possible that most of the activity in a pooled assay is contributed by only a small fraction of the population, and hence, correlations between population behaviour and pooled activity measurements must be viewed with caution (Still and Bradford, 1997).

Obviously, biological variability occurs across multiple scales, ranging from populations to individuals and from cells to the subcellular and molecular components within cells, adding extra complexity to the study of biological systems (Trewavas, 2012). By adopting the terms 'intrinsic' and 'extrinsic', describing processes inherent to the individual and external influences from the environment, respectively, a picture can be drawn of the sources of variability in seeds (Fig. 4; Mitchell et al., 2017).

The plasticity that biological variability provides to seed populations confers a higher probability of survival and progeny. However, this bet-hedging strategy has a negative influence on the association between seed variability and (agricultural) yield potential. In practice, this means that a seed lot will not germinate 100% in the field and not all seeds will germinate at the same speed. Besides sub-optimal yield, this poses serious problems for crop management.

Seed producers have developed technology to synchronize seed populations by pre-germination treatments, allowing seeds to germinate up to a point just prior to radicle protrusion, called 'priming'. This enables the 'slow' seeds to catch up with the faster ones. Whether seed lot synchronization can be achieved by using genetic techniques to reduce the various sources of variability in seeds (Fig. 4) remains to be seen. As seed lots are produced at locations all over the world, in highly variable environments, on an annual basis, it appears that controlling the variability generated by the maternal environment would be most advantageous.

8 Future trends in research

Like other fields of plant science, seed biology has entered the genomics era and many studies employing -omics approaches have been published over the past two decades. Arabidopsis has been the predominant model species to study dormancy and germination mechanisms and their regulation. Many genes, proteins and metabolites have been assigned to the various seed traits. However, it is only during the past 5 years or so that systems approaches have been used to attempt to integrate the molecular, physiological and ecological aspects of seed performance. This is by no means a simple undertaking as seeds comprise a very dense and complex chemical and physical matrix and can hardly ever be studied as single individuals. Furthermore, biological systems in general are very complex with multiple layers of regulation. Will we ever be able to grasp such complexity?

The future of seed biology will be dependent on wider technical and theoretical developments in plant science but will need to account for the inherent variability of studying populations. As mentioned at the beginning of this chapter, it will be pivotal to develop the technology to image seeds and seed processes in a non-destructive manner so as to correlate image parameters of single seeds with seed viability, dormancy and longevity.

9 Where to look for further information

The following works provide an excellent basis to the subject:

- Bewley, J. D., Bradford, K. J., Hilhorst, H. W. M. and Nonogaki, H. (2013). *Seeds. Physiology of Development, Germination and Dormancy.* 3rd Edition. New York, Springer.
- Baskin, C. C. and Baskin, J. M. (2014). *Seeds: ecology, biogeography, and evolution of dormancy and germination*, 2nd Ed. San Diego, CA, USA: Academic/Elsevier.
- Nonogaki, H. (2018) Seed germination and dormancy: The classic story, new puzzles, and evolution. *J. Int. Plant Biol.* 61, 541–563. https://doi.org /10.1111/jipb.12762.

10 References

Ali-Rachedi, S., Bouinot, D., Wagner, M. H., Bonnet, M., Sotta, B., Grappin, P. and Jullien, M. (2004). Changes in endogenous abscisic acid levels during dormancy release and maintenance of mature seeds: studies with the Cape Verde islands ecotype, the dormant model of *Arabidopsis thaliana*. *Planta* 219(3), 479-488. http://dx.doi.org. /https://doi.org/10.1007/s00425-004-1251-4.

Arc, E., Galland, M., Cueff, G., Godin, B., Lounifi, I., Job, D. and Rajjou, L. (2011). Reboot the system thanks to protein post-translational modifications and proteome diversity: how quiescent seeds restart their metabolism to prepare seedling establishment. *Proteomics* 11(9), 1606–1618. https://doi-org./10.1002/pmic.201000641.

Argyris, J., Dahal, P., Hayashi, E., Still, D. W. and Bradford, K. J. (2008). Genetic variation for lettuce seed thermoinhibition is associated with temperature-sensitive expression of abscisic acid, gibberellin, and ethylene biosynthesis, metabolism, and response genes. *Plant Physiol.* 148(2), 926–947. https://doi.org/10.1104/pp.108.125807.

Ashikawa, I., Abe, F. and Nakamura, S. (2014). A transgenic approach to controlling wheat seed dormancy level by using Triticeae DOG1-like genes. *Transgen. Res.* 23, 621–629.

Auge, G. A., Blair, L. K., Karediya, A. and Donohue, K. (2018). The autonomous flowering-time pathway pleiotropically regulates seed germination in *Arabidopsis thaliana*. *Ann. Bot.* 121(1), 183–191. https://doi.org/10.1093/aob/mcx132.

Bai, B., Peviani, A., van der Horst, S., Gamm, M., Snel, B., Bentsink, L. and Hanson, J. (2017). Extensive translational regulation during seed germination revealed by polysomal profiling. *New Phytol.* 214(1), 233–244.

Ballesteros, D., Pritchard, H. W. and Walters, C. (2020). Dry architecture: towards the understanding of the variation of longevity in desiccation-tolerant germplasm. *Seed Sci. Res.* 30(2), 142–155.

Ballesteros, D. and Walters, C. (2019). Solid-state biology and seed longevity: a mechanical analysis of glasses in pea and soybean embryonic axes. *Front. Plant Sci.* 10, 920. https://doi.org/10.3389/fpls.2019.00920.

Basbouss-Serhal, I., Soubigou-Taconnat, L., Bailly, C. and Leymarie, J. (2015). Germination potential of dormant and nondormant Arabidopsis seeds is driven by distinct recruitment of messenger RNAs to polysomes. *Plant Physiol.* 168(3), 1049–1065. https://doi-org.ezproxy.library.wur.nl/10.1104/pp.15.00510.

Baskin, J. M. and Baskin, C. C. (2004). A classification system for seed dormancy. *Seed Sci. Res.* 14(1), 1–16.

Baskin, C. C. and Baskin, J. M. (2014). *Seeds: Ecology, Biogeography, and Evolution of Dormancy and Germination* (2nd edn.). Academic Press/Elsevier, San Diego, CA.

Baskin, J. M., Baskin, C. C. and Li, X. (2000). Taxonomy, anatomy and evolution of physical dormancy in seeds. *Plant Spec. Biol.* 15(2), 139–152.

Bassel, G. W., Lan, H., Glaab, E., Gibbs, D. J., Gerjets, T., Krasnogor, N., Bonner, A. L., Holdsworth, M. J. and Provart, N. J. (2011). Genome-wide network model capturing seed germination reveals coordinated regulation of plant cellular phase transitions. *Proc. Natl. Acad. Sci. U. S. A.* 108 (23), 9709–9714.

Bazin, J., Langlade, N., Vincourt, P., Arribat, S., Balzergue, S., El-Maarouf-Bouteau, H. and Bailly, C. (2011). Targeted mRNA oxidation regulates sunflower seed dormancy alleviation during dry after-ripening. *Plant Cell* 23(6), 2196–2208.

Belmonte, M. F., Kirkbridea, R. C., Stone, S. L., Pelletier, J. M., Buib, A. Q., Yeung, E. C., Hashimoto, M., Feia, J., Harada, C. M., Munoza, M. D., Le, B. H., Drews, G. N., Brady, S. M., Goldberg, R. B. and Harada, J. J. (2013). Comprehensive developmental profiles of gene activity in regions and subregions of the Arabidopsis seed. *Proc. Natl. Acad. Sci. U. S. A.* 110(5), E435–E444. Available at: www.pnas.org/cgi/doi/10.1073/pnas.1222061110.

Belott, C., Janis, B. and Menze, M. A. (2020). Liquid-liquid phase separation promotes animal desiccation tolerance. *Proc. Natl. Acad. Sci. U. S. A.* 117(44), 27676–27684.

Bentsink, L., Hanson, J., Hanhart, C. J., Blankestijn-de Vries, H., Coltrane, C., Keizer, P., El-Lithy, M., Alonso-Blanco, C., de Andrés, M. T., Reymond, M., van Eeuwijk, F., Smeekens, S. and Koornneef, M. (2010). Natural variation for seed dormancy in

Arabidopsis is regulated by additive genetic and molecular pathways. *Proc. Natl. Acad. Sci. U. S. A.* 107(9), 4264–4269.

Bentsink, L., Jowett, J., Hanhart, C. J. and Koornneef, M (2006). Cloning of DOG1, a quantitative trait locus controlling seed dormancy in Arabidopsis. *Proc. Natl. Acad. Sci. U. S. A.* 103(45), 17042–17047.

Bethke, P. C., Libourel, I. G., Reinöhl, V. and Jones, R. L. (2006). Sodium nitroprusside, cyanide, nitrite, and nitrate break Arabidopsis seed dormancy in a nitric oxide-dependent manner. *Planta* 223(4), 805–812.

Bewley, J. D., Bradford, K. J., Hilhorst, H. W. M. and Nonogaki, H. (2013). *Seeds. Physiology of Development, Germination and Dormancy* (3rd edn.). Springer, New York.

Bizouerne, E., Buitink, J., Vu, B. L., Vu, J. L., Esteban, E., Pasha, A., Provart, N., Verdier, J. and Leprince, O. (2021). Gene co-expression analysis of tomato seed maturation reveals tissue-specific regulatory networks and hubs associated with the acquisition of desiccation tolerance and seed vigour. *BMC Plant Biol.* 21(1), 124. https://doi.org/10.1186/s12870-021-02889-8.

Buijs, G., Vogelzang, A., Nijveen, H. and Bentsink, L. (2020). Dormancy cycling: translation-related transcripts are the main difference between dormant and non-dormant seeds in the field. *Plant J.* 102(2), 327–339. https://doi.org/10.1111/tpj.14626.

Carbonero, P., Iglesias-Fernández, R. and Vicente-Carbajosa, J. (2017). The AFL subfamily of B3 transcription factors: evolution and function in angiosperm seeds. *J. Exp. Bot.* 68(4), 871–880. https://doi:10.1093/jxb/erw458.

Carniel, F. C., Fernandez-Marín, B., Arc, E., Craighero, T., Laza, J. M., Incerti, G., Tretiach, M. and Kranner, I. (2021). How dry is dry? molecular mobility in relation to thallus water content in a lichen. *J. Exp. Bot.* 72(5), 1576–1588. https://doi-org.ezproxy.library.wur.nl/10.1093/jxb/eraa521.

Carrera-Castaño, G., Calleja-Cabrera, J., Pernas, M., Gómez, L. and Oñate-Sánchez, L. (2020). An updated overview on the regulation of seed germination. *Plants (Basel)* 9(6), 703. https://doi:10.3390/plants9060703.

Chahtane, H., Kim, W. and Lopez-Molina, L. (2017). Primary seed dormancy: a temporally multilayered riddle waiting to be unlocked. *J. Exp. Bot.* 68(4), 857–869.

Chai, M., Zhou, C., Molina, I., Fu, C., Nakashima, J., Li, G., Zhang, W., Park, J., Tang, Y., Jiang, Q. and Wang, Z. Y. (2016). A class II KNOX gene, KNOX4, controls seed physical dormancy. *Proc. Natl. Acad. Sci. U. S. A.* 113(25), 6997–7002.

Chen, F., Dahal, P. and Bradford, K. J. (2001). Two tomato expansin genes show divergent expression and localization in embryos during seed development and germination. *Plant Physiol.* 127, 929–936.

Corbineau, F., Xia, Q., Bailly, C. and El-Maarouf-Bouteau, H. (2014). Ethylene, a key factor in the regulation of seed dormancy. *Front. Plant Sci.* 5, 539. https://doi.org/10.3389/fpls.2014.00539.

Dalling, J. W., Davis, A. S., Schutte, B. J. and Arnold, A. E. (2011). Seed survival in soil: interacting effects of predation, dormancy and the soil microbial community. *J. Ecol.* 99(1), 89–95.

Dekkers, B. J. W., Pearce, S., van Bolderen-Veldkamp, R. P., Marshall, A., Widera, P., Gilbert, J., Drost, H. G., Bassel, G. W., Müller, K., King, J. R., Wood, A. T., Grosse, I., Quint, M., Krasnogor, N., Leubner-Metzger, G., Holdsworth, M. J. and Bentsink, L. (2013). Transcriptional dynamics of two seed compartments with opposing roles in Arabidopsis seed germination. *Plant Physiol.* 163(1), 205–215.

Dure, L. and Waters, L. (1965). Long-lived messenger RNA: evidence from cotton seed germination. *Science* 147(3656), 410–412.

Finch-Savage, W. E., Cadman, C. S. C., Toorop, P. E., Lynn, J. R. and Hilhorst, H. W. M. (2007). Seed dormancy release in Arabidopsis Cvi by dry after-ripening, low temperature, nitrate and light shows common quantitative patterns of gene expression directed by environment specific sensing. *Plant J.* 51(1), 60–78.

Finch-Savage, W. E. and Footitt, S. (2012). To germinate or not to germinate: a question of dormancy relief not germination stimulation. *Seed Sci Res* 22, 243–248.

Finch-Savage, W. E. and Footitt, S. (2017). Seed dormancy cycling and the regulation of dormancy mechanisms to time germination in variable field environments. *J. Exp. Bot.* 68(4), 843–856. https://doi:10.1093/jxb/erw477.

Finch-Savage, W. E. and Leubner-Metzger, G. (2006). Seed dormancy and the control of germination. *New Phytol.* 171(3), 501–523.

Fleming, M. B., Patterson, E. L., Reeves, P. A., Richards, C. M., Gaines, T. A. and Walters, C. (2018). Exploring the fate of mRNA in aging seeds: protection, destruction, or slow decay? *J. Exp. Bot.* 69(18), 4309–4321. https://doi.org/10.1093/jxb/ery215.

Footitt, S., Walley, P. G., Lynn, J. R., Hambidge, A. J., Penfield, S. and Finch-Savage, W. E. (2019). Trait analysis reveals DOG1 determines initial depth of seed dormancy, but not changes during dormancy cycling that result in seedling emergence timing. *New Phytol.* 225(5), 2035–2047. https://doi.org/10.1111/nph.16081.

Gama-Arachchige, N. S., Baskin, J. M., Geneve, R. L. and Baskin, C. C. (2013). Identification and characterization of ten new water gaps in seeds and fruits with physical dormancy and classification of water-gap complexes. *Ann. Bot.* 112(1), 69–84. https://doi.org/10.1093/aob/mct094.

Gao, F., Rampitsch, C., Chitnis, V. R., Humphreys, G. D., Jordan, M. C. and Ayele, B. T. (2013). Integrated analysis of seed proteome and mRNA oxidation reveals distinct post-transcriptional features regulating dormancy in wheat (*Triticum aestivum* L.). *Plant Biotechnol. J.* 11(8), 921–932.

Graeber, K., Linkies, A., Müller, K., Wunchova, A., Rott, A. and Leubner-Metzger, G. (2010). Cross-species approaches to seed dormancy and germination: conservation and biodiversity of ABA-regulated mechanisms and the Brassicaceae *DOG1* genes. *Plant Mol. Biol.* 73(1–2), 67–87.

Graeber, K. A. I., Nakabayashi, K., Miatton, E., Leubner-Metzger, G. and Soppe, W. J. (2012). Molecular mechanisms of seed dormancy. *Plant Cell Environ.* 35(10), 1769–1786. https://doi.org/10.1111/j.1365-3040.2012.02542.x.

Hilhorst, H. W. M. (1995). A critical update on seed dormancy. I: primary dormancy. *Seed Sci. Res.* 5(2), 61–73.

Hilhorst, H. W. M. (1998). The regulation of secondary dormancy: the membrane hypothesis revisited. *Seed Sci. Res.* 8(2), 77–90.

Hilhorst, H. W. M. (2007). Definition and hypotheses of seed dormancy. In: Bradford, K. J. and Nonogaki, H. (Eds) *Seed Development, Dormancy and Germination*. Annual Plant Reviews Vol. 27. Blackwell Publishing, Sheffield, UK, pp. 50–71.

Hilhorst, H. W. M. and Downie, B. (1996). Primary dormancy in tomato (Lycopersicon esculentum cv. Moneymaker): studies with the sitiens mutant. *J. Exp. Bot.* 47(1), 89–97.

Hubbard, K. E., Nishimura, N., Hitomi, K., Getzoff, E. D. and Schroeder, J. I. (2010). Early abscisic acid signal transduction mechanisms: newly discovered components and newly emerging questions. *Genes Dev.* 24(16), 1695–1708.

Hudson, A. R., Ayre, D. J. and Ooi, M. K. J. (2015). Physical dormancy in a changing climate. *Seed Sci. Res.* 25(2), 66–81. https://doi.org/10.1017/S0960258514000403.

Huo, H., Wei, S. and Bradford, K. J. (2016). Delay of Germination 1 (DOG1) regulates both seed dormancy and flowering through microRNA pathway. *Proc. Natl. Acad. Sci. U. S. A.* 113(15), E2199–E2206.

Jang, S. J., Sato, M., Sato, K., Jitsuyama, Y., Fujino, K., Mori, H., Takahashi, R., Benitez, E. R., Liu, B., Yamada, T. and Abe, J. (2015). A single-nucleotide polymorphism in an endo-1,4-β-glucanase gene controls seed coat permeability in soybean. *PLoS ONE* 10(6), e0128527.

Kanno, Y., Jikumaru, Y., Hanada, A., Nambara, E., Abrams. S. R., Kamiya, Y. and Seo, M. (2010). Comprehensive hormone profiling in developing Arabidopsis seeds: examination of the site of ABA biosynthesis, ABA transport and hormone interactions. *Plant Cell Physiol.* 51(12), 1988–2001.

Law, S. R., Narsai, R., Taylor, N. L., Delannoy, E., Carrie, C., Giraud, E., Millar, A. H., Small, I. and Whelan, J. (2012). Nucleotide and RNA metabolism prime translational initiation in the earliest events of mitochondrial biogenesis during Arabidopsis germination. *Plant Physiol.* 158(4), 1610–1627.

Layat, E., Leymarie, J., El-Maarouf-Bouteau, H., Caius, J., Langlade, N. and Bailly, C. (2014). Translatome profiling in dormant and nondormant sunflower (*Helianthus annuus*) seeds highlights post-transcriptional regulation of germination. *New Phytol.* 204(4), 864–872.

Lepiniec, L., Devic, M., Roscoe, T. J., Bouyer, D., Zhou, D. X., Boulard, C., Baud, S. and Dubreucq, B. (2018). Molecular and epigenetic regulations and functions of the LAFL transcriptional regulators that control seed development. *Plant Reprod.* 31(3), 291–307.

Lin, J. Y., Le, B. H., Chen, M., Henry, K. F., Hur, J., Hsieh, T. F., Chen, P. Y., Pelletier, J. M., Pellegrini, M., Fischer, R. L., Harada, J. J. and Goldberg, R. B. (2017). Similarity between soybean and Arabidopsis seed methylomes and loss of non-CG methylation does not affect seed development. *Proc. Natl. Acad. Sci. U. S. A.* 114(45), E9730–E9739.

Liu, X., Zhang, H., Zhao, Y., Feng, Z., Li, Q., Yang, H. Q., Luan, S., Li, J. and He, Z. H. (2013). Auxin controls seed dormancy through stimulation of abscisic acid signaling by inducing ARF-mediated ABI3 activation in Arabidopsis. *Proc. Natl. Ac. Sci. U.S.A.* 110(38), 15485–15490. https://doi.org/10.1073/pnas.1304651110.

Liu, Y., Shi, L., Ye, N., Liu, R., Jia, W. and Zhang, J. (2009). Nitric oxide-induced rapid decrease of abscisic acid concentration is required in breaking seed dormancy in Arabidopsis. *New Phytol.* 183(4), 1030–1042.

Livne, S., Lor, V. S., Nir, I., Eliaz, N., Aharoni, A., Olszewski, N. E., Eshed, Y. and Weiss, D. (2015). Uncovering DELLA-independent gibberellin responses by characterizing new tomato *procera* mutants. *Plant Cell* 27(6), 1579–1594.

Luckwill, L. C. (1953). Growth-inhibiting and growth-promoting substances in relation to the dormancy and after-ripening of apple seeds. *J. Hortic. Sci.* 27(1), 53–67.

Mapes, G., Rothwell, G. W. and Haworth, M. T. (1989). Evolution of seed dormancy. *Nature* 337(6208), 645–646.

Martínez-Andújar, C., Pluskota, W. E., Bassel, G. W., Asahina, M., Pupel, P., Nguyen, T. T., Takeda-Kamiya, N., Toubiana, D., Bai, B., Górecki, R. J., Fait, A., Yamaguchi, S. and Nonogaki, H. (2012). Mechanisms of hormonal regulation of endosperm cap-specific gene expression in tomato seeds. *Plant J.* 71(4), 575–586.

Meinke, D. W., Franzmann, L. H., Nickle, T. C. and Yeung, E. C. (1994). Leafy cotyledon mutants of Arabidopsis. *Plant Cell* 6(8), 1049–1064.

Mitchell, J., Johnston, I. G. and Bassel, G. W. (2017). Variability in seeds: biological, ecological, and agricultural implications. *J. Exp. Bot.* 68(4), 809–817. https://doi-org.ezproxy.library.wur.nl/10.1093/jxb/erw397.

Mönke, G., Seifert, M., Keilwagen, J., Mohr, M., Grosse, I., Hähnel, U., Junker, A., Weisshaar, B., Conrad, U., Baümlein, H. and Altschmied, L. (2012). Toward the identification and regulation of the Arabidopsis thaliana ABI3 regulon. *Nucleic Acids Res.* 40(17), 8240–8254.

Munz, E., Rolletschek, H., Oeltze-Jafra, S., Fuchs, J., Guendel, A., Neuberger, T., Ortleb, S., Jakob, P. M. and Borisjuk, L. (2017). A functional imaging study of germinating oilseed rape seed. *New Phytol.* 216(4), 1181–1190. https://doi.org/10.1111/nph.14736.

Nakabayashi, K., Okamoto, M., Koshiba, T., Kamiya, Y. and Nambara, E. (2005). Genome-wide profiling of stored mRNA in *Arabidopsis thaliana* seed germination: epigenetic and genetic regulation of transcription in seed. *Plant J.* 41(5), 697–709.

Nikolaeva, M. G. (1999). Patterns of seed dormancy and germination as related to plant phylogeny and ecological and geographical conditions of their habitat. *Rus. J. Plant Physiol.* 46, 369–373.

Nonogaki, H. (2018). Seed germination and dormancy: the classic story, new puzzles, and evolution. *J. Integr. Plant Biol.* 61(5), 541–563.

Nonogaki, H., Chen, F. and Bradford, K. J. (2007). Mechanisms and genes involved in germination *sensu stricto*. In: Bradford, K. J. and Nonogaki, H. (Eds) *Seed Development, Dormancy and Germination*. Annual Plant Reviews Vol. 27. Blackwell Publishing, Sheffield, UK, pp. 264–304.

Nurse, P. (2008). Life, logic and information. *Nature* 454(7203), 424–426.

Oracz, K., El-Maarouf Bouteau, H., Farrant, J. M., Cooper, K., Belghazi, M., Job, C., Job, D., Corbineau, F. and Bailly, C. (2007). ROS production and protein oxidation as a novel mechanism for seed dormancy alleviation. *Plant J.* 50(3), 452–465.

Pelletier, J. M., Kwong, R. W., Park, S., Le, B. H., Baden, R., Cagliari, A., Hashimoto, M., Munoz, M. D., Fischer, R. L., Goldberg, R. B. and Haradaa, J. J. (2017). LEC1 sequentially regulates the transcription of genes involved in diverse developmental processes during seed development. *Proc. Natl. Acad. Sci. U. S. A.* 114(32), E6710–E6719.

Penfield, S. (2017). Seed dormancy and germination. *Curr. Biol.* 27, R853–R855.

Probert, R. J. (2000). The role of temperature in the regulation of seed dormancy and germination. In: Fenner, M. (Ed.) *Seeds: The Ecology of Regeneration in Plant Communities*. CAB International, Wallingford, Oxon, pp. 261–292.

Ritland, K. (1983). The joint evolution of seed dormancy and flowering time in annual plants living in variable environments. *Theor. Popul. Biol.* 24(3), 213–243.

Sajeev, N., Bai, B. and Bentsink, L. (2019). Seeds: a unique system to study translational regulation. *Trends Plant Sci.* 24(6), 487–495. https://doi.org/10.1016/j.tplants.2019.03.011.

Sánchez-Montesino, R., Bouza-Morcillo, L., Marquez, J., Ghita, M., Duran-Nebreda, S., Gómez, L., Holdsworth, M. J., Bassel, G. and Oñate-Sánchez, L. (2019). A regulatory module controlling GA-mediated endosperm cell expansion is critical for seed germination in Arabidopsis. *Mol. Plant* 12(1), 71–85.

Sano, N., Permana, H., Kumada, R., Shinozaki, Y., Tanabata, T., Yamada, T., Hirasawa, T. and Kanekatsu, M. (2012). Proteomic analysis of embryonic proteins synthesized from long-lived mRNAs during germination of rice seeds. *Plant Cell Physiol.* 53(4), 687–698.

Sano, N., Rajjou, L. and North, H. M. (2020). Lost in translation: physiological roles of stored mRNAs in seed germination. *Plants* 9(3), 347. https://doi:10.3390/plants9030347.

Serin, E. A., Nijveen, H., Hilhorst, H. W. and Ligterink, W. (2016). Learning from co-expression network: possibilities and challenges. *Front. Plant Sci.* 7, 444. https://doi:10.3389/fpls.2016.00444.

Silva, A. T., Ligterink, W. and Hilhorst, H. W. M. (2017). Metabolite profiling reveals two metabolic shifts during the seed-to-seedling transition in Arabidopsis thaliana. *Plant Mol. Biol.* 95(4–5), 481–496. https://doi.org/10.1007/s11103-017-0665-x.

Silva, A. T., Ribone, P. A., Chan, R. L., Ligterink, W. and Hilhorst, H. W. M. (2016). A predictive coexpression network identifies novel genes controlling the seed-to-seedling phase transition in *Arabidopsis thaliana*. *Plant Physiol.* 170(4), 2218–2231.

Sliwinska, E., Bassel, G. W. and Bewley, J. D. (2009). Germination of Arabidopsis thaliana seeds is not completed as a result of elongation of the radicle but of the adjacent transition zone and lower hypocotyl. *J. Exp. Bot.* 60(12), 3587–3594.

Soppe, W. J. J. and Bentsink, L. (2020). Seed dormancy back on track; its definition and regulation by DOG1. *New Phytol.* 228(3), 816–819.

Sperber, K., Steinbrecher, T., Graeber, K., Scherer, G., Clausing, S., Wiegand, N., Hourston, J. E., Kurre, R., Leubner-Metzger, G. and Mummenhoff, K. (2017). Fruit fracture biomechanics and the release of *Lepidium didymum* pericarp-imposed mechanical dormancy by fungi. *Nat. Commun.* 8(1), 1868. https://doi.org/10.1038/s41467-017-02051-9.

Steinbrecher, T. and Leubner-Metzger, G. (2017). The biomechanics of seed germination. *J. Exp. Bot.* 68(4), 765–783. https://doi:10.1093/jxb/erw428.

Still, D. W. and Bradford, K. J. (1997). Endo-[beta]-mannanase activity from individual tomato endosperm caps and radicle tips in relation to germination rates. *Plant Physiol.* 113(1), 21–29. https://doi:10.1104/pp.113.1.21.

Sun, L., Miao, Z., Cai, C., Zhang, D., Zhao, M., Wu, Y., Zhang, X., Swarm, S. A., Zhou, L., Zhang, Z. J., Nelson, R. L. and Ma, J. (2015). GmHs1-1, encoding a calcineurin-like protein, controls hard-seededness in soybean. *Nat. Genet.* 47(8), 939–943.

Tao, Z., Shen, L., Gu, X., Wang, Y., Yu, H. and He, Y. (2017). Embryonic epigenetic reprogramming by a pioneer transcription factor in plants. *Nature* 551(7678), 124–128. https://doi:10.1038/nature24300.

Tian, R., Wang, F., Zheng, Q., Niza, V. M. A. G. E., Downie, A. B. and Perry, S. E. (2020). Direct and indirect targets of the Arabidopsis seed transcription factor abscisic acid INSENSITIVE3. *Plant J.* 103(5), 1679–1694.

To, A., Valon, C., Savino, G., Guilleminot, J., Devic, M., Giraudat, J. and Parcy, F. (2006). A network of local and redundant gene regulation governs Arabidopsis seed maturation. *Plant Cell* 18(7), 1642–1651. https://doi.org/10.1105/tpc.105.039925.

Topham, A. T., Taylor, R. E., Yan, D., Nambara, E., Johnston, I. G. and Bassela, G. W. (2017). Temperature variability is integrated by a spatially embedded decision-making center to break dormancy in Arabidopsis seeds. *Proc. Natl. Acad. Sci. U. S. A.* 114(25), 6629–6634. Available at: www.pnas.org/cgi/doi/10.1073/pnas.1704745114.

Trewavas, A. (2012). Information, noise and communication: thresholds as controlling elements in development. In: Witzany, G. and Baluska, F. (Eds) *Biocommunication of Plants*. Springer, Berlin, pp. 11–35.

Tuan, P. A., Kumar, R., Rehal, P. K., Toora, P. K. and Ayele, B. T. (2018). Molecular mechanisms underlying abscisic acid/gibberellin balance in the control of seed dormancy and germination in cereals. *Front. Plant Sci.* 9, 668. https://doi.org/10.3389/fpls.2018.00668.

Verdier, J., Lalanne, D., Pelletier, S., Torres-Jerez, I., Righetti, K., Bandyopadhyay, K., Leprince, O., Chatelain, E., Ly Vu, B. L., Gouzy, J., Gamas, P., Udvardi, M. K. and Buitink, J. (2013). A regulatory network-based approach dissects late maturation processes related to the acquisition of desiccation tolerance and longevity of *Medicago truncatula* seeds. *Plant Physiol.* 163(2), 757–774.

Willis, C. G., Baskin, C. C., Baskin, J. M., Auld, J. R., Venable, D. L., Cavender-Bares, J., Donohue, K., Rubio de Casas, R. and NESCent Germination Working Group (2014). The evolution of seed dormancy: environmental cues, evolutionary hubs, and diversification of the seed plants. *New Phytol.* 203(1), 300–309. https://doi.10.1111/nph.12782.

Xia, Q., Ponnaiah, M., Thanikathansubramanian, K., Corbineau, F., Bailly, C., Nambara, E., Meimoun, P. and El-Maarouf-Bouteau, H. (2019). Re-localization of hormone effectors is associated with dormancy alleviation by temperature and after-ripening in sunflower seeds. *Sci. Rep.* 9(1), 4861. https://doi-org./10.1038/s41598-019-40494-w.

Xu, H., Lantzouni, O., Bruggink, T., Benjamins, R., Lanfermeijer, F., Denby, K., Schwechheimer, C. and Bassel, G. W. (2020). A molecular signal integration network underpinning Arabidopsis seed germination. *Curr. Biol.* 30(19), 3703–3712.e4. https://doi.org/10.1016/j.cub.2020.07.012.

Xu, X., Legay, S., Sergeant, K., Zorzan, S., Leclercq, C. C., Charton, S., Giarola, V., Liu, X., Challabathula, D., Renaut, J., Hausman, J. F., Bartels, D. and Guerriero, G. (2021). Molecular insights into plant desiccation tolerance: transcriptomics, proteomics and targeted metabolite profiling in *Craterostigma plantagineum*. *Plant J.* 107(2), 377–398. https://doi.org/10.1111/tpj.15294.

Ytterberg, A. J. and Jensen, O. N. (2010). Modification-specific proteomics in plant biology. *J. Proteomics* 73(11), 2249–2266.

Zinsmeister, J., Berriri, S., Basso, D. P., Ly-Vu, B., Dang, T. T., Lalanne, D., da Silva, E. A. A., Leprince, O. and Buitink, J. (2020). The seed-specific heat shock factor A9 regulates the depth of dormancy in *Medicago truncatula* seeds via ABA signalling. *Plant Cell Environ.* 43(10), 2508–2522.

Chapter 2

Understanding the effects of maternal environment in controlling seed dormancy

Roberto L. Benech-Arnold, Cátedra de Cultivos Industriales, IFEVA-Facultad de Agronomía (CONICET/UBA), Argentina; M. Verónica Rodríguez, Cátedra de Fisiología Vegetal, IFEVA-Facultad de Agronomía (CONICET/UBA), Argentina; Diego Batlla, Cátedra de Cerealicultura, IFEVA-Facultad de Agronomía (CONICET/UBA), Argentina; and Rocío Fernández Farnocchia, Cátedra de Cultivos Industriales, IFEVA-Facultad de Agronomía (CONICET/UBA), Argentina

1 Introduction

Variation in an individual's phenotype may be determined not only by the genotype and environment of that individual but also by maternal effects, that is, the contribution of the maternal parent to the phenotype of its offspring beyond the equal chromosomal contribution expected from each parent (Roach and Wulff, 1987). Maternal effects in plants can be classified into three different categories: (i) cytoplasmic genetic, (ii) endosperm nuclear and (iii) maternal phenotypic (Roach and Wulff, 1987).

http://dx.doi.org/10.19103/AS.2022.0105.04

Cytoplasmic genetic maternal effects result from the fact that, in eukaryotes, some plastids and mitochondria have DNA (i.e. some genes had been retained in these organelles after endo-symbiosis, instead of being transferred to the nucleus) which is directly transferred from the maternal plant to the progeny during ovule formation and development. This type of maternal effect contributes to heritable variation in numerous traits in plants.

Double fertilization in angiosperms derives in a 3N endosperm with two nuclei from the maternal and only one from the paternal parent. This means more doses of maternal than paternal genes which leads to the second category of maternal effects: endosperm nuclear effects. Beyond the fact that the endosperm is the site of action of enzymes important for seedling growth until autotrophy is achieved, the female gametophyte, which derives also in endosperm after double fertilization, is where maternal imprinting processes (i.e. a form of epigenetic regulation through which methylation and de-methylation of target DNA sequences are determined in the central cell of the female gametophyte) take place. Maternal imprinting occurs during pre- and post-fertilization stages and affects methylation patterns and, in consequence, expression patterns of genes in the developing endosperm (Kohler et al., 2012). But in turn, endosperm imprinting regulates gene methylation and expression in the embryo through the production of nt siRNA which are translocated to the embryo, thus influencing the phenotype (Kohler et al., 2012).

Maternal phenotypic effects result from the environment and/or the genotype of the mother plant (Roach and Wulff, 1987). The tissues surrounding the endosperm and the embryo are maternal since they derive from the integuments of the ovule (the testa) and from the wall of the ovary (the pericarp). Hence, the influence of maternal phenotypic effects on traits that affect seed behavior may occur via structure (Karssen, 1970; Pourrat and Jacques, 1975; Ceccato et al., 2015; Penfield and MacGregor, 2017) or physiology (Kendall et al., 2011).

Maternal effects have consequences at many levels. They are considered important sources of error in selection studies because of their non-Mendelian nature which reduces the precision of genetic work. However, it is only recently that the adaptive significance of maternal effects has started to be recognized. It has become evident that many maternal effects have been shaped by the action of natural selection to act as a mechanism for adaptive phenotypic response to environmental heterogeneity (Mousseau and Fox, 1998). But, in addition, there is evidence that understanding and even quantifying maternal effects can yield tools of decisive importance for phenotype prediction which can be used in agriculture.

Seed dormancy (i.e. an internal condition of the seed that impedes its germination under otherwise adequate hydric, thermal and gaseous conditions; Benech-Arnold et al., 2000) is one trait prone to be controlled by maternal effects belonging to any of the three different categories mentioned above. In this chapter, however, mostly maternal phenotypic effects on seed dormancy will be dealt with, unless stated otherwise. Seed dormancy is a trait with enormous importance both for plant adaptation to the environment and for agriculture. Indeed, it is well known the role of (physiological) dormancy in the seasonal adjustment of emergence and establishment of temperate species. On the other hand, the literature is abundant on examples of persistent dormancy that interferes with grain industrialization when it depends on seed germination (i.e. malting industry) or the seed industry, or the short-lived dormancy that makes crops prone to suffer pre-harvest sprouting (PHS).

This chapter discusses the effects of maternal environment in controlling seed dormancy. We will start by describing some dormancy response patterns to the different environmental factors to which the mother plant can be exposed during seed development. Second, we will go through the known physiological and molecular mechanisms underlying such response patterns. The chapter will end by illustrating with examples the ecological and agricultural implications of dormancy modulation by the maternal environment.

2 Dormancy response patterns to the effect of the maternal environment in seeds displaying physiological and physical dormancy

The effect of the maternal environment on germination has been studied in several species, and the significant influence of a wide range of environmental factors has been determined (reviewed in Roach and Wulff, 1987; Fenner, 1991; Gutterman, 2000). This section will give a brief but actualized overview of experimental studies that led to the identification of patterns in dormancy responses to the different factors that make the maternal environment.

Although limited because of the lack of characterization of individual causal factors, field experiments have yielded important evidence about the responsibility of the environment experienced by the seed while still on the mother plant, in the phenotypical variation in terms of dormancy between populations growing in contrasting habitats. These experiments consist basically of either (i) using staggered sowing dates to generate variation in the environment experienced by the mother plant during seed development (Rodríguez et al., 2001; Ceccato et al., 2015; Fernández Farnocchia et al., 2019; 2021) or (ii) experiments aimed to separate environmental from genetic effects by reciprocal transplanting material between locations. Through the latter, a strong positive relationship between latitude and dormancy was determined in *Daucus carota* (Lacey, 1984), while Dorne (1981) found that coat thickness and polyphenol content correlate positively with altitude in *Chenopodium bonus-henricus* and showed a positive correlation between these factors and dormancy. Field experiments also provided evidence of year-to-year variation in the dormancy of plant populations (Baskin and Baskin, 1975; Rodríguez et al., 2001) and within-year variations in some species with an extended flowering period (Okunsaya and Ungar, 1983; Venable and Levin, 1985).

The role of the maternal thermal environment during seed development on dormancy has been recognized since the seminal study by Koller (1962). A negative relationship has been found in several species between the temperature experienced by the mother plant during seed development and maturation and the level of primary dormancy they present at the time of dispersal (Sawhney et al., 1985; Peters, 1982; Rodríguez et al., 2001; Steadman et al., 2004; Kendall et al., 2011; Kendall and Penfield, 2012; Chen et al., 2014; MacGregor et al., 2015; Huang et al., 2018). In some cases, however, dormancy level has been found to increase with temperature during maturation: *Aegilops ovata* (Datta et al., 1972), *Syringa vulgaris* (Juntilla, 1973) and sunflower (Bodrone et al., 2017). In turn, when dormancy is imposed mechanically (i.e. physical dormancy), high temperatures during development appear to increase dormancy due to a thickening of the seed coat, as has been found in soybean (Keigley and Mullen, 1986). As will be referred to in further sections, some researchers have stressed the importance of the timing of a certain thermal regime within the seed development period, to exert a particular effect on dormancy (Rodríguez et al., 2001; Bodrone et al., 2017). Some studies show that changes in maternal temperature determine important variations in the ability of seeds to germinate under different water availability during their incubation (Edwards et al., 2016). Furthermore, some works show that the prevailing temperature during the vegetative phase would also affect the level of dormancy of the seeds in several species, such as Arabidopsis and oats (Thomas and Raper, 1975; Sawhney et al., 1985; Chen et al., 2014). In turn, effects of maternal temperature have been detected on the induction dynamics in secondary dormancy, especially in seeds of *A. thaliana*. In this species, seeds that matured at a lower temperature generally have a higher sensitivity to factors that promote induction in secondary dormancy than those seeds that experienced a higher temperature during their formation in the mother plant (Auge et al., 2015; Burghardt et al., 2016; Coughlan et al., 2017).

Daylength experienced during seed maturation is known to affect the dormancy of a vast number of species. As a general rule, the shorter the daylength, the lower the resultant dormancy (Wurzburger and Koller, 1976; Gutterman and Heydecker, 1973; Gutterman, 1974, 1978; Kigel et al., 1977; Munir et al., 2001; Imaizumi et al., 2017; Fernández Farnocchia et al., 2019; 2021). Long days are known to enhance dormancy by thickening the seed coat (Karssen, 1970; Gutterman and Heydecker, 1973; Gutterman, 1978; Fernández Farnocchia et al., 2021). In some species, however, short days can promote dormancy, for example, *Avena fatua* and *Avena sterilis* (Somody et al., 1984) and *Polygonum monspeliensis* (Gutterman, 1982).

The light environment is of paramount importance as a modulator of dormancy during seed development. In this regard, a positive relationship has been found between the level of dormancy of the seeds and the amount of radiation received by the mother plant (Sánchez et al., 1981). But, on the other hand, the R:FR (Red to Far Red) ratio to which the seeds are exposed during their development and maturation has been indicated also as a key factor in determining the light requirements to germinate in mature seeds and, consequently, their dormancy level. The latter was initially investigated by McCullough and Shropshire (1970) who found that the ability to germinate in the dark in *Arabidopsis thaliana* was higher in seeds obtained from environments rich in R light. Following this line of evidence, later studies revealed that these effects were the result of variations in R:FR incident on the seed during its formation in the mother plant (Cresswell and Grime, 1981; Schmitt and Wulff, 1993; Orozco-Segovia et al., 2000; Vayda et al., 2018). Based on these results, it had been widely accepted that the effect of the presence of a canopy would consist in increasing seed dormancy level through the reduction in the R:FR ratio of the light environment that the canopy produces. However, recent studies suggest that the influence of a canopy would include the participation of other factors beyond the R:FR ratio alteration (Leverett et al., 2016; Vayda et al., 2018). Supporting this conclusion, a recent work by Fernández Farnocchia (2020) which discriminated the effects of changes in R:FR from those of light intensity (PAR) on the dormancy level of *Polygonum aviculare* fruits determined an antagonistic participation of these factors: low R:FR increased dormancy and light requirements for germination of *P. aviculare* fruits as has been reported for many species (Hayes and Klein, 1974; Cresswell and Grime, 1981; Orozco-Segovia et al., 1993; Benvenuti and Macchia, 1997; Hinsberg, 1998; Contreras et al., 2009), but a low PAR generated opposite effects, as has also been verified in numerous wild and cultivated species (Kigel et al., 1977; Sánchez et al., 1981; Jha et al., 2010). Since, contrary to expectations from a reduction of the R:FR ratio, the environment generated by a wheat canopy reduced seed dormancy. Fernández Farnocchia (2020) concluded that that light quantity, and not the R:FR (nor the temperature), modulates canopy effects on dormancy level in *P. aviculare* fruits. Experiments with other species are needed to determine whether this could be considered a general pattern of behavior.

Experimental evidence suggests that the direction in which dormancy is modified by water availability during seed development depends on the dormancy mechanism that is being influenced (Fenner, 1991). When dormancy is imposed mechanically by the presence of a thick coat, drought during seed development increases dormancy by thickening the seed coat. This was shown in soybean by Hill et al. (1986) who found that seeds from water-stressed plants had, as a percentage, more dry weight allocated to the seed coat than control seeds and presented a reduced permeability to water. In contrast, in species whose seeds display physiological dormancy, water stress during seed development normally results in reduced dormancy. Early work by Sánchez et al. (1981) showed that the reduced dormancy in *Datura ferox* seeds resulting from intermittent drought imposed during seed maturation was accompanied by a lower endogenous level of an unidentified germination inhibitor. Later on, Benech-Arnold et al. (1991) found that *Sorghum bicolor* seeds produced under water stress during development had low dormancy, and this was correlated with reduced embryo sensitivity to abscisic acid (ABA). Water stress was also found to reduce dormancy of *Sorghum halepense* seeds (Benech-Arnold et al., 1992) and of wheat and barley grains (Goldbach and Goldbach, 1977; Haeder and Beringer, 1981; Gualano and Benech-Arnold, 2009a).

Few workers have assessed the effect of nutrient availability to the mother plant on seed dormancy. In a pioneer study, Harrington (1960) carried out a wide survey to analyze the effect of reducing the provision of some macronutrients on seed dormancy of several vegetables. Dormancy was not significantly affected by the application of the different treatments on the different species, except for *Capsicum annun* growing under severe potassium deficiency, where sprouting of the immature seeds was observed in the developing fruits. This was presumed to be related to a failure in the synthesis of some germination inhibitors. In contrast, Haeder and Beringer (1981) found that wheat plants growing under low potassium levels produced seeds that during development reached a higher content of ABA than that measured in well-supplied plants, but no attempt was made to link this effect with seed dormancy. However, Benech-Arnold et al. (1995) found that potassium deprivation reduced *S. bicolor* seed dormancy through a marked reduction in embryo sensitivity to ABA and also through decreasing ABA endogenous content. Molybdenum deficiency in soils was found to induce pre-harvest sprouting (PHS) in wheat due to reduced grain dormancy (Cairns and Kritzinger, 1992). Molybdenum acts as a cofactor of the enzyme aldehyde oxidase (AO) which is involved in ABA biosynthesis (Mendel and Haensch, 2002). Therefore, reduced dormancy in wheat grains from the plant growing in molybdenum-deficient soils is due to ABA synthesis partial impairment. Nitrogen supply has been shown to reduce dormancy in seeds of a number of species and this effect will be treated in depth in further sections (Watson and Watson, 1982; Hilhorst and Karssen, 2000; Alboresi et al., 2005; Matakiadis et al., 2009; He et al., 2016; Longás et al., 2021). However, Gualano and Benech-Arnold (2009a) did not find differences in dormancy in barley crops grown in the field under contrasting nitrogen supply.

In contrast to maternal experience of abiotic environmental factors, maternal biotic stress (i.e. herbivory and competition) effects on seed dormancy have remained relatively unexplored. In this regard, Singh et al. (2017) found that maternal herbivory resulted in the accumulation of jasmonic acid-isoleucine and loss of dormancy in seeds of stressed Arabidopsis plants. The authors concluded that herbivory generated a jasmonic acid-dependent reduction in seed dormancy, mediated by alteration of gibberellic and ABA signaling, which can be regarded as a direct maternal effect, operating independently of transgenerational herbivore resistance priming.

In spite of the obvious relevance of all the referred studies, it is important to note that the role of each factor has mostly been analyzed individually but seldom the relative importance that each of them could have on the level of dormancy in a given species has been considered (Donohue, 2013). Similarly, it is not frequent to find studies in which the impact of these factors on the emergence, the population dynamics and the fitness of the plants has been assessed (Fenner, 1991; Donohue et al., 2005; Donohue, 2009; Kendall and Penfield, 2012). Further sections in this chapter will be devoted to discussing these possibilities as well as other practical implications.

3 Physiological and molecular bases underlying the modulation of dormancy by the environment experienced by the mother plant during reproductive development

How are environmental signals perceived by the mother plant and translated to a particular dormancy state in the progeny seed? Dormancy mechanisms controlling the germination process (and repressing embryo growth) can operate in the embryo itself but also on the surrounding structures of the seed or fruit; all these structures can potentially be targets for changes induced by the maternal environment that will end up affecting dormancy. The environment is expected to trigger biochemical and molecular changes that will in turn affect one or several steps of the germination process. These changes may relate to biochemical and physical properties of the seed coats (pericarp, testa) that are non-living in the mature seed, which in turn affect some physiological processes in the embryo (e.g. hormone metabolism or signaling). Also, the maternal environment could trigger direct physiological changes in the living endosperm and in the embryo (e.g. hormone metabolism, transport and/or signaling) that will affect the dormancy status of the seed. Therefore, there are multiple ways through which the maternal environment can potentially affect the dormancy of the progeny. Important progress toward the understanding of these mechanisms has been made in the last decade mostly in model plant species *Arabidopsis thaliana* (*Brassicaceae* family) regarding the most influential environmental cues on seed dormancy in this species (cold and nitrate). A few important advances made in other species will also be discussed in this section. Evidence obtained in Arabidopsis supports that deeper dormancy levels induced by cold temperatures and low nitrate levels experienced by the mother plant are the result of multiple changes involving the embryo, the endosperm, the seed coat and the fruit (silique) tissues. Figure 1 summarizes how different cues of the maternal environment can affect primary dormancy based on information presented in this section.

Sensing of environmental conditions prior to seed development, either in vegetative organs or in the already differentiated floral meristems, may involve specific dormancy signals that are then transmitted to the seeds or may also result from broader changes in physiology (e.g. altered source-sink relations, or carbon and nitrogen metabolism) that will later impact on seed dormancy. Chiang et al. (2009) proposed that germination and flowering are both fundamental developmental transitions that require precise environmental sensing and responses to multiple seasonal cues to accurately match developmental timing to appropriate seasonal conditions. Because both transitions respond to similar seasonal cues, they hypothesized that the genetic pathways of these two transitions would share common elements, with some differences in their regulation as both developmental transitions take place under different seasons. These authors proposed that regulators of floral transition may have a role as a molecular link between environmental conditions previous to seed development and traits manifested in the mature seed.

A few candidates acting as major regulators involved in the control of dormancy by temperature during seed development have been identified in Arabidopsis and wheat and belong to the phosphatidyl ethanolamine-binding protein (PEBP) family. This evolutionarily conserved family includes well-characterized members such as the 'florigen' FLOWERING-LOCUS (FT), TERMINAL-FLOWER1 (TFL1) and MOTHER-OF-FT-ANDTFL1 (MFT). Interestingly, cold temperatures affecting the mother plant before anthesis can induce deep-dormant states in Arabidopsis seeds, and this involves regulation of FT (Chen et al., 2014) and also FLOWERING LOCUS C (FLC), another canonical component of the flowering pathway (Chen and Penfield, 2018). Chen et al. (2014) first observed that cold temperatures promote proanthocyanidin (PA) accumulation in seeds, and this is mediated by decreasing FT protein levels that otherwise inhibit PA synthesis. Because FT inhibits PA synthesis in the maternal fruit tissues (carpels that develop into siliques), these authors proposed that cold-induced inhibition of FT in the silique results in increased synthesis and transport (by phloem) of PA from the silique to the developing seeds, where PA finally accumulate in the seed coat. Chen and Penfield (2018) further dissected the mechanism by which FT controls PA metabolism. Under warm temperatures during pre-anthesis, *FT* is expressed in maternal fruit tissues and causes epigenetic silencing of *FLC* gene (through the expression of antisense '*COOLAIR*' transcripts at the *FLC* locus). Similar to what happens with FT in

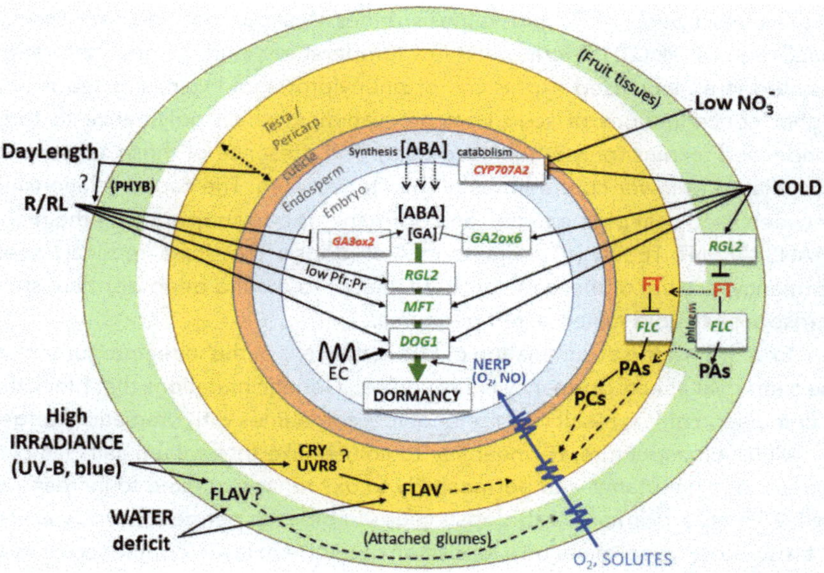

Figure 1 Summarizes current knowledge on how maternal environmental cues (nitrates, temperature, light and water) operate on seed components leading to changes in primary dormancy. Some of the major molecular targets are presented (up-regulated genes, green font in white box; down-regulated genes, red font). Phenylpropanoid compounds may accumulate in seed fruit tissues and/or glumes (depending on species) and in the seed coat. In Arabidopsis, proanthocyanidins (PAs) are synthesized first in the silique and then in the seed coat but are also transported by phloem; procyanidins (PCs or condensed tannins), also derived from PAs, are formed only in the seed coat. The outer walls of endosperm cells may develop a thick waxy cuticle (by deposition of suberin). All these compounds increase resistance to diffusion of gas and solutes when imbibed, and O_2 and NO levels are signaled by the N-END RULE PATHWAY (NERP) to affect ABA signal (hypoxia enhances ABA signal). Dormancy regulators within the zygote include RGL2, DOG1, MFT and are also targets of signaling in response to COLD (DOG1 and MFT) and probably by light cues perceived through phytochromes and integrated with a circadian oscillator (DOG1 is controlled by the evening complex, EC). Hormone metabolism genes are also environmental targets (GA2ox3, regulated by phytochrome-mediated signals and CYP707A2 repressed by cold or low nitrates). Abscisic acid (ABA) synthesis occurs mostly in the endosperm layer, from where it is transported to the embryo. ABA signaling in the embryo involves MFT, RGL2 and DOG1 and many other transcription factors not shown here (e.g. ABI5, ABI3 and ABI4). High irradiance and water deficit also appear to affect the accumulation of flavonoid compounds in the maternal structures/coats. Daylength has also been reported to impact on seed coat development and thickness, another source of variation for permeability to gases and solutes. The NERP is expected to act as a sensor for O_2 or NO levels in the embryo (either during development or after imbibition in the mature seed) and modulate ABA signaling pathway. For more information and citations refer to the main text.

leaves, the FT protein is likely transported from silique to seed where it also silences FLC and represses PA synthesis. On the contrary, cool temperatures repress FT and allow expression of FLC both in silique and seeds, resulting in the transcription of PA biosynthesis genes in seed coat and fruit tissues. MacGregor et al. (2015) found that low temperatures during seed maturation resulted in an increased expression of phenylpropanoid genes in seeds and higher concentrations of seed coat procyanidins, which polymerize to form condensed tannins (only in the seed coat). The presence of these substances was related to lower coat permeability to tetrazolium. The causal relationship between seed coat pigmentation and dormancy is also supported by the many TRANSPARENT TESTA (*tt*) mutants in Arabidopsis that have reduced seed dormancy; several of these mutants are unable to enter a deep-dormant state in response to cold maternal temperatures.

Exactly how the accumulation of PAs in the cells of the inner integument of the seed coat affects dormancy is still unclear. Genetic mutations that block the accumulation of PAs result in a *tt* phenotype, sometimes with dramatic changes in cellular ultrastructure. Permeability to solutes like tetrazolium salts (TTZ) is increased in *tt* mutants, and permeability to oxygen is expected to increase as well. Wild type, dormant Arabidopsis seeds imbibe, so permeability to water is not restrictive for germination. Deposition of a suberin layer (composed of lipid polyesters) in the seed coat also appears to have a similar role in many respects to that of seed coat tannins; both are required for coat-imposed dormancy and reduced permeability of the seed to TTZ and other dyes (Debeaujon et al., 2000; Fedi et al., 2017). Both the accumulation of suberin and tannin layers in the seed coat are important for dormancy imposition and have been shown to be targets of environmental factors during seed development (Fedi et al., 2017). These authors examined the germination response of WT and suberin mutants (*awe1-1*, *gpat5-1* and *tt4-1*) incubating seeds in low-oxygen atmospheres and found that germination at 25°C of the three highly permeable mutants was significantly enhanced compared with the wild type at 5% oxygen. This supports the idea that the presence of suberin in the seed coat limits the oxygen supply into the imbibed seed.

Alternatively, a direct effect of PAs in germination inhibition has been proposed. Jia et al. (2012) observed that soluble PA (extracted from black soybean hulls) inhibited Arabidopsis and *Brassica napus* seed germination; this was related to increased ABA levels by up-regulation of two *NCED* genes. However, it is still uncertain how endogenous PAs that accumulate in the inner integument and pigment strand affect ABA metabolism in the endosperm layer or in the embryo. Using a similar approach, Shah et al. (2018) worked with *Sapium sebiferum* seeds and observed that seed coat extracts containing soluble PA applied to isolated embryos inhibited their germination, and this was related to altered expression of ABA and GA metabolism and signaling genes, and also *DOG1* (see below). A positive association between PAs metabolism and ABA signaling is also supported by Gao et al. (2018), who characterized a tomato gene, *SlAN11*, encoding a WD40 transcriptional regulator that promotes PAs synthesis and expression of ABA signaling genes *SlABI3* and *SlABI5* during seed maturation.

A seed-specific, central regulator of seed dormancy and responsible for intraspecific variability for dormancy among Arabidopsis accessions is DOG1 (DELAY OF GERMINATION 1; Bentsink et al., 2006). In addition to the mentioned changes in seed coat composition, low temperature during seed development has been shown to increase the expression of *DOG1*. Both transcript and protein levels of *DOG1* increase by low temperatures during seed maturation leading to enhanced dormancy, and *dog1* null mutants are unable to respond to cold (Chiang et al., 2011; Kendall et al., 2011; Nakabayashi et al., 2012). Therefore, DOG1 protein has an important role in the translation of seed maturation temperature to dormancy levels (Chiang et al., 2011; Footitt et al., 2011; Kendall et al., 2011). Function of DOG1 in dormancy requires ABA, and *vice versa*. The ABA and DOG1 signaling pathways converge since DOG1 requires protein phosphatase 2C (PP2C) to control primary dormancy. DOG1 enhances ABA signaling by direct binding and suppressing a PP2C-type negative regulator of ABA signaling, AHG1. DOG1 suppresses the AHG1 action and increases ABA sensitivity to impose dormancy (Carrillo-Barral et al., 2020). DOG1 localizes in the cell nucleus, within vascular tissue in the embryo axis. ABA-deficient lines also cannot enter the strongly dormant states after low maturation temperatures. In parallel to modulation of DOG1 levels, low temperature during seed maturation strongly reduces ABA catabolism through transcriptional inhibition of *CYP707A2* encoding an ABA-degrading enzyme, resulting in increased ABA levels in dry seeds. As ABA metabolism occurs primarily in the living endosperm (from where it is secreted and transported into the embryo; Kang et al., 2015), changes in ABA levels as reported in previous works in seeds with varying dormancy (caused by different maternal environments) are expected to occur in this tissue.

In addition to high levels of DOG1 and ABA, a functional GA signaling pathway is also necessary for dormancy promotion by low temperature during seed development (Kendall et al., 2011). The *RGL2* gene encodes a DELLA protein that blocks GA responses when GA levels are low. The interaction between DOG1, RGL2 and FT during dormancy modulation by maternal temperature was investigated by Chen et al. (2014) in Arabidopsis using single and double mutants and measuring seed coat permeability to TTZ and the dormancy phenotypes. Analysis of *rgl2-1 ft-1* double mutants showed that the low seed coat permeability of *ft-1* is epistatic to the high permeability of *rgl2-1* seeds, showing that maternal RGL2 affects the seed coat via FT. However, dormancy assays showed that the low dormancy of *rgl2-1* null mutants was epistatic to the increased dormancy of *ft-1* mutants. These authors concluded that the zygotic activity of RGL2 is required for seed coat properties to affect dormancy, and RGL2 therefore affects dormancy by both maternal and zygotic mechanisms. In contrast, loss or gain of function of DOG1 did not affect seed coat permeability, or the permeability response to maternal temperature experienced, but like *rgl2-1*, *dog1* null mutants could not increase dormancy in response to low temperatures before bolting. Based on these results, Chen et al. (2014) concluded that DOG1 acts exclusively in the zygote to control dormancy. Altogether these results show that maternal-thermal environment induces changes in the maternal tissues (i.e. PA synthesis and seed coat permeability), in the endosperm (i.e. ABA metabolism and transport) and in the zygote (i.e. DOG1, RGL2 activity and ABA signaling) to control primary dormancy levels.

Another member of the PEBP family involved in the flowering pathway (as described above for Arabidopsis) also has a regulatory role in the control of dormancy in response to maternal temperatures. MFT was found to control dormancy in cultivated wheat. Natural alleles in this locus produce intraspecific variability for pre-harvest sprouting susceptibility but are also involved in sensing the maternal-thermal environment. Nakamura et al. (2011) provided evidence that *TaMFT* encodes an inhibitor of seed germination and has a key role in translating temperature effects during the establishment of seed dormancy in the developing grain. Exposure to cold (13°C vs 25°C) during seed development promoted *MFT* expression and seed dormancy. In contrast with *FT* in Arabidopsis (expressed in the fruit tissues but not in the seed), the role of MFT is zygotic, and *TaMFT* expression was observed in the embryo and coleorhiza. In barley, it has been proposed that the coleorhiza has a regulatory role in dormancy similar to the micropylar endosperm cap in tomato and Arabidopsis (Barrero et al., 2009); these authors observed that changes in ABA levels within the coleorhiza precede those in the radicle. Nevertheless, in wheat, it remains unknown if repression of germination by MFT induced by cold maternal temperatures involves changes in hormonal balance ABA/GA or increased ABA signaling (resulting in higher sensitivity) in the embryo or the coleorhiza. In Arabidopsis, *MFT* expression was up-regulated by low seed-maturation temperature (15°C), and expression levels correlated with dormancy release rate (after-ripening time to 50% germination). In this species, zygotic MFT controls seed dormancy through complex interactions with ABA signaling component ABI5, DELLA protein RGL2 and metabolism of OPDA, another germination inhibitor compound required for ABA action (Dave et al., 2016).

Regulation of seed dormancy by nitrate supply to the mother plant has also been investigated at the molecular level in Arabidopsis seeds. Nitrate accumulation in dry seeds negatively correlated with their ABA content (Matakiadis et al., 2009). These authors first observed that high nitrate levels provided to the mother plant (leading to high endogenous nitrate in the seed) produced a transcriptomic profile similar to that of stratified or after-ripened seeds. Regulation of nitrate-dependent gene expression involves *NIN-like protein 8* (*NLP8*) encoding a transcription factor, which is induced by high nitrate levels. One of the direct targets of NLP8 is *CYP707A2*. Therefore, high nitrate levels promote ABA catabolism resulting in low ABA content and reduced dormancy. Using a metabolomic and transcriptomic approach, He et al. (2016) compared seed phenotypes produced under different levels of nitrate, light and temperature. Low nitrogen and cold strongly promoted dormancy. These authors identified a strong overlapping effect for seeds matured at 15°C and those with low nitrate, based on the correlation of metabolites with phenotypes, fold change of metabolites, over-representation analysis of the transcriptome (with many N-metabolism related genes) and overlapping of up-/down-regulated genes. Based on these observations, the authors suggest that low temperature also affected nitrate transport and assimilation and the accumulation of amino acids and N-rich compounds in dry seeds. One of the genes up-regulated in plants exposed to 15°C and low N treatments after flowering was NITRATE TRANSPORTER 1.1 (NRT1.1). These authors discuss the possible role of NRT1.1 as interacting with several signaling pathways and propose that NRT1.1 may sense low temperatures, causing similar effects in nitrate-deficient environments. Another nitrate transporter NRT1.2 is also able to transport ABA (Kanno et al., 2012). Deeper understanding of the interactions between these nitrate transporters and hormonal pathways in seed development is still needed. Modulation of seed dormancy by nitrogen supply to the mother plant probably involves the endosperm layer where ABA synthesis takes place, and other changes in the zygote; differently from regulation by temperature, it does not appear to involve changes in seed coat characteristics.

Perception of the light environment in plants involves at least five types of photoreceptors. Blue and UV-A (320–500 nm) are absorbed by cryptochromes (CRYs), ZEITLUPE and phototropins; red/far-red light (600–750 nm) is sensed by phytochromes (PHY); and UV-B (280–320 nm) light is sensed by UVRESISTANCE LOCUS8 (UVR8) (reviewed by Yang et al., 2020). The role of PHY in the control of germination in photoblastic species has been known for a long time, but a role for PHYB in dormancy imposition in the developing seed, as well as any possible role of other photoreceptors, is much less clear. Mechanistic studies with Arabidopsis using mutants for light signaling components exhibit dormancy/germination phenotypes and alterations in levels of dormancy regulators. These studies therefore support that core regulators of dormancy, that is, DOG1, RGL2 and the ABA/GA, appear as downstream targets of the molecular networks involved in light perception and signaling. PHYTOCHROME B (PHYB) has a main role in the control of germination in many photoblastic species like Arabidopsis and lettuce. A direct role of the fruit covers and the light environment on the Pfr:Pr ratio established in the embryo during late maturation (before dehydration impairs reversibility) can influence the light requirement for germination after imbibition. In addition, signaling through PHYB during seed development may also affect dormancy through regulation of DOG1 and RGL2 and GA metabolism. The REVEILLE1(RVE1) and RVE2 transcription factors repress GA synthesis under low PHYB activation (Jiang et al., 2016). Expression of *RVE1, RVE2* and *DOG1* genes increases during seed development; all three are downregulated by PHYB active form (Pfr) in the endosperm and embryo of imbibed, mature seeds. Yang et al. (2020) reported that RVE proteins and RGL2 are part of a regulatory module controlling seed dormancy and germination by integrating light perception, GA metabolism and GA signaling pathways. In addition to the repression of DOG1 by active PHYB (Pfr) reported by Jiang et al. (2016), *DOG1* also exhibits photoperiodic regulation during early fruit development. Expression of *DOG1* in developing siliques displays a circadian expression pattern and is under the control of the evening complex that transmits circadian signals to regulate plant growth and development (Zha et al., 2019). Relevance of these regulatory networks in the control of dormancy by the light environment under natural conditions in Arabidopsis, as well as in different species, is still unclear. For instance, Arabidopsis seed dormancy is unaffected by day length during seed development (He et al., 2014), but regulation of DOG1 by photoperiod may play a key role in other species where dormancy is strongly influenced by day length.

The reported effects of light intensity on seed dormancy may depend on the range explored by the light treatments. Dormancy promotion by high irradiances in the field (comparing direct sunlight and shading treatments) likely involves changes in fruit/seed coat pigmentation (and permeability), including PA and other flavonoids. No studies have focused on this response so far at the molecular/biochemical level yet partial evidence supports this possibility. Pericarp accumulation of flavonoids in *Sorghum bicolor* is promoted by sunlight; prolonged shadowing in developing panicles reduced anthocyanidin (3-DOA) content in the seed coat at maturity (Pfeiffer and Rooney, 2015). Flavonoids have protective roles in biotic and abiotic stress. High light stress and nitrogen deficiency both increased foliar PA levels in hybrid poplar (James et al., 2017, discussed by Dixon and Sarnala, 2020). These evidences, although incomplete (no dormancy testing in sorghum) or not seen in seeds (for the poplar study), support that flavonoid metabolism is regulated by irradiance level. The specific wavelength signaling for high irradiance may be blue, UV-A and/or UV-B, involving several photo-receptors other than phytochromes. Light in the range of UV-B is known to regulate flavonoid biosynthesis through activation of the UVR8 photoreceptor, as part of acclimation responses to UV-B in vegetative and reproductive organs and is highly expressed in developing siliques and seeds. Activated UVR8 interacts with COP1 and HY5 to positively regulate the expression of UV-B response genes (reviewed by Yin and Ulm, 2017). The role of this signaling network in developing fruit and seed tissues remains unknown. So far, the responses to light intensity in Arabidopsis have been explored within a range where light is well below natural light conditions (i.e. between 75 and 300 umol photons.m^{-2}.s^{-1}). Indeed, combinations of such low light intensities and long days are so unnatural that responses to these cues as could happen in the field may have been overlooked.

A wide range of components in the light signaling networks are expressed during seed development with distinctive temporal patterns (Yang et al., 2020). Nevertheless, understanding how these signaling networks operate to modulate dormancy in response to the light environment 'in the field' is still very limited. Also, the potential of the different attributes of the light environment to impact dormancy may occur at different times during fruit/seed development. For example, PHYB signaling may be relevant in imposing dormancy to the early maturing embryo underneath the green (red absorbing) covers, while high irradiance may promote flavonoid accumulation in the seed coats during the late maturation stages.

Among the maternal-environmental cues affecting primary dormancy, temperature, light and possibly water regime have been reported to produce changes in seed coat/covers (Benech-Arnold et al., 1992) characteristics, in addition to direct effects on the zygote. The presence of seed coat layers impregnated with flavonoids, suberin or cutin (which forms a thick cuticle on the outer side of the endosperm layer; De Giorgi et al., 2015) seems to function in an additive way to increase overall seed coat resistance to the diffusion of solutes and gases; alterations in any of these layers result in dormancy/germination phenotypes in Arabidopsis. Their abundance, composition and distribution (within the seed coats) vary greatly among species and are strongly affected by the environment. Because synthesis and transport of these substances are incredibly complex and diverse among species, the molecular targets involved in this regulation may also differ greatly. Nevertheless, sensing of different cues of the maternal environment may converge in central regulators such as key flowering-time genes (*FT, FLC* and *MFT*), as observed in response to temperature in Arabidopsis and wheat. These flowering-time genes have pleiotropic effects on seed dormancy and germination, and vice versa. The temporal adjustment of both key transitions –germination and flowering – to a changing environment involves the same central regulators, leading to pleiotropic effects. This idea and its implications are further discussed by Auge et al. (2019), who have reviewed examples of pleiotropy of flowering-time genes and highlighted those that also influence seed germination.

4 Ecological implications of dormancy modulation by the maternal environment

Seedling emergence at the right time and in the right place is essential for ensuring wild species persistence in a given ecological scenario (Donohue et al., 2010). As plants are sessile organisms, the seasonal timing of germination establishes the environment they are going to experience in subsequent life stages (Donohue et al., 2005; Burghardt et al., 2016), establishing the chances of survival, flowering and dispersal times, their competitive ability in relation to neighbors and their reproductive output. Thus, germination has been under high selective pressure (Penfield, 2017), and sophisticated mechanisms have evolved to ensure seeds germinate in the right context.

Many wild species (i.e. weeds) form seed banks in the soil, from which a fraction of the seed population emerges each year in a seasonal pattern. In most species forming seed banks, dormancy is a common feature, and the timing of seedling emergence in the field depends on the dormancy state of the seed bank (Batlla et al., 2020). Each year, new seeds produced by plants of the population are dispersed, entering the seed bank. For these seeds, the timing of seedling emergence in the field is dependent on the level of dormancy at dispersal (primary dormancy) and the regulation of dormancy level by the post-dispersal environment. The relative influence of the pre- and post-dispersal environment on the timing of emergence is variable and has been shown to depend on how predictable or unpredictable is the environment (Vayda et al., 2018). Maternal environmental effects are usually stronger than the effect of the post-dispersal environment under stable conditions or in those cases in which environments vary predictably. In these scenarios, conditions explored by mother plants during seed development and maturation can affect seed primary dormancy level at dispersal provoking changes in how dispersed seeds respond to the post-dispersal environment and, consequently, in emergence timing in the field.

Most of the studies in relation to the effect of the maternal environment on seed dormancy have been carried out under laboratory-controlled conditions. Thus, the ecological consequences of the maternal regulation of seed dormancy in the field are not totally understood. However, the few studies that tried to understand the ecological reason for dormancy regulation by the maternal environment suggest that this regulation is extremely relevant for species fitness. For example, Galloway and Etterson (2007) observed that a higher proportion of seeds from *Campanula americana* plants (an understory forest herb) grown in tree-fall light gaps germinates during fall, survives and behaves as annuals, while most seeds from plants grown in the forest understory, which receives ten times less irradiance, germinate in the spring, survive and behave as biennials. Projections of population growth rate for annuals and biennials showed that in the light gap environment more than 90% of the projected population growth is due to annuals, while population growth is due to biannual in the light-limited forest understory environment. Moreover, when projections were calculated considering that offspring did not grow in the same environment as mother plants (where seeds are mainly dispersed), the population growth rate was less than a third of that when the mother's environment and offspring's environment matched. These results show that seasonal germination plasticity in response to the maternal light environment enhances the fitness of offspring in their own mother environment.

Leverett et al. (2016) and Vayda et al. (2018) showed that Arabidopsis plants that were grown under a simulated canopy (green filters) produced less dormant seeds, able to germinate in a wider environmental range than those seeds that matured under neutral filters. This effect of maternal light environment on seed dormancy has been interpreted to enhance the population fitness by allowing offspring to emerge before competitive conditions intensify, providing an advantage to seedlings over their future competitors. Similar results have been reported recently by Fernández Farnocchia (2020) for *P. aviculare* seeds matured under a wheat canopy.

Quantitative models, for example population threshold models, have been widely used to investigate dormancy changes in seed populations and are excellent tools to interpret the effect of the maternal environment in relation to its adaptive significance. Indeed, Fernández Farnocchia et al. (2021) applied this approach to quantify the effect of different environments during seed maturation as the result of differences in mother plant sowing date (i.e. different cohorts of emergence) on *Amaranthus hybridus* seed dormancy level. Dormancy at dispersal and during post-harvest storage of seeds matured under the different environments was quantified through changes in the thermal range permissive for seed germination, which is flanked by a low and a high limit temperature (Fig. 2A–D). The lower the seed dormancy level, the wider the thermal range for seed germination (i.e. a larger difference between the low and the high limit temperature), and vice versa. Results showed that a delay in the sowing date (i.e. emergence date) and, consequently, in the maturation environment produced significant less dormant seeds at dispersal (see changes in the width of the thermal range permissive for seed germination at dispersal (0 days) from Fig. 1A–D) but did not affect dormancy release rate after dispersal (slopes of lines describing changes in the low and high limit temperature between seeds from plants sown on different dates – Fig. 2A–D – were similar; $P > .05$).

Fernández Farnocchia et al. (2021) also examined the emergence patterns of *A. hybridus* seedlings under field conditions, by burying in the field recently dispersed seeds from mother plants that were sown at contrasting dates (late

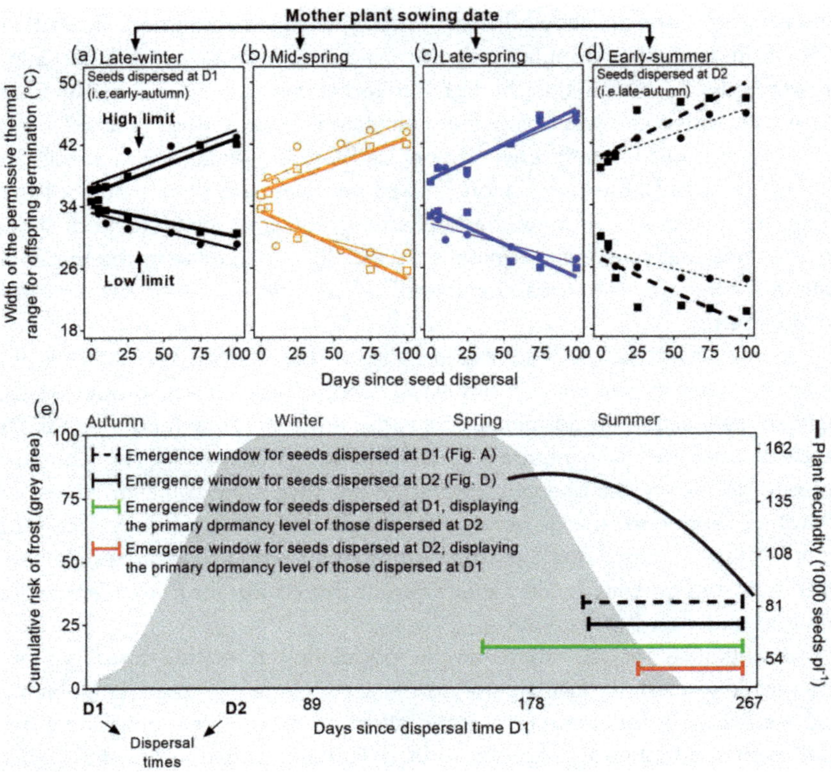

Figure 2 Upper panels a–d: Dormancy level of *Amaranthus hybridus* seeds that matured on mother plants that were sown in (a) late winter, (b) mid-spring, (c) late spring and (d) early summer, quantified as changes in the width of the permissive thermal range for seed germination (which is flanked by a low and a high limit temperature, as shown by the vertical arrows in panel (a)) in function of the storage time since seed dispersal. Fine lines and thick lines correspond to the seed after ripening conditions (dry storage at 25°C and stratification at 5°C, respectively). In all cases, the value of the thermal limits differed among sowing dates when measured at the time of seed dispersal (i.e. at 0 days since seed dispersal), while the slopes of the linear regressions (i.e. the rate at which dormancy release occurs) were not statistically different ($P > .05$). Lower panel E: Cumulative risk of frost damage (grey area) and simulated plant fecundity (grey curve) for *A. hybridus* plants from the north of Buenos Aires, Argentina. Horizontal bars show the simulated emergence window under different scenarios. Black bars refer to seeds dispersed in different moments (D1 - dotted line and D2 - full line - times) from plants that were sown in late winter (figure panel a) and early -summer (figure panel d). These simulations showed that differences in primary dormancy level observed at dispersal allow next generation to synchronize the temporal emergence window in the field. Green bar shows the location of the emergence window for seeds dispersed early (i.e. at D1) but displaying the same primary dormancy level at dispersal of those dispersed late in the season (i.e. at D2), whereas the red bar shows the location of the emergence window for the opposite scenario, seeds dispersed late (i.e. at D2) but displaying the same primary dormancy level at dispersal of those dispersed early (i.e. at D1). These scenarios illustrate what would happen whether variations in mother plant sowing date did not affect offspring dormancy level: an anticipation of emergence to those periods when the risk of frost damage is high (green bar) or a delay of emergence with a concomitant decrease in plants reproductive output. Adapted from Fernández Farnocchia et al. (2021).

winter and early summer, in Fig. 2a and d, respectively). Strikingly, the differences observed in the primary dormancy level under laboratory-controlled conditions did not provoke changes in the timing of germination in the field, where almost identical temporal emergence patterns were observed (see the horizontal full and dashed black bars indicating the temporal location of emergence windows in panel e of Fig. 2). Model simulations showed that the synchronization observed in the temporality of emergence in the field was a direct consequence of the maternal effects on primary dormancy level: changes in primary dormancy level (panels a and d in Fig. 2) allowed compensating differences in dispersal timings (D1 and D2 in Fig. 2e) to synchronize emergence windows. Reciprocal models assignments (i.e. simulating emergence for seeds dispersed in D1 with a primary dormancy level of those dispersed in D2 and vice versa) were developed in order to analyze the ecological implications of the maternal regulation of seed dormancy level (Fig. 2e). Results showed that such synchronization is crucial for optimizing next-generation fitness, by allowing the new generation to emerge not so early when the risk of frost damage is high (as would happen if seeds from an early maturation and dispersal environment (Fig. 2a) have the same dormancy level than those matured and dispersed later (Fig. 2c), as shown by the horizontal green-bar in Fig. 2e), or late enough to provoke a reduction in the reproductive output of the plant population (which would occur if seeds from a late maturation and dispersal environment (Fig. 2d) display the same dormancy level than those matured and dispersed early (Fig. 2a), as shown by the horizontal red-bar in Fig. 2e). Additional experiments showed that longer photoperiods during seed maturation (associated to early sowings) enhanced dormancy, suggesting that this factor, which represents a clear seasonal cue, is involved in the above-described regulation. The same modeling approach revealed a similar ecological significance (locating the temporal window of emergence to maximize fitness) of dormancy regulation by the maternal environment in *P. aviculare* seeds (Fernández Farnocchia et al., 2019). Quantifying the effects of the main factors (i.e. temperature) involved in the control of seed dormancy by the maternal environment would allow us to predict seed banks' dynamics in changing climatic scenarios, such as global warming. Moreover, developed models could be used for predicting possible responses of plant populations, by combining the effect of the maternal environment on seed banks' dormancy dynamics with additional models that allow for predicting other phenological events (Burghardt et al., 2015, 2016; Donohue et al., 2015).

5 Agricultural implications of dormancy modulation by the maternal environment: managing crops for dormancy

Although the adaptive significance of dormancy is evident for plants living in the wild, it has been always seen as a troublesome trait in crop production.

Since uncertainty on the emergence of a crop because of the possible existence of remnant dormancy in the seed lot that is used to generate a new crop cannot be tolerated, selection has gone heavily against dormancy during crop domestication (Rodríguez et al., 2015). This process has been particularly severe in crops that, like malting barley, are subjected to an industrialization procedure that relies on grain germination. The undesired consequence of taking too far this selection against dormancy is to have made many crops susceptible to suffer pre-harvest sprouting (PHS, i.e. the untimely germination in the mother plant as a result of damp conditions prior to harvest). Other crops, in contrast, have a shorter history of domestication and, in consequence, usually display persistent dormancy that could last for many months or even years. A conspicuous example of the latter is sunflower. Understanding the effects of the maternal environment on dormancy in these situations brings an opportunity to develop tools for managing crops for dormancy.

5.1 Malting barley

Barley, when its end use is malting for brewery, is typically the crop in which the statement 'not too much dormancy, but not so little' applies tightly (Rodríguez et al., 2015). Indeed, a long-lasting dormancy impairs the rapid industrialization of seed lots and, in consequence, increases costs and losses due to the necessity of storage until dormancy is terminated and the lot is ready for industrialization (Gualano et al., 2014). But, on the other hand, an anticipated exit from dormancy to permit immediate malting upon harvest exposes the crop to suffer pre-germination or PHS (Rodríguez et al., 2001). The barley grain is a 'dressed' grain, with the lemma and the palea strongly adhered to the pericarp of the caryopsis. Rarely the barley grain is able to germinate prior to physiological maturity (PM) and, as in many other cereals, this is due to coat-imposed dormancy (Benech-Arnold et al., 1999). Three classes of barley cultivars exist in relation to the timing with which the grains are released from dormancy after PM: (i) those whose grains are rapidly released from dormancy after PM; these cultivars are very prone to suffer PHS and are very difficult to manage in environments where rains occur prior to crop harvest; (ii) those with long-lasting dormancy after harvest maturity (HM); these cultivars are usually resisted by the industry but they display no risk of suffering PHS; (iii) those whose grains are released from dormancy at different velocity from year to year; these cultivars behave as sprouting-resistant in some years, and as resistant in others. The latter type is the most widespread cultivars nowadays and their different behavior between years is modulated by the environment experienced by the mother plant during seed development. Among the different factors that might control this variation, temperature during seed development has been reported to have a strong influence. Rodríguez et al. (2001) found that it was not temperature throughout the whole developmental period, but during a specific 'sensitivity thermal time window' located close to the end of the grain-filling period. This window was found to go from 300°C days to 350°C days after heading (accumulated over a base temperature of 5.5°C) for one barley cultivar whose PM was attained 420°C days after anthesis (Rodríguez et al., 2001). Identifying this window as a thermal time interval allowed to anchor it within the grain-filling period, making its location independent of the thermal environment for a particular year. Further work with other barley cultivars confirmed the relative location of this sensitivity window within the grain-filling period (Gualano and Benech-Arnold, 2009b). A positive linear relationship was established between the average temperature perceived by the crop during this thermal time window and the germination index of the grains either 12 days after PM or when the grains have between 10% and 20% humidity (this is approximately halfway between PM and HM and is a good estimator of the velocity with which the seeds are being released from dormancy after PM) (Fig. 3). With high temperatures sensed during the sensitivity window, rapid dormancy release might be expected after PM which, combined with a forecast of heavy rains prior HM, would indicate an almost certain risk of sprouting. After such a prediction the farmer can take management decisions as, for example, applying a desiccant after PM and harvesting before the rains. Conversely, low temperatures during the sensitivity window would result in grains with high dormancy prior to harvest thus making the crop resistant to sprouting. The relationship between dormancy after PM and thermal conditions during the sensitivity windows was successfully validated against data collected 600 km away from where the relationship was established experimentally. From this validation, it was also evident that temperature explains only one dimension of the variability observed in dormancy and that other environmental factors might be also responsible for dormancy modulation (Rodríguez et al., 2001). Therefore, water and nitrogen availability were incorporated into the analysis; the results showed that while nitrogen supply did not affect the rate of dormancy release after PM, drought during seed development accelerated dormancy release to a similar extent than high temperatures during the sensitivity window (Gualano and Benech-Arnold, 2009b). In this way, the low susceptibility to suffer PHS that might have been expected in a year with cool temperatures during the sensitivity window would be reverted by a drought during grain filling (Gualano and Benech-Arnold, 2009b).

5.2 Sunflower

The short history of domestication of sunflower (<4000 years), as compared to cereals like wheat and barley with more than 10000 years of domestication, has probably aided to the fact that this crop has retained some dormancy

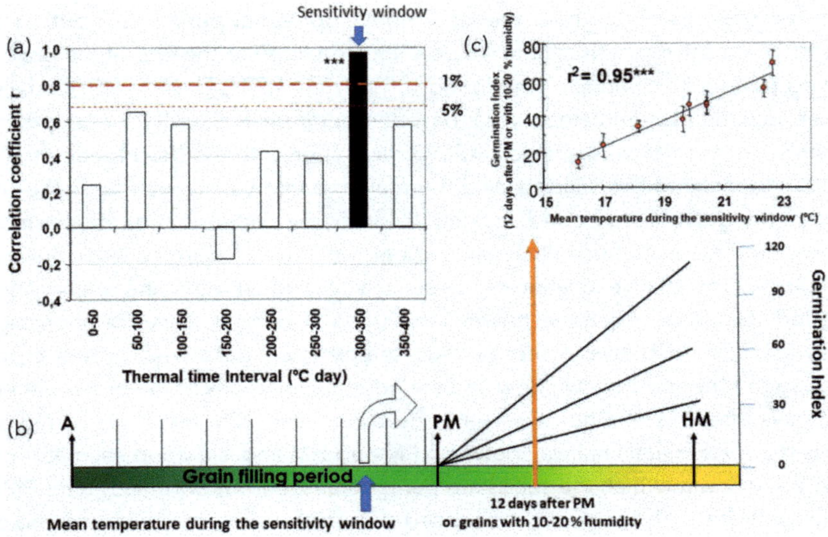

Figure 3 Schematic description of the identification of sensitivity windows to temperature throughout grain filling, for the prediction of pre-harvest sprouting (PHS) susceptibility in barley crops. (a) The grain-filling period (420°C days from anthesis (A) to physiological maturity (PM) accumulated over a base temperature (Tb) of 5.5°C for this example) is divided into intervals of 50°C days for the analysis, and correlation analysis is performed between temperature experienced during each interval and the germination index (GI) of the grains 12 days after PM or with 10–20% humidity, which is an indicator of PHS susceptibility. A sensitivity window is detected when the correlation coefficient (r) is significant (the interval of 300–350°C days in the example). (b) Temperature experienced during the sensitivity window influences the velocity with which the grains are released from dormancy after PM and determines their GI 12 days after PM or when having 10–20% humidity, prior to harvest maturity (HM). (c) The linear relationship between temperature experienced by the crop during the sensitivity window (300-350°C days after A in the example) and the GI 12 days after PM or in grains with humidity between 10% and 20%. This relationship is used to predict PHS susceptibility as a function of temperature experienced during the sensitivity window.

which becomes big trouble to the seed industry in some years. Indeed, seed commercialization is permanently subjected to the uncertainty of having or not non-dormant seeds in time for the next-crop implantation.

The various structures that make up the sunflower achene contribute to dormancy and, their relative contribution, varies during fruit development. The embryo is surrounded by a single layer of living endospermatic cells intimately coalesced with the seed coat which is dead at maturity, forming the true sunflower 'seed' (Seiler, 1997; Szemruch et al., 2014). This 'seed' is further enclosed by the pericarp, forming the fruit. Dormancy is imposed mainly by the seed coat at the early stages of development (Corbineau et al., 1990). Dormancy of the embryo increases gradually as development progresses until it reaches a maximum at about 20–22 days after anthesis (DAA). Approaching

PM, embryo dormancy decreases gradually and achene dormancy is mostly due to the seed coat. This coat-imposed dormancy is usually persistent and may require several weeks or months of dry storage to disappear completely (Corbineau et al., 1990).

Contrary to most examples given in this chapter, an extended dormancy in sunflower has been associated with high temperatures during seed development (Bodrone et al., 2017). Conducting controlled (greenhouse) and field experiments (sowing dates and plastic tents), Bodrone et al. (2017) also found that temperatures during later stages of achene development significantly increased the level of dormancy at harvest and its persistence during post-harvest storage. These authors considered the impact of the maternal thermal environment on embryo and coat-imposed dormancy and found that, although maturation under warm environments reduced embryo dormancy, this effect was overcompensated for by the enhancement of inhibition imposed by the pericarp and the seed coat (Bodrone et al., 2017). These results stress the importance of avoiding warm environments during later stages of seed development in crops grown for seed production; this can be managed through sowing dates that place the grain-filling period under cooler conditions or avoiding production sites characterized by high temperatures.

This enhancement of dormancy by high temperatures during achene maturation has been proposed as one of the reasons for the low primary dormancy level usually observed in sunflower fruits harvested from late sown crops, which are exposed to lower temperatures during late stages of achene maturation (Riveira et al., 2017). However, recent results showed that sunflower fruit dormancy level at harvest also depends on the level of radiation to which plants are exposed during seed maturation. Low incident radiation, which corresponds to late season maturation, also determined a low dormancy level (Pizzorno et al., 2012).

6 Summary

The maternal environment explored by the developing seed has a decisive importance in shaping the seed dormancy phenotype. Although these effects have started to be recognized several decades ago, only in recent years we began to understand the underlying mechanisms through which the environmental signals perceived by the mother plant are translated into a particular dormancy phenotype. On the other hand, there is an increasing body of evidence suggesting the adaptive significance of maternal effects on dormancy. Recent work strongly suggests that these effects have been shaped by natural selection to act as precise devices to maximize phenotypic fitness in response to environmental uncertainty or heterogeneity. Maternal effects on dormancy have obvious implications also for agricultural production.

Taking into account these effects to design agricultural practices that minimize dormancy adverse effects on crop production should be part of a package of strategies toward a more sustainable agriculture.

7 Future trends in research

In spite of the significant progress made in recent years on the molecular mechanisms behind maternal effects on dormancy, there are still enormous holes in knowledge that need to be filled before we gain a comprehensive understanding of how the maternal environment modulates dormancy. Given the velocity with which new techniques and tools are developed in the field of plant biology, a bulk of exciting revelations might be expected in years to come. Similarly, ecological implications of maternal effects on dormancy have just begun to be disclosed. New insights in this area will be of great value to further understanding plant population and community dynamics. Finally, we intended to depict with examples how understanding and quantifying these maternal effects on dormancy could yield powerful tools to manage the different constraints to production resulting from either a short- or a long-lasting dormancy. We hope that these examples will inspire new experiments to assess maternal effects with practical ends, since these dormancy constraints are transversal to many crops.

8 Where to look for further information

Further information on the subject can be found in the following text:

Bewley, J. D., Bradford, K. J., Hilhorst, H. and Nonogaki, H. (2013) Seeds: physiology of development, germination and dormancy. Third Edition. Springer. New York. 392 pp.

Key research in this area can be found at the following organizations:

- Instituto de Investigaciones Fisiológicas y Ecológicas aplicadas a la Agricultura (IFEVA) (UBA/CONICET). Seed Biology Laboratory. https://agro.uba.ar/laboratorios/semillas.
- John Innes Centre. Steven Penfield's Group. https://www.jic.ac.uk/people/steven-penfield/.
- Duke University. The Donohue Lab. https://sites.duke.edu/donohuelab/.

9 References

Alboresi, A., Gestin, C., Leydecker, M. T., Bedu, M., Meyer, C. and Truong, H. N. (2005). Nitrate, a signal relieving seed dormancy in Arabidopsis. *Plant, Cell and Environment* 28(4), 500–512.

Auge, G. A., Blair, L. K., Burghardt, L. T., Coughlan, J., Edwards, B., Leverett, L. D. and Donohue, K. (2015). Secondary dormancy dynamics depends on primary dormancy status in Arabidopsis thaliana. *Seed Science Research* 25(2), 230–246.

Auge, G. A., Penfield, S. and Donohue, K. (2019). Pleiotropy in developmental regulation by flowering-pathway genes: Is it an evolutionary constraint? *New Phytologist* 224(1), 55–70.

Barrero, J. M., Talbot, M. J., White, R. G., Jacobsen, J. V. and Gubler, F. (2009). Anatomical and transcriptomic studies of the coleorhiza reveal the importance of this tissue in regulating dormancy in barley. *Plant Physiology* 150(2), 1006–1021.

Baskin, J. M. and Baskin, C. C. (1975). Year-to-year variation in the germination of freshly-harvested seeds of *Arenaria patula* var. *robusta* from the same site. *Journal of the Tennessee Academy of Sciences* 50, 106–108.

Batlla, D., Malavert, C., Fernández Farnocchia, R. B. and Benech-Arnold, R. (2020). Modelling weed Seedbank dormancy and germination. In: *Decision Support Systems for Weed Management*. Springer, Cham, pp. 61–83.

Benech-Arnold, R. L., Fenner, M. and Edwards, P. J. (1991). Changes in germinability, ABA levels and ABA embryonic sensitivity in developing seeds of *Sorghum bicolor* induced by water stress during grain filling. *New Phytologist* 118(2), 339–347.

Benech-Arnold, R. L., Fenner, M. and Edwards, P. J. (1992). Changes in dormancy level in *Sorghum halepense* seeds induced by water stress during seed development. *Functional Ecology* 6, 460–468.

Benech-Arnold, R. L., Fenner, M. and Edwards, P. J. (1995). Influence of potassium nutrition on germinability, ABA content and embryonic sensitivity to ABA of developing seeds of *Sorghum bicolor* (L.) Moench. *New Phytologist* 130, 207–216.

Benech-Arnold, R. L., Giallorenzi, M. C., Frank, J. and Rodriguez, V. (1999). Termination of hull-imposed dormancy in barley is correlated with changes in embryonic ABA content and sensitivity. *Seed Science Research* 9, 39–47.

Benech-Arnold, R. L., Sánchez, R. A., Forcella, F., Kruk, B. C. and Ghersa, C. M. (2000). Environmental control of dormancy in weed seed soil banks. *Field Crops Research* 67, 105–122.

Bentsink, L., Jowett, J., Hanhart, C. J. and Koornneef, M. (2006). Cloning of DOG1, a quantitative trait locus controlling seed dormancy in Arabidopsis. *Proceedings of the National Academy of Sciences of the United States of America* 103(45), 17042–17047.

Benvenuti, S. and Macchia, M. (1997). Light environment, phytochrome and germination of Datura stramonium L. seeds. *Environmental and Experimental Botany* 38(1), 61–71.

Bodrone, M. P., Rodríguez, M. V., Arisnabarreta, S. and Batlla, D. (2017). Maternal environment and dormancy in sunflower: The effect of temperature during fruit development. *European Journal of Agronomy* 82, 93–103.

Burghardt, L. T., Edwards, B. R. and Donohue, K. (2016). Multiple paths to similar germination behavior in Arabidopsis thaliana. *New Phytologist* 209(3), 1301–1312.

Burghardt, L. T., Metcalf, C. J. E., Wilczek, A. M., Schmitt, J. and Donohue, K. (2015). Modeling the influence of genetic and environmental variation on the expression of plant life cycles across landscapes. *The American Naturalist* 185(2), 212–227.

Cairns, A. L. P. and Kritzinger, J. H. (1992). The effect of molybdenum on seed dormancy in wheat. *Plant and Soil* 145(2), 295–297.

Carrillo-Barral, N., Rodríguez-Gacio, M. D. C. and Matilla, A. J. (2020). Delay of Germination-1 (DOG1): A key to understanding seed dormancy. *Plants* 9(4), 480.

Ceccato, D., Bertero, D., Batlla, D. and Galati, B. (2015). Structural aspects of dormancy in quinoa (Chenopodium quinoa): Importance and possible action mechanisms of the seed coat. *Seed Science Research* 25(3), 267–275.

Chen, M., MacGregor, D. R., Dave, A., Florance, H., Moore, K., Paszkiewicz, K., Smirnoff, N., Graham, I. A. and Penfield, S. (2014). Maternal temperature history activates Flowering Locus T in fruits to control progeny dormancy according to time of year. *Proceedings of the National Academy of Sciences* 111(52), 18787–18792.

Chen, M. and Penfield, S. (2018). Feedback regulation of COOLAIR expression controls seed dormancy and flowering time. *Science* 360(6392), 1014–1017.

Chiang, G. C., Bartsch, M., Barua, D., Nakabayashi, K., Debieu, M., Kronholm, I., Koornneef, M., Soppe, W. J., Donohue, K. and De Meaux, J. (2011). DOG1 expression is predicted by the seed-maturation environment and contributes to geographical variation in germination in Arabidopsis thaliana. *Molecular Ecology* 20(16), 3336–3349.

Chiang, G. C., Barua, D., Kramer, E. M., Amasino, R. M. and Donohue, K. (2009). Major flowering time gene, flowering locus C, regulates seed germination in Arabidopsis thaliana. *Proceedings of the National Academy of Sciences of the United States of America* 106(28), 11661–11666.

Contreras, S., Bennett, M. A., Metzger, J. D., Tay, D. and Nerson, H. (2009). Red to far-red ratio during seed development affects lettuce seed germinability and longevity. *Hortscience* 44(1), 130–134.

Corbineau, F., Bagniol, S. and Côme, D. (1990). Sunflower (Helianthus annuus L.) seed dormancy and its regulation by ethylene. *Israel Journal of Botany* 39, 313–325.

Coughlan, J. M., Saha, A. and Donohue, K. (2017). Effects of pre- and post-dispersal temperature on primary and secondary dormancy dynamics in contrasting genotypes of Arabidopsis thaliana (Brassicaceae). *Plant Species Biology* 32(3), 210–222.

Cresswell, E. G. and Grime, J. P. (1981). Induction of a light requirement during seed development and its ecological consequences. *Nature* 291, 585–586.

Datta, S. C., Gutterman, Y. and Evenari, M. (1972). The influence of the origin of the mother plant on yield and germination of their caryopses in *Aegilops ovata*. *Planta* 105(2), 155–164.

Dave, A., Vaistij, F. E., Gilday, A. D., Penfield, S. D. and Graham, I. A. (2016). Regulation of Arabidopsis thaliana seed dormancy and germination by 12-oxo-phytodienoic acid. *Journal of Experimental Botany* 67(8), 2277–2284.

Debeaujon, I., Leon-Kloosterziel, K. M. and Koornneef, M. (2000). Influence of the testa on seed dormancy, germination, and longevity in Arabidopsis. *Plant Physiology* 122(2), 403–414.

De Giorgi, J., Piskurewicz, U., Loubery, S., Utz-Pugin, A., Bailly, C., Méne-Saffrané, L. and López-Molina, L. (2015). An endosperm-associated cuticle is required for arabidopsis seed viability, dormancy and early control of germination. *PLoS Genetics* 11(12), e1005708. doi:10.1371/journal.pgen.1005708

Dixon, R. A. and Sarnala, S. (2020). Proanthocyanidin biosynthesis-a matter of protection. *Plant Physiology* 184(2), 579–591. doi: 10.1104/pp.20.00973.

Donohue, K. (2009). Completing the cycle: Maternal effects as the missing link in plant life histories. *Philosophical Transactions of the Royal Society of London. Series B, Biological Sciences* 364(1520), 1059–1074.

Donohue, K. (2013). Why ontogeny matters during adaptation: Developmental niche construction and pleiotoypy across the life cycle in Arabidopsis thaliana. *Evolution; International Journal of Organic Evolution* 68(1), 32–47.

Donohue, K., Burghardt, L. T., Runcie, D., Bradford, K. J. and Schmitt, J. (2015). Applying developmental threshold models to evolutionary ecology. *Trends in Ecology and Evolution* 30(2), 66–77.

Donohue, K., Dorn, L., Griffith, C., Kim, E., Aguilera, A., Polisetty, C. R. and Schmitt, J. (2005). Environmental and genetic influences on the germination of Arabidopsis thaliana in the field. *Evolution; International Journal of Organic Evolution* 59(4), 740–757.

Donohue, K., Rubio de Casas, R., Burghardt, L., Kovach, K. and Willis, C. G. (2010). Germination, postgermination adaptation, and species ecological ranges. *Annual Review of Ecology, Evolution, and Systematics* 41(1), 293–319.

Dorne, A. J. (1981). Variation in seed germination inhibition of *Chenopodium bonus-henricus* in relation to altitude of plant growth. *Canadian Journal of Botany* 59(10), 1893–1901.

Edwards, B. R., Burghardt, L. T., Zapata-Garcia, M. and Donohue, K. (2016). Maternal temperature effects on dormancy influence germination responses to water availability in Arabidopsis thaliana. *Environmental and Experimental Botany* 126, 55–67.

Fedi, F., O'Neill, C. M., Menard, G., Trick, M., Dechirico, S., Corbineau, F., Bailly, C., Eastmond, P. J. and Penfield, S. (2017). Awake1, an ABC-type transporter, reveals an essential role for suberin in the control of seed dormancy. *Plant Physiology* 174(1), 276–283. doi: 10.1104/pp.16.01556.

Fenner, M. (1991). The effects of the parent environment on seed germinability. *Seed Science Research* 1(2), 75–84.

Fernández Farnocchia, R. B. (2020). *Efecto del ambiente materno sobre el nivel de dormición y los patrones temporales de emergencia de dos especies maleza primavero-estivales: Polygonum aviculare L. y Amaranthus hybridus L. Tesis de Doctorado en Ciencias Agropecuarias.* EPG Alberto Soriano, Universidad de Buenos Aires. 200 p.

Fernández Farnocchia, R. B., Benech-Arnold, R. L. and Batlla, D. (2019). Regulation of seed dormancy by the maternal environment is instrumental for maximizing plant fitness in *Polygonum aviculare*. *Journal of Experimental Botany* 70(18), 4793–4806.

Fernández Farnocchia, R. B., Benech-Arnold, R. L., Mantese, A. and Batlla, D. (2021). Optimization of next-generation emergence timing of *Amaranthus hybridus* is determined through seed dormancy modulation by the maternal environment. *Journal of Experimental Botany* 72(12), 4283–4297, doi: 10.1093/jxb/erab141.

Footitt, S., Douterelo-Soler, I., Clay, H. and Finch-Savage, W. E. (2011). Dormancy cycling in *Arabidopsis* seeds is controlled by seasonally distinct hormone-signaling pathways. *Proceedings of the National Academy of Sciences of the United States of America* 108(50), 20236–20241.

Galloway, L. F. and Etterson, J. R. (2007). Transgenerational plasticity is adaptive in the wild. *Science* 318(5853), 1134–1136.

Gao, Y., Liu, J., Chen, Y., Tang, H., Wang, Y., He, Y., Ou, Y., Sun, X., Wang, S. and Yao, Y. (2018). Tomato SlAN11 regulates flavonoid biosynthesis and seed dormancy by interaction with bHLH proteins but not with MYB proteins. *Horticulture Research* 5, 27.

Goldbach, H. and Goldbach, E. (1977). Abscisic acid translocation and influence of water stress on grain abscisic acid content. *Journal of Experimental Botany* 28(6), 1342–1350.

Gualano, N. A. and Benech-Arnold, R. L. (2009a). The effect of water and nitrogen availability during grain filling on the timing of dormancy release in malting barley crops. *Euphytica* 168(3), 291–301.

Gualano, N. A. and Benech-Arnold, R. L. (2009b). Predicting pre-harvest sprouting susceptibility in barley: Looking for "sensitivity windows" to temperature throughout grain filling in various commercial cultivars. *Field Crops Research* 114(1), 35–44.

Gualano, N. A., Del Fueyo, P. A. and Benech-Arnold, R. L. (2014). Potential longevity (Ki) of malting barley (Hordeum vulgare L.) grain lots relates to their degree of pre-germination assessed through different industrial quality parameters. *Journal of Cereal Science* 60(1), 222–228.

Gutterman, Y. (1974). The influence of the photoperiodic regime and red-far red light treatments of *Portulaca oleracea* L. plants on the germinability of their seeds. *Oecologia* 17(1), 27–38.

Gutterman, Y. (1978). Seed coat permeability as a function of photoperiodical treatments of the mother plant during seed maturation in the desert annual plant *Trigonella arabica* del. *Journal of Arid Environments* 1(2), 141–144.

Gutterman, Y. (1982). Phenotypic maternal effect of photoperiod on seed germination. In: Kahn, A. A. (Ed.), *The Physiology and Biochemistry of Seed Development, Dormancy and Germination*. Elsevier, Amsterdam, pp. 67–79.

Gutterman, Y. (2000). Maternal effects on seeds during development. In: Fenner, M. (Ed.), *Seeds: The Ecology of Regeneration in Plant Communities* (2nd edn.). CAB International, Wallingford, pp. 59–83.

Gutterman, Y. and Heydecker, W. (1973). Studies of the surfaces of desert plants. I. Effect of daylength upon maturation and seed coat on *Ononis sicula* Guss. *Annals of Botany* 37(5), 1049–1050.

Haeder, H. E. and Beringer, H. (1981). Influence of potassium nutrition and water stress on the content of absicic acid in grains and flag leaves of wheat during grain development. *Journal of the Science of Food and Agriculture* 32(6), 552–556.

Harrington, J. F. (1960). Germination of seeds from carrot, lettuce and pepper plants grown under severe nutrient deficinecies. *Hilgardia* 30(7), 219–235.

Hayes, R. G. and Klein, W. H. (1974). Spectral quality influence of light during development of *Arabidopsis thaliana* plants in regulating seed germination. *Plant and Cell Physiology* 15(4), 643–653.

He, H., de Souza Vidigal, D., Snoek, L. B., Schnabel, S., Nijveen, H., Hilhorst, H. and Bentsink, L. (2014). Interaction between parental environment and genotype affects plant and seed performance in Arabidopsis. *Journal of Experimental Botany* 65(22), 6603–6615.

He, H., Willems, L. A., Batushansky, A., Fait, A., Hanson, J., Nijveen, H., Hilhorst, H. W. and Bentsink, L. (2016). Effects of parental temperature and nitrate on seed performance are reflected by partly overlapping genetic and metabolic pathways. *Plant and Cell Physiology* 57(3), 473–487. doi: 10.1093/pcp/pcv207.

Hilhorst, H. W. and Karssen, C. M. (2000). Effect of chemical environment on seed germination. In: Fenner, M. (Ed.), *Seeds: The Ecology of Regeneration in Plant Communities* (2nd edn.). CAB International, Wallingford, pp. 293–309.

Hill, H. J., West, S. H. and Hinkson, K. (1986). Effect of water stress during seedfill on impermeable seed expression in soybean 1. *Crop Science* 26(4), 807–812.

Hinsberg, A. Van (1998). Maternal and ambient environmental effects of light on germination in Plantago lanceolata: Correlated responses to selection on leaf length. *Functional Ecology* 12(5), 825–833.

Huang, Z., Footitt, S., Tang, A. and Finch-Savage, W. E. (2018). Predicted global warming scenarios impact on the mother plant to alter seed dormancy and germination behaviour in Arabidopsis. *Plant, Cell and Environment* 41(1), 187–197.

Imaizumi, T., Auge, G. and Donohue, K. (2017). Photoperiod throughout the maternal life cycle, not photoperiod during seed imbibition, influences germination in Arabidopsis thaliana. *American Journal of Botany* 104(4), 516–526.

James, A. M., Ma, D., Mellway, R., Gesell, A., Yoshida, K., Walker, V., Tran, L., Stewart, D., Reichelt, M., Suvanto, J., Salminen, J. P., Gershenzon, J., Séguin, A. and Constabel, C. P. (2017). Poplar MYB115 and MYB134 transcription factors regulate proanthocyanidin synthesis and structure. *Plant Physiology* 174(1), 154–171.

Jha, P., Norsworthy, J. K., Riley, M. B. and Bridges, W. (2010). Shade and plant location effects on germination and hormone content of Palmer amaranth (Amaranthus palmeri) seed. *Weed Science* 58(1), 16–21.

Jia, L., Wu, Q., Ye, N., Liu, R., Shi, L., Xu, W., Zhi, H., Rubaiyath Bin Rahman, A. N. M., Xia, Y. and Zhang, J. (2012). Proanthocyanidins inhibit seed germination by maintaining a high level of abscisic acid in *Arabidopsis thaliana*. *Journal of Integrative Plant Biology* 54(9): 663–673.

Jiang, Z. M., Xu, G., Jing, Y. J., Tang, W. J. and Lin, R. C. (2016). Phytochrome B and REVEILLE1/2-mediated signalling controls seed dormancy and germination in Arabidopsis. *Nature Communications* 7, 12377.

Juntilla, O. (1973). Seed and embryo germination in *Syringa vulgaris* and *S. retroflexa* as affected by temperature during seed development. *Physiologia Plantarum* 29(2), 264–268.

Kang, J., Yim, S., Choi, H., Kim, A., Lee, K. P., Lopez-Molina, L., Martinoia, E. and Lee, Y. (2015). Abscisic acid transporters cooperate to control seed germination. *Nature Communications* 6, 8113.

Kanno, Y., Hanada, A., Chiba, Y., Ichikawa, T., Nakazawa, M., Matsui, M., Koshiba, T., Kamiya, Y. and Seo, M. (2012). Identification of an abscisic acid transporter by functional screening using the receptor complex as a sensor. *Proceedings of the National Academy of Sciences of the United States of America* 109(24), 9653–9658.

Karssen, C. M. (1970). The light promoted germination of the seeds of Chenopodium album L. III. Effect of the photoperiod during growth and development of the plants on the dormancy of the produced seeds. *Acta Botanica Neerlandica* 19(1), 81–94.

Keigley, P. J. and Mullen, R. E. (1986). Changes in soybean seed quality from high temperature during seed fill and maturation 1. *Crop Science* 26(6), 1212–1216.

Kendall, S. and Penfield, S. (2012). Maternal and zygotic temperature signalling in the control of seed dormancy and germination. *Seed Science Research* 22(S1), S23–S29.

Kendall, S. L., Hellwege, A., Marriot, P., Whalley, C., Graham, I. A. and Penfield, S. (2011). Induction of dormancy in Arabidopsis summer annuals requires parallel regulation of DOG1 and hormone metabolism by low temperature and CBF transcription factors. *The Plant Cell* 23(7), 2568–2580.

Kigel, J., Ofir, M. and Koller, D. (1977). Control of the germination responses of Amaranthus retroflexus L. seeds by their parental photothermal environment. *Journal of Experimental Botany* 28(5), 1125–1136.

Kohler, C., Wolff, P. and Spillane, C. (2012). Epigenetic mechanisms underlying genomic imprinting in plants. *Annual Review of Plant Biology* 63, 331–352.

Koller, D. (1962). Preconditioning of germination in lettuce at time of fruit ripening. *American Journal of Botany* 49(8), 841–844.

Lacey, E. P. (1984). Seed mortality in *Daucus carota*: Latitudinal effects. *American Journal of Botany* 71(9), 1175–1182.

Leverett, L. D., Auge, G. A., Bali, A. and Donohue, K. (2016). Contrasting germination responses to vegetative canopies experienced in pre-vs. post-dispersal environments. *Annals of Botany* 118(6), 1175–1186.

Longás, MdlM., Sabbatini, M. R., Batlla, D. and Chantre, G. R. (2021). Effect of contrasting maternal nitrogen environments on Buglossoides arvensis seed germination response to gibberellic and abscisic acids. *Weed Research* 61(3), 221–230.

MacGregor, D. R., Kendall, S. L., Florance, H., Fedi, F., Moore, K., Paszkiewicz, K., Smirnoff, N. and Penfield, S. (2015). Seed production temperature regulation of primary dormancy occurs through control of seed coat phenylpropanoid metabolism. *New Phytologist* 205(2), 642–652.

Matakiadis, T., Alboresi, A., Jikumaru, Y., Tatematsu, K., Pichon, O., Renou, J. P., Kamiya, Y., Nambara, E. and Truong, H. N. (2009). The Arabidopsis abscisic acid catabolic gene CYP707A2 plays a key role in nitrate control of seed dormancy. *Plant Physiology* 149(2), 949–960.

McCullough, J. M. and Shropshire, W. (1970). Physiological predetermination of germination responses in Arabidopsis thaliana (L.) Heynh. *Plant and Cell Physiology* 11(1), 139–148.

Mendel, R. R. and Haensch, R. (2002). Molybdoenzymes and molybdenum cofactor in plants. *Journal of Experimental Botany* 53(375), 1689–1698.

Mousseau, T. A. and Fox, C. W. (1998). The adaptive significance of maternal effects. *Trends in Ecology and Evolution* 13(10), 403–407.

Munir, J., Dorn, L. A., Donohue, K. and Schmitt, J. (2001). The effect of maternal photoperiod on seasonal dormancy in Arabidopsis thaliana (Brassicaceae). *American Journal of Botany* 88(7), 1240–1249.

Nakabayashi, K., Bartsch, M., Xiang, Y., Miatton, E., Pellengahr, S., Yano, R., Seo, M. and Soppe, W. J. J. (2012). The time required for dormancy release in Arabidopsis is determined by DELAY OF GERMINATION1 protein levels in freshly harvested seeds. *The Plant Cell* 24(7), 2826–2838.

Nakamura, S., Abe, F., Kawahigashi, H., Nakazono, K., Tagiri, A., Matsumoto, T., Utsugi, S., Ogawa, T., Handa, H., Ishida, H., Mori, M., Kawaura, K., Ogihara, Y. and Miura, H. (2011). A wheat homolog of MOTHER OF FT AND TFL1 acts in the regulation of germination. *Plant Cell* 23(9), 3215–3229.

Okunsaya, O. T. and Ungar, I. A. (1983). The effects of time of seed production on the germination response of *Spergularia marina*. *Physiologia Plantarum* 59(3), 335–342.

Orozco-Segovia, A., Brechú-Franco, A. E., Zambrano-Polanco, L., Osuna-Fernández, R., Laguna-Hernández, G. and Sánchez-Coronado, M. E. (2000). Effects of maternal light environment on germination and morphological characteristics of Sicyos deppei seeds. *Weed Research* 40(6), 495–506.

Orozco-Segovia, A., Sanchez-Coronado, M. E. and Vazquez-Yanes, C. (1993). Effect of maternal light environment on seed germination in *Piper auritum*. *Functional Ecology* 7(4), 395–402.

Penfield, S. (2017). Seed dormancy and germination. *Current Biology: CB* 27(17), R874–R878.

Penfield, S. and MacGregor, D. R. (2017). Effects of environmental variation during seed production on seed dormancy and germination. *Journal of Experimental Botany* 68(4), 819–825.

Peters, N. C. B. (1982). The dormancy of wild oat seed (*Avena fatua* L.) from plants grown under various temperature and soil moisture conditions. *Weed Research* 22(4), 205–212.

Pfeiffer, B. K. and Rooney, W. L. (2015). Sunlight induces black color and increases flavonoid levels in the grain of sorghum line Tx3362. *Crop Science* 55(4), 1703–1711. doi: 10.2135/cropsci2014.11.0757.

Pizzorno, L., Batlla, D. and Lopez Pereira, M. (2012). Effect of maternal environment on sunflower fruit dormancy (*Helianthus annuus*): Photoperiod and incident radiation. *Proceddings of the 18th International Sunflower Conference*, Mar del Plata, Argentina.

Pourrat, Y. and Jacques, R. (1975). The influence of photoperiodic conditions received by the mother plant on morphological and physiological characteristics of Chenopodium polyspermum L. seeds. *Plant Science Letters* 4(4), 273–279.

Riveira, M., Arata, G., Rodriguez, M. V. and Batlla, D. (2017). Intraespecific variability and maternal environment effects on sunflower (*Helianthus annuus*) dormancy traits. *Proceedings of the 12th Triennial Conference of the International Society for Seed Science*, Monterey, CA.

Roach, D. A. and Wulff, R. D. (1987). Maternal effects in plants. *Annual Review of Ecology and Systematics* 18(1), 209–235.

Rodríguez, M. V., Barrero, J. M., Corbineau, F., Gubler, F. and Benech-Arnold, R. L. (2015). Dormancy in cereals (not too much, not so little): About the mechanisms behind this trait. *Seed Science Research* 25(2), 99–119.

Rodríguez, M. V., Margineda, M., González-Martín, J. F., Insausti, P. and Benech-Arnold, R. L. (2001). Predicting pre-harvest sprouting susceptibility in barley: A model based on temperature during grain filling. *Agronomy Journal* 93(5), 1071–1079.

Sánchez, R. A., Eyherabide, G. and de Miguel, L. (1981). The influence of irradiance and water deficit during fruit development on seed dormancy in *Datura ferox* L. *Weed Research* 21(3–4), 127–132.

Sawhney, R., Quick, W. A. and Hsiao, A. I. (1985). The effect of temperature during parental vegetative growth on seed germination of wild oats (*Avena fatua*). *Annals of Botany* 55(1), 25–28.

Schmitt, J. and Wulff, R. D. (1993). Light spectral quality, phytochrome and plant competition. *Trends in Ecology and Evolution* 8(2), 47–51.

Seiler, G. J. (1997). Anatomy and morphology of sunflower. In: Schneiter, A. A. (Ed.), *Sunflower Technology and Production*. American Society of Agronomy, Crop Science Society of America, Soil Science Society of America, Madison, WI, pp. 67–111.

Shah, F. A., Ni, J., Chen, J., Wang, Q., Liu, W., Chen, X., Tang, C., Fu, S. and Wu, L. (2018). Proanthocyanidins in seed coat tegmen and endospermic cap inhibit seed germination in *Sapium sebiferum*. *PeerJ* 6, e4690. doi: 10.7717/peerj.4690.

Singh, P., Dave, A., Vaistij, F. E., Worrall, D., Holroyd, G. H., Wells, J. G., Kaminski, F., Graham, I. A. and Roberts, M. R. (2017). Jasmonic acid-dependent regulation of seed dormancy following maternal herbivory in Arabidopsis. *New Phytologist* 214(4), 1702–1711.

Somody, C. N., Nalewaja, J. D. and Miller, S. D. (1984). The response of wild oat (*Avena fatua*) and *Avena sterilis* accessions to photoperiod and temperature. *Weed Science* 32(2), 206–213.

Steadman, K. J., Ellery, A. J., Chapman, R., Moore, A. and Turner, N. C. (2004). Maturation temperature and rainfall influence seed dormancy characteristics of annual ryegrass (Lolium rigidum). *Australian Journal of Agricultural Research* 55(10), 1047–1057.

Szemruch, C. L., Renteria, S. J., Moreira, F., Cantamutto, M. A., Ferrari, L. and Rondanini, D. P. (2014). Germination, vigour and dormancy of sunflower seeds following chemical desiccation of female plants. *Seed Science and Technology* 42(3), 454–460.

Thomas, J. F. and Raper, C. D. (1975). Seed germinability as affected by the environmental temperature of the mother plant. *Tobacco Science* 19, 98–100.

Vayda, K., Donohue, K. and Auge, G. A. (2018). Within- and trans-generational plasticity: Seed germination responses to light quantity and quality. *AoB Plants* 10(3), ply023.

Venable, D. L. and Levin, D. A. (1985). Ecology of achene dimorphism in *Heterotheca latifolia*. I. Achene structure, germination and dispersal. *Journal of Ecology* 73(1), 133–145.

Watson, C. E. and Watson, V. H. (1982). Nitrogen and date of defoliation effects on seed yield and seed quality of tall fescue 1. *Agronomy Journal* 74(5), 891–893.

Wurzburger, J. and Koller, D. (1976). Differential effects of the parental photothermal environment on development of dormancy in caryopses of *Aegilops kotschyi*. *Journal of Experimental Botany* 27(1), 43–48.

Yang, L., Jiang, Z., Liu, S. and Lin, R. (2020). Interplay between REVEILLE1 and RGA-LIKE2 regulates seed dormancy and germination in Arabidopsis. *New Phytologist* 225(4), 1593–1605.

Yin, R. and Ulm, R. (2017). How plants cope with UV-B: From perception to response. *Current Opinion in Plant Biology* 37, 42–48.

Zha, P., Liu, S., Li, Y., Ma, T., Yang, L., Jing, Y. and Lin, R. (2019). The evening complex and the chromatin-remodeling factor PICKLE coordinately control seed dormancy by directly repressing DOG1 in Arabidopsis. *Plant Communications* 1(2), 100011.

Chapter 3

Applying population-based threshold models to quantify and improve seed quality attributes

Kent J. Bradford and Pedro Bello, University of California-Davis, USA

> It has been suggested that seeds are unpredictable things and consequently it would be a waste of time to try to perceive laws of behavior among such erratic individuals. . . . But other disciplines have shown us quite clearly that, although the behavior of any individual in a population may be quite unpredictable, the behavior of populations of individuals can often be defined very accurately.
>
> *Eric Roberts (1973)*

1 Introduction

Seeds are the remarkable achievement of evolution to convey most terrestrial plant species from one generation to the next. In some cases, this happens immediately after the seeds are released from the mother plant, while in others, seeds are the vehicle to transport species from one satisfactory growing

http://dx.doi.org/10.19103/AS.2022.0105.05

environment to the next, for example, through winter freezes or summer droughts, or to a new location entirely. Increasingly, as human impacts on the natural environment mount, we will be attempting to aid species in adapting to changing conditions by replanting them or establishing them in locations more amenable to their future success. On the other hand, some species will be able to take advantage of the changing conditions and become invasive in new locations through the assistance of their seeds. In addition, critically for feeding humans and their domesticated animals, seeds are the delivery system for transferring improvements incorporated by breeders into new varieties into the field for crop production. If the seeds do not fulfill their role by germinating and producing viable seedlings, that value is not transferred and productivity is less than it could be. Efficiency in crop production will be essential for reducing the expansion of agriculture's footprint and maintaining natural ecosystems (Folberth et al., 2020). At the same time, biological approaches to controlling weeds in crop production are increasingly focusing on managing or eliminating soil weed seed banks (Ramesh, 2015). Thus, understanding what seeds pay attention to and how they respond to environmental signals is critical to sustaining plants that we wish to promote (crops and native species) and restricting those that we wish to suppress (weeds and invasive species) (Long et al., 2015; Klupczyńska and Pawłowski, 2021).

This chapter will focus on approaches to understand seed behavior by modeling germination responses to key factors in their environment, particularly temperature (T) and water potential (ψ), that are being impacted by climate change. Seeds' responses to these environmental signals are also influenced by their physiological state, such as dormancy to time germination to coincide with appropriate conditions for seedling success and seed age or vigor that limits the useful lifetime of a given seed. Our attention will be primarily on what we have learned from the application of these population-based threshold (PBT) models to quantify and predict seed behavior. It is not expected that farmers need to become mathematicians or statisticians in order to optimally manage their planting dates or seedbed preparation. However, they do need to understand intuitively how seeds respond to their environments and how to extract the best performance possible from the seeds that they plant. Similarly, seed companies focus primarily on developing new varieties, but in the end, their product is a seed and its performance in the greenhouse or field determines the likelihood of a repeat customer. In addition, as some herbicides that have been relied on for weed control become less effective due to plant adaptation through genetic and metabolic resistance mechanisms (Perotti et al., 2020), managing weed seed production and persistence in the soil seed bank is assuming greater importance. Ecologists seeking to maintain or reestablish species in natural habitats must be able to characterize seed dormancy and germination patterns in order to schedule planting dates for successful plant establishment and regeneration (Larson and Funk, 2016; ten Brink et al., 2020). Thus, we will also emphasize how models quantifying seed behavior, particularly as a result of its population nature, can be used to assess, improve and maintain seed quality, and we will introduce new tools to facilitate the application of these models for practical purposes.

2 The importance of seed quality and behavior in crop production and weed control

It is hard to overemphasize the impact of seed quality and seedling emergence on crop production and management. In some highly domesticated and self-pollinating crops, such as wheat, rice or beans, seed production largely resembles crop production, with some additional precautions to maintain the identity of the variety and of the seeds produced. The seeds retain little dormancy from their pre-domestication ancestors and generally have relatively few problems in stand establishment that cannot be overcome by planting a bit more seed or adjusting planting dates. In addition, these crops can tiller or branch and fill spaces left by the failure of emergence. The situation is less simple for maize or many vegetables, including tomatoes, peppers, melons, onions, carrots and Brassicas, as almost all modern highly productive varieties in these crops are F1 hybrids, requiring specialized seed production methods to produce separate male and female parent plants and then enable them to cross specifically to produce the F1 hybrid seeds. In addition, each seed produces one plant in the crop field, and modern varieties are bred to perform best at specific plant populations per hectare. Thus, emergence failure means a missing production unit, and even delayed emergence can handicap the tardy plant due to shading by its neighbors, making it essentially a weed by using resources (water, nutrients) but not producing a harvestable plant.

These issues are particularly critical for vegetable crops. In lettuce or cabbage, for example, each plant produces one harvestable head, so if seeds are planted at the desired population density, emergence failure means lower yield per hectare. In addition, the optimal harvest time is dependent upon plant development to a specific state, so delay in emergence by even 2 or 3 days may mean additional costly harvest operations. Overseeding followed by thinning can assure an optimal plant population but with additional costs for seeds and labor for thinning at an early stage. Transplanting seedlings directly into the field solves these problems for the farmer but transfers them to the transplant producer. Flats of transplants must be fully filled, and all seedlings should be at an identical stage of development to satisfy customers. If seed germination is uniform and complete, this is simple. But if the seed lot germinates less than 100% or is not uniform in its emergence, this creates major and costly problems for the transplant producer to sort seedlings and assemble new trays of usable transplants. The requirements and demands of both field and transplant customers are therefore transferred to the company producing the seeds to not only provide a variety that grows well in the intended climate and meets consumers' desires and demands but also deliver that variety in seed lots of extremely high quality and uniformity. This is in addition to the fact that most of these crops are also hybrids and may require hand pollination and greenhouse or off-season production in diverse locations and may take 2 years to produce one seed crop, as in the biennials such as cabbage, onion or carrot. We will therefore briefly identify some of these issues and indicate how understanding seed behavior can contribute to assuring the high seed performance that is expected in modern agriculture.

2.1 Seed production, conditioning and enhancement

Seed quality begins with seed production and, in particular, with the seed maturity at harvest. In a number of crops, it was once thought that when the seeds had attained maximum dry weight, seed development was completed and harvest could begin. However, it has become clear now that the final stages of seed maturation after full dry weight accumulation but prior to desiccation are important for maximum seed quality, particularly with respect to the development of seed longevity (Egli, 2017; Leprince et al., 2017; Ellis, 2019). This creates a particular problem for indeterminate crops in which flowering and seed development continue sequentially so that a range of seed maturities are present simultaneously (Still and Bradford, 1998). This issue is exacerbated by the fact that many of these same crops retain the seed-shattering trait of their undomesticated ancestors, which has been eliminated in most domesticated grain and pulse crops (Maity et al., 2021). When seed shattering is present, the more mature seeds will be released from the fruit and lost before the later seeds are mature (Steiner and Akintobi, 1986). Thus, the timing of harvest is always a trade-off between optimal quality and highest yield, and the inevitable mix of maturities creates additional problems for seed conditioning and performance. PBT models have been used to quantify seed germination characteristics of seeds of different maturities in cabbage and identify features that could be useful for guiding decisions on when to harvest such crops (Still and Bradford, 1998). New developments in seed conditioning and image analysis are creating opportunities to identify and separate seeds on a number of criteria, potentially enabling the upgrading of such lots (Galletti et al., 2020; Mortensen et al., 2021). Correlating these performance indices with physical features of seeds can be applied to guide seed conditioning equipment to remove lower-quality fractions from the seed lot (Bruggink and van Duijn, 2017; Bello and Bradford, 2021; https://www.seed-x.com/product/sorter/; https://videometer.com/videometerlab/; https://www.vibeia.com/qm3-seeds-phenotyping-analyzer). Physiological enhancement of seed performance via priming treatments can also be optimized with the use of PBT models (see Section 4.5).

2.2 Crop management and yield

As indicated above, it is difficult to recover during the growing season from the consequences of poor stand establishment. Some crops can spread to fill empty spaces, but others cannot, while weeds consider such gaps as opportunities. The lack of uniformity among plants creates problems for the timing of management activities, such as pest control, pollination, growth regulator applications or harvest dates. The most critical factor in stand establishment in the field is also the least controllable: the weather immediately after planting. In most crops, guidelines are well established with respect to the minimum soil temperatures at planting. Crop management is also now largely guided by degree-day (or thermal time) thresholds associated with different developmental stages. This enables the management of plants or pests in relation to actual growth and development rates rather than calendar days. The importance of a well-prepared seedbed, precision control of planting depth and optimal water management are also widely appreciated, even if achieving such goals can be thwarted by soil conditions or unexpected changes in weather.

Understanding and modeling seed behavior can have multiple applications for optimizing stand establishment. The degree-day models mentioned above are based on the cardinal temperatures (minimum, optimum and maximum) for seed germination, and the models for determining these values and the associated germination rates will be explored in detail in Section 4.1. A key parameter of germination models is related to the uniformity of germination among the seeds in the lot. As standard germination tests are often conducted under optimal conditions, the variation observed sets an upper limit for the uniformity that can be expected in the field. However, standard germination tests focus on total viability and seldom quantify uniformity, so this information is generally not available. It is also useful to understand and appreciate the sensitivity of seeds to the T and ψ conditions of the seedbed. Even relatively slight reductions in water availability can have considerable negative effects on germination speed and uniformity. As climates are changing, the ability to match seed lots with marketing environments in which they will be successful is a considerable advantage for a seed company and its customers (https://climate.ai). The speed of germination is the first component of seed vigor that decreases in association with seed aging, and it can be quantified by PBT models (Section 5.2). Knowing where a seed lot is in its progression from high vigor and viability toward death is critical for managing seed inventories for sale or for germplasm conservation. In addition to focusing on the crop, the management of weed populations is a critical component of efficient agriculture. Germination and dormancy modeling of weed seed banks and their behavior in response to weather patterns can enable the prediction of weed emergence patterns and target management actions to control them (Section 5.1).

2.3 Transplant production

Given the difficulties and uncertainties noted above for field planting, it is no surprise that transplanting is becoming more popular for crop stand establishment, particularly for high-value crops. In many cases, seeded flats are placed in climate-controlled rooms so that optimal temperature and moisture conditions can be maintained during germination and emergence to increase uniformity. Even so, problems can persist, such as in coordinating the emergence of rootstock seedlings of wild species accessions for grafting with hybrid scion varieties in tomatoes (Mahmoud, 2020). Machines have been developed specifically to image young plants and sort them into trays with uniform seedlings for automated grafting (https://www.tta.eu/equipment/selecting/flexsorter). However, given the imperatives of scheduling and the financial risks of failure to achieve a required standard of usable plants, it is not uncommon to do pretests of seed lots to confirm the seeds' response to the planned conditions and assess uniformity before large-scale plantings. Doing these pretests in such a way that they can be assessed by germination models would improve both their accuracy and utility. For example, doing a test at two temperatures enables estimation of the germination rates across the entire temperature range likely to be employed, enabling the selection of a temperature that assures complete germination in the shortest time possible. The same approach can be applied to testing seeds prior to applying priming treatments that can speed germination and improve the uniformity of emergence (Section 4.5). Appreciation of the high sensitivity of seeds to water availability will encourage the use of irrigation systems that can apply water very uniformly and accurately to minimize variation among individual seeds or trays. Using transplants has reduced the risks of stand establishment for the farmer, and it has moved the critical germination and seedling emergence phase into more controlled conditions. However, it has also increased the expectations for seed performance.

These considerations indicate why it would be beneficial for seed companies and seed testing organizations to more fully adopt population-based approaches to assessing and managing seed quality. The models discussed below are based on knowing the time courses of germination, on every seed tested if possible; this will require further automation of seed testing procedures. The labor required to do repeated observations (at least daily, often more frequently) to capture adequately the full germination time

Published by Burleigh Dodds Science Publishing Limited, 2022.

course is unsupportable using human analysts. However, various methods are now available to collect such information robotically and process it in a form usable for model fitting and evaluation (Ducournau et al., 2005; Bello and Bradford, 2016; Colmer et al., 2020). Thus, while using germination-based models may have been applicable only for research in the past due to the labor requirements, advancements in imaging and robotics soon will routinely and accurately acquire data for complete germination time courses, enabling full use of the capabilities of the models described in the following sections.

2.4 Seed dormancy and germination characteristics in weed control and plant population ecology

Our primary focus in this chapter will be on crop seeds, but understanding and characterizing seed dormancy and germination behavior are also critical for managing unwanted plants (weeds) and encouraging the survival or reestablishment of plants in natural locations. Weed management entails the study of the soil seed bank (Schwartz-Lazaro and Copes, 2019), which persists as a result of seed dormancy mechanisms (Soltani et al., 2019). PBT models have been applied effectively to understand weed seed bank persistence and emergence in relation to environmental factors (Boddy et al., 2012; Batlla and Agostinelli, 2017). In ecological contexts, PBT models have been utilized in combating invasive species (Meyer and Allen, 2009; Hawkins et al., 2017; Liu et al., 2019), enabling revegetation (Larson and Funk, 2016) and characterizing species demographics over time (Gremer et al., 2016; Liu et al., 2020). Given the wide diversity of seed dormancy types and plant life cycles (Saatkamp et al., 2019; Fernández-Pascual et al., 2021), we can only cover a few examples of the possible applications of PBT approaches in seed ecology. However, we note that the population-based understanding of seed biology we describe is especially useful in ecological and plant diversity contexts due to the more prominent dormancy mechanisms present in wild species compared to domesticated plants and their importance in bet-hedging adaptations (Rubio de Casas et al., 2015; Gianella et al., 2021).

3 Understanding the population-based behavior of seeds

3.1 Basis of population-based approaches to seed biology

It is a fundamental fact of seed biology that seeds can be regarded simultaneously as individuals and as members of a larger population or cohort. Seeds are designed for dispersal, with each seed on its own to assess the environment in which it finds itself and to determine whether and when to commit itself to germination (Bassel, 2016). That commitment is a life-or-death decision for the individual, as once a seed has initiated germination to the point of protrusion of the embryo from its enclosing tissues, it cannot go back. If it has sensed and interpreted its external and internal conditions correctly, its embryo/seedling has a higher probability of success in growing to maturity. If it misinterprets the conditions or fails to heed negative signals, the risk of failure for that seedling is increased. The evolutionary pressures on seeds to make the correct choices have resulted in the myriad of strategies embodied in seed dormancy and environmental signaling systems represented among the diversity of plant species and climates around the globe (Baskin and Baskin, 2014, 2021).

However, while it is individual seeds that face the dilemma of whether to germinate now or wait until later, it is the collective action of populations of seeds that ultimately determines the success of the species or crop, whether in nature or in agricultural fields (Mitchell et al., 2017; Gremer et al., 2020). In nature, species whose seeds can more accurately predict the optimal time to germinate in their environment have a greater probability of long-term success in that environment (Larson and Funk, 2016; ten Brink et al., 2020). There are still risks involved, however, due to variability in environmental conditions, leading to bet-hedging strategies in which the seed population as a whole consists of subpopulations responding differently in variable environments (Gremer et al., 2016; Gianella et al., 2021). Some seeds in the population will be less dormant and will germinate immediately when adequate conditions (e.g., temperature, water) are present, possibly reaping the advantage of early emergence ahead of competitors, but also risking death if the favorable environmental conditions do not persist. Other seeds in the same population may take the opposite strategy, bypassing apparently adequate opportunities for years and representing a bank account of seeds invested in the future. Success for the species incorporates the likelihood of failure for most seeds but sufficient spreading of risk to assure that a sufficient fraction of individuals establish and grow to reproductive maturity.

Once humans began collecting seeds and purposefully replanting them, the selective pressures changed and traits such as retention of mature seeds on the mother plant (nonshattering) and absence of dormancy were associated strongly with successful propagation. Rather than the wide diversity in individual responses often characterizing seed populations in native species (e.g., bet-hedging strategies), farmers prefer seed populations to germinate immediately and simultaneously after planting, taking the responsibility on themselves to select the optimal timing for successful emergence and subsequent growth and yield (Finch-Savage and Bassel, 2016). At the same time, cultivation activities changed the selective pressures on non-crop species, resulting in the selection of weeds whose seed dormancy characteristics enable them to both persist for long periods in the soil and to emerge at times that enhance the likelihood of their seedlings to both compete with the crop and avoid removal (Malavert et al., 2021a).

3.2 Characterizing germination timing of seed populations

When studying seeds, whether in natural or agricultural systems, we must appreciate the roles of both the individuals and the populations that they collectively constitute. The range of seed germination behavior observed in a given population is not an indication of measurement error, stochastic noise or uncertainty but simply the reality of their inherent biological diversity. The most common measure of seed performance is the percentage of germination (or viability) in the population after being exposed to a suitable condition of hydration and temperature for a period of time. Here, we will be interested also in the cumulative increase in the percentage of seeds germinated with time after initiation or the germination time course (Fig. 1). Whenever a measurement is expressed as a percentage, it immediately indicates that there are individuals in the population that can vary in that particular measure. Thus, the total population is composed of individuals or subpopulations that can differ in their characteristics or responses, which themselves are often of a quantal or digital nature (i.e., living or dead, dormant or nondormant, germinated or not, etc.).

Despite its persistent usage in the scientific literature, the term 'germination rate' is incorrect when applied to data measured in percentages. In physics, a 'rate' is a measure of a change over time, or a velocity, and it should be also in seed biology. A germination 'rate' could therefore be either the increase in percentage germination per unit of time or the time it takes a given percentage of seeds to germinate after imbibition. The former term is equivalent to the slope of the cumulative germination time course, which approaches linearity around 50%, but is actually sigmoid (Fig. 1). However, rather than representing a rate, the slope is actually determined by the variation among seeds in their individual times required to achieve embryo (generally radicle) emergence. Therefore, the term 'germination rate' should refer to the time required for a given percentage of the seed population to germinate and will therefore be defined at a specific percentage, for example, 50% to indicate the median time to germination in the population (Fig. 1). As the germination time course is not symmetric around 50%, that percentage should be termed the 'median', that is, half of the seeds germinating faster and half slower. The mean (average) time to germination is shifted to longer times due to the asymmetry of the time courses. In a symmetric distribution (such as the normal distribution), the mean and the median are equivalent.

The inverse of this time (*t*) to a specific germination percentage (*g*) (i.e., $1/t_g$) is considered to be the germination rate for that percentage (GR_g) with

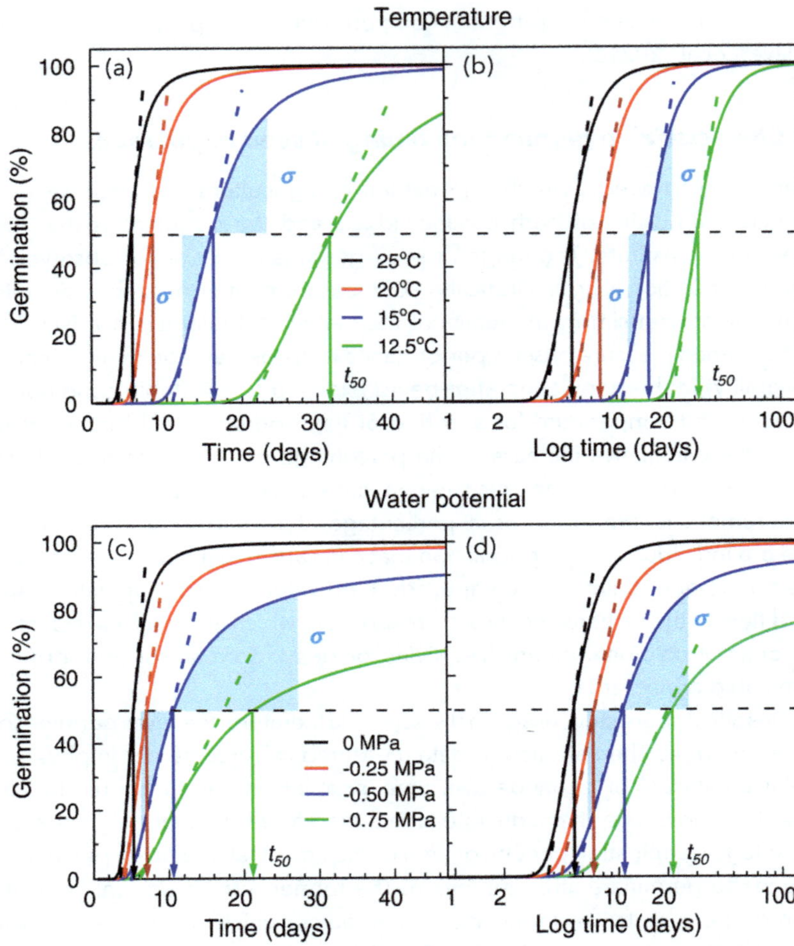

Figure 1 General features of seed germination time courses in response to temperature (T) (a, b) and water potential (ψ) (c, d) on a linear time scale (a, c) and a logarithmic time scale (b, d). (a) The black curve shows a standard germination time course on water (0 MPa) at 25°C with a base temperature (T_b) of 10°C. The red, blue and green curves show the time courses for the same seed lot at 20°C, 15°C and 12.5°C on water. The sides of the shaded areas on an example time course indicate the time from 16% to 50% or 50% to 84% germination, which is ±1 standard deviation (σ) in the seed population, assuming a normal distribution. Note that the period on the time axis is not symmetrical around the median (50%) but rather is much shorter for 16–50% compared to 50–84%, indicating that the time courses are not symmetric but rather are right skewed. The dashed lines indicate the initial slopes of the curves, which decrease as the time to germination increases. The times to 50% germination (t_{50}) are also shown on each curve (colored arrows). (b) Plotting these time courses at different temperatures on a $\log_{10}$ scale results in parallel slopes for different temperatures and more symmetric sigmoidal curves. Note also that at suboptimal temperatures, it is generally observed that all seeds can eventually germinate even at lower temperatures (above T_b) when the incubation period is long enough. (c) The black curve shown is under the same initial conditions as in (a), but the temperature

units of hour^{-1} or day^{-1}. The slope of the cumulative germination curve is a result of the variation in germination times in the population, with highly uniform populations having steep germination time course curves while less-uniform populations have shallower curves with larger times between successive germination events. The slopes of the time courses are also sensitive to environmental conditions, particularly T and ψ (Fig. 1). It is evident that on a population basis, the key descriptive parameters in relation to germination are the percentages of seeds that can complete radicle emergence, also termed germination *sensu stricto* (Nonogaki et al., 2007), the times when individual seeds did so (or their inverses, the germination rates for a specific cumulative percentage) and variation in those times among individual seeds in the population (Fig. 1).

The analysis of seed behavior in population terms has a long history, but Eric Roberts was a key innovator in applying population-based approaches to quantify seed characteristics (Roberts, 1960). In 1961, he demonstrated that the increase in germination percentage as rice seeds lost dormancy during storage could be described by a cumulative normal distribution or a sigmoidal curve (Roberts, 1961). This showed that the after-ripening period required to satisfy the dormancy requirements varied among seeds, with the highest frequency around the mean duration and lower frequencies earlier and later. The required after-ripening periods were distributed normally among seeds in the population, which resulted in a symmetrical sigmoidal curve of increasing germination percentage with increasing storage duration. Roberts illustrated that plotting this data on a probability (or probit) scale based on standard deviation units (Bliss, 1934) resulted in a straight line, the slope of which is inversely proportional to the standard deviation of the distribution of individual after-ripening requirement times in the seed population. Consistent with the terminology discussed above for germination rates, he also noted that 'the gradient of the curve showing percentage germination plotted against time does not indicate the rate of breaking dormancy, as has sometimes been implied in the literature. A steep gradient shows that the spread or standard deviation of the distribution of dormancy periods among the seeds of a population is small, whereas a shallow gradient represents a wide spread of dormancy periods' (Roberts, 1961).

Figure 1 (*Continued*)

is maintained at 25°C while the ψ is reduced to −0.25 (red), −0.5 (blue) or −0.75 MPa (green) relative to a median base water potential (ψ_b*(50)*) of −1 MPa with a standard deviation (σ_{ψ_b}) of 0.3 MPa. The skewness in the time courses is even more evident as ψ decreases as shown by the shaded areas indicating apparent σ values. (d) Plotting germination time courses at different ψ on a log scale does not completely normalize the slopes or the skewness of the time courses. See Sections 4.1 and 4.2 for further discussion of the features of these time courses.

Richard Ellis and Eric Roberts subsequently extended this approach to quantify seed longevity in relation to the temperature and seed moisture content (or water activity) at which seeds were stored (Ellis and Roberts, 1981; Roberts and Ellis, 1989). After an initial lag period of variable duration, the percentage of seeds retaining viability over storage time also closely matched a normal distribution, with the (negative) slope of the regression line on a probability scale indicating the standard deviation of the variation in individual seed lifetimes within the population under the specified conditions. This parameter is sensitive to the temperature and seed moisture content during storage, enabling Ellis and Roberts to develop their seed longevity equation (Ellis and Roberts, 1980; see Section 5.2). The parameters of this equation (seed viability constants) can be used to estimate the potential seed longevity under combinations of temperature and moisture content (Ellis, 1988) and are compiled (when known for a species) at the Kew Seed Information Database (Royal Botanic Gardens, 2021). The Ellis–Roberts viability equation and other models to predict seed longevity in storage or under natural conditions (Walters et al., 2005; Long et al., 2015; Solberg et al., 2020) represent critical insights and achievements in the development of mathematical models in seed science but will not be addressed in detail in this chapter. In general, these models of seed longevity or persistence address only final percentages of germination/viability in a seed population rather than the kinetics of germination over time following imbibition, which will be the primary focus here due to its greater relevance to seed performance in agriculture. We will discuss the effects of seed storage and aging on germination rates in Section 5.2.

A final general point can be made before examining the details of seed responses to different factors. The timing of seed germination for a seed lot is often calculated using mean germination times (MGT) according to a commonly used formula (ISTA, 2020), which multiplies the number of newly germinated seeds (n) at a given time by that time (t), sums those values, and divides by the total number of germinated seeds, or

$$\mathrm{MGT} = \Sigma(nt)/\Sigma(n). \tag{1}$$

While these MGT values match fairly closely with the median time to germination when germination is rapid and complete in the population, it becomes increasingly skewed to longer times when germination is delayed (Fig. 1). The exponential increase in time to germination of the slower seeds increasingly shifts the MGT values to longer times relative to the median when germination is delayed by environmental conditions. This issue is compounded when total germination is less than 100%, as the population (n) is often defined as only the number of seeds that actually germinated, regardless of the number of seeds tested. This error is not too problematic when most of the

Published by Burleigh Dodds Science Publishing Limited, 2022.

seeds germinate (e.g., 90% or more), but as the total germination percentage goes down, this measure becomes increasingly unreliable as a description of the total population (Bewley et al., 2013; Gianinetti, 2020). In the extreme, if only two seeds in a seed population germinated, the MGT calculation based on germinated seeds would be the average of the times to germination of those two seeds. If this MGT value was compared to that of a seed lot that germinated 100%, it would be comparing the best two seeds in one population to the average for the other, which is clearly a misleading comparison. It is true that an MGT value based only on the viable seeds will give an estimate of when those seeds will germinate, but it must be paired with the germination percentage to give an accurate description of the seed lot performance. Thus, in general, for modeling germination, the complete seed population should be considered, and comparisons between seed lots should be based on the same fractions in the entire population, not upon values for only the fraction of the population capable of germinating under a given condition. For example, the times at which seed lots achieved a given germination percentage that they all exceeded, say 25%, could be used to compare across those lots. The importance of comparing seed lots with these population principles in mind will become evident as we explore the patterns that seed populations exhibit when germinating in response to different environmental and physiological factors.

3.3 Population-based threshold models describing seed germination behavior

In addition to approaching the analysis of seed behavior through population-based concepts, another key feature is the seeds' responses to different levels or dosages of a given factor influencing germination. The approach here has been to utilize thresholds, which imply that there is a minimum dosage or factor level required to initiate the response. This concept has a long history in biological responses to T, for example, as characterized by a minimum or base T (termed T_b) below which the response does not occur. Above this T_b, the process or response progresses at a rate dependent upon the amount by which the current T exceeds T_b, as will be explored in detail in Section 4.1. This is the basis for the degree-days or thermal time approach to normalizing developmental processes across variation in T noted previously. This combination of a minimum threshold sensitivity and an increasing rate of progress as the dosage increases above this threshold is fundamental for describing seed behavior in response to many factors. However, while in thermal time calculations, the minimum thresholds are often considered to be constant among individuals, in the case of seeds, these thresholds can vary among individuals. As shown earlier for seed dormancy loss and seed aging, the variation in these thresholds is often normally distributed among seeds in the population. Thus, instead of a single threshold, there is a population distribution of thresholds that can be characterized by its median and standard deviation. Each seed has a specific sensitivity threshold, and once it has been exceeded by the dosage level of the factor of interest, progress toward germination can begin (or will be prevented in the case of inhibitors), and it will proceed at a rate proportional to the amount by which the factor level exceeds the threshold (negatively in the case of inhibitors, i.e., slowing germination). It turns out that this fundamentally very simple PBT model can describe seed behavior in response to essentially all of the factors that influence germination, as will be illustrated in detail in the following sections.

3.4 General features of germination time courses and their characterization for seed populations

The remainder of this chapter will focus on the description and application of these PBT models that characterize, quantify and predict seed germination rates and percentages across a wide range of conditions (Bradford, 1995). They are based on the concept that seeds require particular environmental (e.g., T, ψ, O_2) or physiological (e.g., after-ripening, chilling, light, hormones) factors to exceed a certain minimum level or threshold in order to proceed toward germination. In addition, the PBT models explicitly incorporate the concept that as the amount by which a given factor level exceeds its threshold (or base) level required for a response, the time to exhibit the response will be shortened proportionately. That is, the time to germination is inversely proportional to the difference between the current factor level and the threshold for response to it. This can be generalized by the following equation for a given factor X that promotes germination:

$$\theta_X = \left(X - X_b\right)t \quad (2)$$

$$\text{or, } GR = 1/t = \left(X - X_b\right)/\theta_X \quad (3)$$

where X is the level or concentration of factor X, X_b is the base (or threshold) level for a response to X, θ_X is a time constant, and t is the time to germination. Since θ_X is a constant, this means that the difference between X and X_b is inversely proportional to the time required for the response to occur (t) or directly proportional to the germination rate ($GR = 1/t$). If X is a germination-promoting factor, then once X exceeds X_b, germination can occur, and the greater the amount by which X exceeds X_b, the shorter the time required to complete germination, so that θ_X remains constant. If X is a germination-inhibiting factor, then X_b refers to the concentration or level that will prevent germination, and the difference term can be reversed (i.e., $X_b - X$) so that as the level of X increases toward X_b, the difference between them decreases and the value of t increases until germination is prevented (i.e., $t = \infty$) at $X = X_b$:

$$\theta_X = \left(X_b - X\right)t. \tag{4}$$

As an intuitive understanding of how the components of this model interact is helpful for appreciating how different factors influence germination, we present a simple analogy based on a ball rolling down a hill and then up and over a second hill onto a flat surface (Fig. 2). In this model, the height of the second hill represents the threshold that the ball (representing the seed) must climb before it can germinate (i.e., reach the flat surface beyond). The factor levels are represented as the starting heights of the ball on the first hill; when $X \leq X_b$, the ball will not have sufficient momentum to climb the second hill, and germination is prevented. As X (starting height) increases above X_b for germination promoters, the ball will have sufficient momentum to overcome the threshold hill and proceed further distances past it, the latter representing the germination rate (*GR*) or 1/*t* (Fig. 2a). The more momentum gained (higher factor level), the farther the ball can roll past the threshold hill (higher *GR* or faster germination). These relationships are reversed for inhibitors: *GR* is reduced as increasing levels (downward on the scale in this case) approach X_b, the level that prevents germination completely (Fig. 2b). The time constant (θ_X) is represented by the slope of the first hill; at the same factor level, the final momentum will be the same, so the threshold hill can be overcome, but the speed of the ball will be slowed (*GR* reduced) if the first hill is less steep (θ_X is increased) (Fig. 2c). A change in θ_X will proportionately affect all factor levels. This analogy aspires to make the relationships among the three major parameters of the PBT models clear. However, if the ball and hill analogy is not intuitive for the reader, an alternative hydraulic analogy has also been presented (Donohue et al., 2015).

Equations 2 and 4 describe the response of an individual seed to promoters or inhibitors relative to its own X_b or threshold. To extend this to a population basis, either the time constant (θ_X in the case where X is temperature) or the threshold values (X_b for other factors) can vary among individual seeds (models for specific factors influencing germination will be discussed in detail below). In the ball and hill model, this is represented as a range of X_b thresholds, each representing a specific germination fraction within the total population (Fig. 2d). For a given factor level (starting height), how far the ball will eventually roll is determined by how much momentum is required to climb the second hill (threshold). As these thresholds vary among individual seeds (i.e., $X_b(g)$, where g refers to a specific germination fraction in the population), they must be described in population terms. In most cases, this variation among seeds in their individual thresholds will occur in a normal distribution, which can be defined by its mean and standard deviation (Fig. 3a). The larger the standard deviation in sensitivity thresholds among seeds, the larger the range of dosages

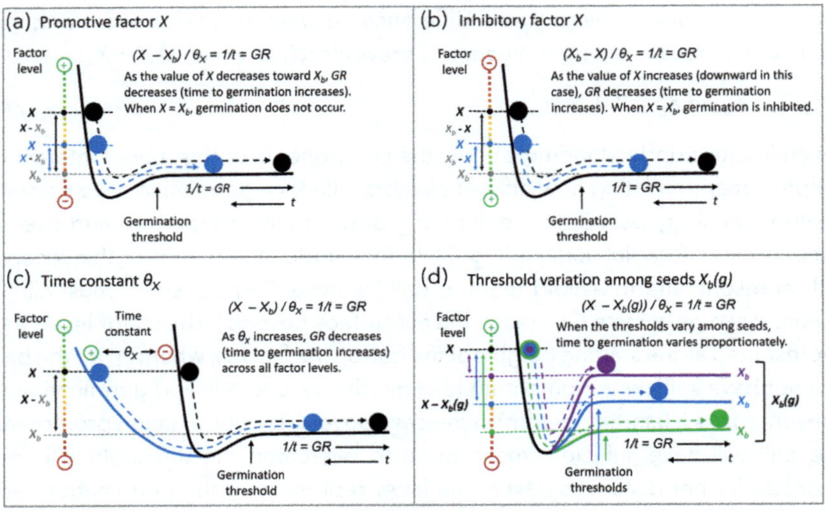

Figure 2 Rolling ball analogy to population-based threshold models. To conceptually illustrate the features and parameters of population-based threshold (PBT) models, consider the case of a ball rolling down an incline and then rolling up over a subsequent hill (threshold). The distance that the ball would roll on the flat after surmounting the hill represents the germination rate (*GR*, or 1/*t*). (a) In the case of a factor promoting germination (e.g., temperature or GA), decreasing the factor level would lower the starting height of the blue ball, giving it less momentum to overcome the hill (threshold), and it would subsequently roll a shorter distance, indicating a slower *GR*, or longer time to germination; increasing the factor level would have the opposite effect, speeding germination. If the promotive factor level is reduced to equal X_b, germination is prevented as the ball will not have sufficient momentum to overcome the threshold. (b) To illustrate the effect of germination inhibitors (e.g., more negative ψ or ABA), we reverse the scale on the factor level; that is, lower factor levels increase the height of the ball and *GR* is increased. As the inhibitory factor level increases (downward on the scale; blue ball) it approaches X_b, reducing *GR* and preventing germination when $X = X_b$. (c) The time constant (θ_X) has the effect of speeding or slowing germination at all factor levels; a shallower slope of the incline illustrates the effect of increasing θ_X. It does not affect the final momentum, so the ball can still overcome the threshold (second hill), but it will delay germination proportionately across factor levels. In the case of suboptimal temperatures, the value of θ_T varies among seeds in the population (see Fig. 4). (d) For most factors, it is the threshold that varies among seeds, illustrated as variation in the height of the second hill that the ball must climb. Thus, for a given factor level *X* (starting height on the first hill), the ball will use increasing amounts of momentum to scale the second hill, leaving it with less to proceed on the flat, or reducing *GR*. When the value of X_b increases to equal *X*, germination is prevented, that is, the ball would not have sufficient momentum to climb the second hill. To view an animated version of panels in this figure, see www.animation.pbtmodels.org.

of a given factor that is required to permit (or prevent) germination of the entire population. A consequence of this model is that the effect of a given change in the level of factor *X* will not have an equal effect on all seeds in a population.

This is why germination time courses are not symmetrical, with the time from the beginning of germination to 50% being shorter than the time from 50% to 100%, or right skewed in time (Fig. 1a and c). This becomes increasingly evident as the time to germination is extended, such as by lowering the *T* or ψ. In the case of suboptimal temperatures, seeds often share a common T_b, so a given change in *T* has the same proportional effect on all seeds, but its effect is still more pronounced on the slower germinating seeds. This is illustrated in Figure 1a, where a reduction in temperature slows germination of all seeds, but has a greater effect on the slower seeds in the population, resulting in the shift in the median slope of the latter curve on a linear scale. When these curves are plotted on a logarithmic time scale, however, the median slopes of these time courses are essentially parallel (Fig. 1b). Thus, the effect of lowering the temperature has the same proportionate effect on all seeds, but as the differences in times to germination increase logarithmically within the population, the differences become progressively larger with time on a linear scale.

If those seeds were exposed to reduced ψ rather than lower *T*, times to germination are also delayed; the differences between the early and late germinating seeds are increased and are not fully compensated by plotting on a log scale (Fig. 1c and d). This is because rather than sharing a common base water potential threshold (ψ_b), seeds vary in their thresholds with respect to ψ. This results in a much greater effect of a given change in ψ for some seeds compared to others, depending upon their position in the sensitivity distribution (Fig. 3a). When the promotive factor level (e.g., high ψ) exceeds the response thresholds of all seeds, germination of all seeds is rapid and the asymmetry in the germination time course is hardly noticeable (Fig. 3b, black curve). As the level of the promotive factor is reduced (or the inhibitor level increases, e.g., ψ becomes lower or more negative), seeds with higher thresholds (right side of the distribution) are more strongly affected than the seeds with lower thresholds (left side of the distribution). This is reflected in the greater delays in germination at higher percentages compared to lower percentages (Fig. 3b). As the factor level is further reduced, an increasing fraction of seeds are unable to germinate as the factor level falls below their individual thresholds, and the remaining seeds take proportionately longer to germinate. It is also important to note that as the model is driven by the difference between *X* and X_b, the same effect occurs if the factor level is held constant and the threshold distribution is increased or decreased. Thus, as we will see, dormancy can be alleviated by reducing inhibitors, increasing promoters, or by shifting the cellular sensitivity thresholds to either or both of these regulators. This basic pattern has been identified in seed responses to essentially all environmental and hormonal factors that influence germination, indicating that population variation in

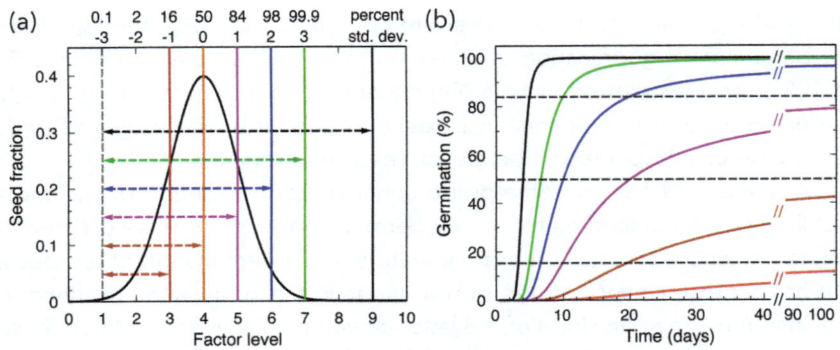

Figure 3 Sensitivity distribution in a seed population and consequent effects on germination time courses. (a) The distribution of sensitivities to a given factor affecting germination can be described in most cases as a normal distribution (black curve). This distribution is centered at the median (same as the mean for a normal distribution) or the 50th percentile in the population (upper scale). Percentages can be converted to probit values, which are in standard deviation units from the mean (lower scale above panel). The percentage values corresponding to increments of one standard deviation are shown, which is nonlinear; 68% of the population lies within one standard deviation below and above the median. The colored vertical lines indicate different levels of a factor that promotes germination (e.g., high water potential or GA). The dashed arrows indicate the difference between the factor levels and the minimum level that must be exceeded for any seeds to germinate (dashed vertical line). It is apparent that a given reduction in the factor level (e.g., from level 9 to level 6) has a much greater effect on seeds with higher thresholds (right side of the distribution) than on seeds with lower thresholds (left side of the distribution). Also, as the factor level falls below the threshold of a given fraction of the population, those seeds are unable to germinate. (b) The germination time courses predicted at each factor level shown in panel (a). Due to the relationship described above, the effect of a given reduction in factor level is much greater for the higher percentages (which have higher response thresholds) than for the lower percentages (with lower thresholds), although all fractions will respond proportionately to their own thresholds. This results in the right skewing of germination time courses and in the reduction in final germination percentages as the level of the required promotive factor is reduced. The time axis is broken at the right to indicate the extended time required for the slowest germinating seeds to approach the final percentages predicted by the model.

response thresholds is a fundamental feature of seed biology and likely of many other aspects of biology (Bradford, 2018).

3.5 Implications of population-based models for the management of stand establishment

Understanding these population-based features of germination time courses and thinking of seeds in population terms is essential in managing them to achieve optimal performance in agricultural or ecological contexts. From an agricultural point of view, we want to minimize the variation in the time of germination and maximize the emergence percentage. Already from the

Published by Burleigh Dodds Science Publishing Limited, 2022.

simple examples shown, it is clear that T and ψ are critical factors for uniform germination and that the farther they are from optimal, the greater the impact on uniformity. Thus, providing optimal conditions during the initial imbibition and germination phase after planting is critical for achieving maximum uniformity. In addition, reducing the variation in thresholds within the seed population is an important criterion for seed quality. A fundamental physiological feature of a seed population (i.e., a given seed lot) is its sensitivity distribution, as characterized by its median and standard deviation (σ). It is this inherent variation within the seed lot that we need to focus on, as its effects are magnified under suboptimal environmental conditions. In addition, these population features can change during seed maturation and storage, so optimizing them during harvest, storage and delivery to the grower is critical for the highest performance. As seed companies have tools to improve seed uniformity (e.g., grading, sorting, priming, etc.) (Halmer, 2004; Bruggink and van Duijn, 2017; Mortensen et al., 2021), understanding these basic behavioral principles of seed populations and using population-based models to characterize these enhancement operations provide opportunities for advancements in seed technology.

In ecological applications, seed lots are often collected in nature and are much less uniform than agricultural seeds. They also often exhibit diverse types of dormancy that may require exposure to specific conditions to enable germination (Baskin and Baskin, 2020). On the other hand, uniformity in germination after planting may be less critical or even detrimental, depending upon the planting conditions and potential lifetime of seeds in the soil (Burghardt et al., 2015). For example, the dormant fraction of a seed lot planted for restoration purposes may be able to survive in the seed bank and emerge in a subsequent season. However, the low success rates of such ecological interventions (Commander et al., 2020) indicate that better knowledge about and prediction of native seed behavior is sorely needed.

4 Population-based models for environmental factors affecting seed germination

4.1 Temperature (thermal time model)

The population-based models for describing seed germination time courses are based upon the threshold concept described above (Bradford, 1995). In general, a threshold indicates the magnitude or intensity of some requirement or input that must be exceeded for a certain reaction, phenomenon, result or condition to occur. In the case of a seed, a threshold is the minimum effective level or dosage of the relevant factor required for the seed to respond by initiating germination (or inhibiting germination for repressors). This is the case for thermal time models, for example, in which there is a base (or minimum) temperature below which germination will not occur (or will take infinitely long). Above this T_b, germination is possible, and its rate (inverse of time to a specific percentage germination percentage) will increase linearly with temperature (Fig. 4b). This is the thermal time model, which states that the thermal time required for germination (θ_T) of a specific percentage or fraction (g) of the seed population ($\theta_T(g)$) is equal to the degrees in excess of T_b (or $T - T_b$) multiplied by the time to germination of percentage g (t_g) (Bierhuizen and Wagenvoort, 1974; Alvarado and Bradford, 2002).

$$\theta_T(g) = (T - T_b)t_g. \tag{5}$$

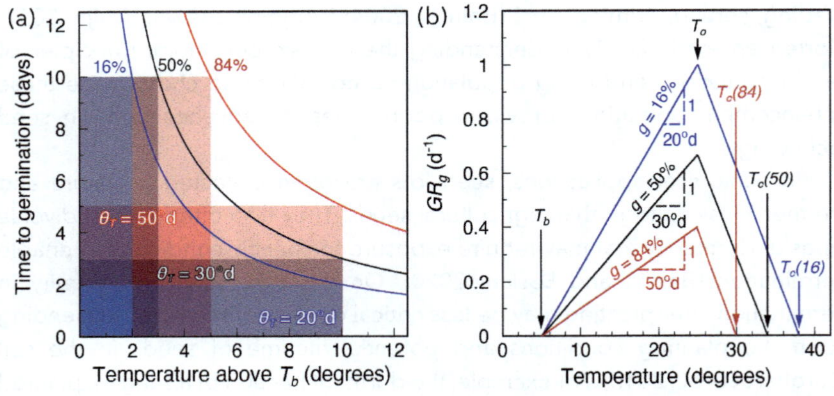

Figure 4 Temperature and thermal time relationships of seed germination. (a) The thermal time concept is based upon the fact that the time to germination for a given percentage fraction of the seed population is inversely proportional to the degrees above the base temperature (T_b). The blue, black and red curves show this relationship for the 16%, 50% and 84% fractions or one standard deviation below and above the median. As is indicated by the shaded rectangles, any such rectangle constructed below these curves encloses the same area, for example, 20, 30 and 50 degree-days (°d) for 16%, 50% and 84%, respectively. Thus, as the temperature increases above T_b, the time to germination reduces proportionately so that their product remains constant. This fundamental feature of thermal time enables developmental time to be normalized across temperatures by putting time and temperature on an accumulated thermal time basis, i.e., $(T - T_b)\,t$. Note that the differences between the curves (and between the areas of the thermal time boxes) are not equal, due to the skewing of germination time courses as percentage increases. (b) Plotting the inverse of the times to germination from panel (a) (i.e., the germination rates or GR_g) results in linear relationships with temperature. The accumulated thermal time ($\theta_T(g)$) is now represented by the inverse of the slopes of the lines for each fraction (g), and T_b (arbitrarily set at 5°C) is their common intercept on the x-axis. This provides a convenient method to determine T_b. These linear relationships only hold up to the optimum temperature (T_o, at which germination is most rapid). Increasing temperature above T_o generally results in parallel decreasing GR values with different ceiling temperatures ($T_c(g)$) at which germination is prevented (intercepts on the x-axis).

The thermal time (θ_T) is in degree-day (°C d) or degree-hour (°C h) units, indicating that it is the product of the temperature and the accumulated time. As there is an inverse relationship between T in excess of T_b and t_g (see curves in Fig. 4a), and $\theta_T(g)$ is a constant for a given fraction g, this means that any combination of T and t_g will result in the same rectangular area under the T versus t_g curve for fraction g (see shaded rectangles in Fig. 4a), which is equal to $\theta_T(g)$. In seed germination tests, it is generally possible to control the temperature to a specific value for the duration of the test, so that $T - T_b$ multiplied by the time to germination of a given percentage (t_g) equals the thermal time for that percentage ($\theta_T(g)$). However, in the field, the temperature varies daily during the germination and emergence period, so the thermal time can be accumulated for each day (i), and germination will occur when the total thermal time requirement is achieved.

$$\theta_T(g) = \Sigma_i (T - T_b)_i (t_g)_i. \tag{6}$$

As temperature varies diurnally, the daily mean temperature is often used for thermal time calculations. This also works for seed germination, with the caveat that some seeds require alternating temperatures for germination, so germination performance will not be the same if the seeds are held at the mean temperature constantly (Batlla et al., 2003). In general, developmental timing varies across temperatures for most plants, insects and poikilothermic (cold-blooded) animals and can be normalized on a thermal time scale, resulting in its widespread use in modeling and predicting crop growth and maturity, pest management and many other biological processes in addition to seed germination (Trudgill et al., 2005).

It is often, but not always (see Section 5.1), the case that all seeds in the population will have the same (or very similar) value for T_b. In this case, the value of $\theta_T(g)$ varies depending upon the physiological variation in t_g in the seed population. When plotted as germination rates ($1/t_g$) versus time, the lines representing different fractions of the population will vary in slope but with a common intercept at T_b (Fig. 4b). The inverses of the slopes of these lines will be equal to $\theta_T(g)$, as seen by rearranging Equation 5:

$$GR_g = 1/t_g = (T - T_b)/\theta_T(g), \tag{7}$$

where GR_g refers to the germination rate (GR) for fraction g. Between the base and optimal (T_o) temperatures, this equation generally works well to describe the rates of germination for a given seed population across temperatures (as in Fig. 1a). Recommendations for many crops will include minimum soil temperatures at planting and the expected times to seedling emergence based upon the thermal time requirements for germination and subsequent seedling growth to the soil surface.

Equations 5 and 7 predict that the time to germination for a given fraction g would continue to decrease indefinitely as $T - T_b$ increased. However, this generally occurs only until an optimum temperature (T_o) is reached; with further increases in T, the time to germination (or other biological process) slows down and eventually stops or is prevented when a maximum or ceiling (T_c) temperature is reached (Fig. 4b). In many cases, this T_c varies among different seeds in the population ($T_c(g)$), resulting in the characteristic triangular shapes of complete GR temperature responses shown in Fig. 4b (Covell et al., 1986; Alvarado and Bradford, 2002; Bakhshandeh et al., 2020). This figure reemphasizes that each seed in a population represents a particular fraction g in that population and will have a unique time to germination (t_g) or rate (GR_g) determined by its physiological state (potential speed of germination) and its base, optimum and ceiling temperatures. These latter temperatures are known as the cardinal temperatures for seed germination and are critical for understanding seed performance in relation to temperature (Alvarado and Bradford, 2002; Rowse and Finch-Savage, 2003). Figure 4 also illustrates that for a given factor (T in this case), seeds can share a common threshold within a population (as for T_b) or can vary in their thresholds (as for $T_c(g)$). It is also evident that the earliest germinating seeds have the greatest GR values at suboptimal temperatures and the highest T_c values, or more generally, the most rapidly germinating seeds are also the most vigorous and stress tolerant.

The simplest description of germination in response to temperature therefore has two components, one for temperatures between T_b and T_o (suboptimal region) (Equations 5 and 7) and another for temperatures between T_o and T_c (supraoptimal region) (Equations 8 and 9) (Ellis et al., 1986):

$$\theta_T = \left(T_c(g) - T\right)t_g, \tag{8}$$

$$\text{or, } GR_g = 1/t_g = \left(T_c(g) - T\right)/\theta_T. \tag{9}$$

For this supraoptimal region, it is the T_c that varies with g ($T_c(g)$), while θ_T (with a value distinct from $\theta_T(g)$) is a constant for all seeds. Above T_o, increasing temperature inhibits rather than promotes germination rates, so the terms are reversed on the right side of the equation; as T increases above T_o toward T_c, the difference between them decreases and the time to germination increases proportionately to maintain θ_T constant. This results in the parallel linear relationships with decreasing GR_g between T_o and $T_c(g)$ (Fig. 4b). There also are modifications to this bilinear model that use continuous variables to describe this change in direction of germination speed rather than two linear components (Grundy et al., 2000; Rowse and Finch-Savage, 2003), which is appropriate if the germination behavior exhibits a plateau or curve near T_o rather than a sharp reversal (Watt and Bloomberg, 2012).

The models in Equations 8 and 9 also assume that thermal time would continue to accumulate as T increased above T_o. However, Alvarado and Bradford (2002) found that the models fit their data better if thermal time did not continue to accumulate above T_o, that is, if Equation 8 was modified in the supraoptimal temperature region to Equation 10, capping thermal time accumulation to that achieved at T_o.

$$\theta_T = \left(T_c(g) - T_o\right)t_g. \tag{10}$$

On the other hand, Rowse and Finch-Savage (2003) proposed that thermal time should continue to accumulate but decreasing germination rates should start to occur below T_o. This will be discussed further in Section 4.3 after the introduction of the hydrotime and hydrothermal time (HTT) models.

4.2 Water (hydrotime model)

In addition to a temperature between its T_b and T_c, a seed also requires water in order to germinate. Pure water represents the highest water potential (ψ) value (0 MPa) at a given temperature, so reducing ψ (to more negative values) has a proportional effect in delaying germination, just as reducing temperature toward T_b slows germination (Fig. 1) (Bradford, 1990, 1995). At the minimum or base threshold ψ (or ψ_b), the time to germination becomes infinite, or germination is prevented. A critical development in modeling seed germination was based on the insight of Gummerson (1986) that individual seeds could differ in the ψ_b thresholds that must be exceeded to allow them to proceed toward germination. Combining these two components resulted in the hydrotime model for germination:

$$\theta_H = \left(\psi - \psi_b(g)\right)t_g, \tag{11}$$

$$\text{or, } GR_g = 1/t_g = \left(\psi - \psi_b(g)\right)/\theta_H, \tag{12}$$

where θ_H is the hydrotime constant (MPa h or MPa d), $\psi_b(g)$ represents the base water potential of fraction g of the seed population (MPa), and t_g is the time to germination of fraction g (h or d) (Bradford, 1990). This pattern is displayed in Fig. 3b if we use ψ as the factor, that is, factor level 9 represents pure water and the other factor levels represent progressively lower ψ values. All of the features of the germination time courses at different ψ values, the delays in germination, the skewness of the curves, the final germination percentages, and so on, are generated by this simple model that requires only three parameters: the value of θ_H, the mean ($\psi_b(50)$) and standard deviation (σ_{ψ_b}) of the $\psi_b(g)$ distribution among seeds in the population.

In contrast to the common T_b in the suboptimal temperature region (Fig. 4b), ψ_b values vary for different fractions of the population (Figs. 2d and

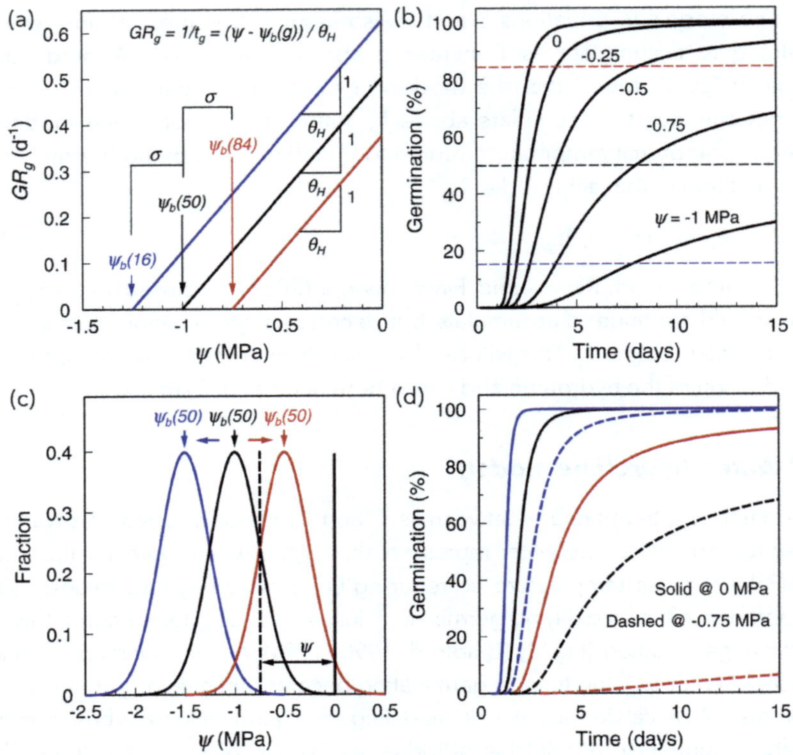

Figure 5 Key characteristics of the hydrotime model of seed germination. (a) Plotting the germination rates to different percentages (GR_g or $1/t_g$) shows that while the hydrotime constant (θ_H or the inverse of the slope of the lines) is the same across percentages, the intercepts on the *x*-axis (the base water potentials or $\psi_b(g)$) differ among percentages. When the ambient water potential (ψ) is 0 MPa, GR_g is at its maximum for each percentage of the population. As ψ decreases, the GR_g values fall and become zero when $\psi = \psi_b(g)$ for that fraction of the population. Values are plotted for 16%, 50% and 84%, representing the population mean and one standard deviation (σ) above and below it. (b) Germination time courses are illustrated for the seed population in panel (a) at different ψ values between 0 and −1 MPa (the latter is the $\psi_b(50)$ used in this illustration along with σ_{ψ_b} = 0.25 MPa and θ_H = 2 MPa d). The horizontal dashed lines indicate the 16%, 50% and 84% germination percentiles for comparison with panel (a). Note that relatively slight reductions in ψ notably delay germination, particularly in the slower fractions of the population. The curves for −0.75 and −1 MPa will asymptotically approach 84% and 50% germination, as indicated by the $\psi_b(50)$ values for these percentages (panel a), after very long times (more than 200 days). (c) This panel illustrates shifts in the $\psi_b(g)$ distributions from the values used in panels (a) and (b) (black curve) to 0.5 MPa higher (red curve) or lower (blue curve), without changing the σ_{ψ_b} or θ_H values. The vertical solid line indicates the relationship of the environmental ψ to these distributions when ψ = 0 MPa, and the vertical dashed line indicates when ψ = −0.75 MPa. (d) Germination time courses for the seed populations indicated in panel (c) at ψ = 0 MPa (solid lines) or −0.75 MPa (dashed lines). Reductions in environmental ψ have a much greater effect in delaying germination in populations with higher $\psi_b(g)$ distributions (red lines) than in those with lower $\psi_b(g)$ distributions (blue lines).

5a). As ψ decreases toward ψ_b, GR_g decreases (or t_g increases) until germination is prevented when ψ is equal to or less than $\psi_b(g)$. The parallel slopes of the GR_g versus ψ lines indicate that the hydrotime constant (θ_H, or the inverse of the slope; Eqn. 12) is the same across all seeds in the population. This is similar to the situation for germination at supraoptimal temperatures (Eqn. 8; compare Fig. 5a with Fig. 4b); the reason for this will become clear in the next section. Depending upon the mean and spread of the ψ_b values in the seed population, germination will occur over time in a characteristic pattern at different water potentials (Fig. 5b). When the $\psi_b(50)$ is fairly high, for example, −1 MPa in the example shown, the time to germination increases markedly with relatively slight reductions in ψ, particularly in the slower fractions of the seed population. For example, −0.03 MPa is an approximate value for the water potential of soil at field capacity, when bulk water has drained from its pores. For seed germination, this is essentially the same as the 0 MPa curve shown. However, depending upon the soil type and weather conditions, the soil ψ near the surface can drop considerably lower over a few days. A reduction even to −0.25 MPa considerably slows germination and increases the spread in germination times for a population with a $\psi_b(50)$ of −1 MPa (Fig. 5b). In addition, although half of the seed population should be able to complete germination at −1 MPa, it would take approximately 300 days for the germination percentage to approach 50% at that ψ due to the exponential increase in the time to germination as the ψ approaches the ψ_b for a particular seed fraction, 50% in this example (Fig. 5b; c.f., Fig. 3b showing extended time axis). Seeds in the population with $\psi_b(g)$ values more positive than the ψ of the environment are prevented from germinating, that is, t_g becomes infinitely long.

In addition to the environmental conditions (primarily T and ψ) of the seedbed, the physiological status of the seeds can also dramatically affect their response to a given environment. For example, seed lots can differ in their hydrotime parameters. If the population has a $\psi_b(50)$ of either −1.5 or −0.5 MPa, two standard deviations below and above the $\psi_b(50)$ in the example above (Fig. 5c), the germination performance at a given ψ will be markedly affected (Fig. 5d). If the $\psi_b(50)$ is lower, germination will be more rapid and uniform, while if it is higher, germination will be delayed and more dispersed, even at 0 MPa (Fig. 5d, solid lines). If the soil ψ decreases, such as to −0.75 MPa (Fig. 5c), the effect on germination will be much greater on the population with the higher $\psi_b(50)$ (Fig. 5d, dashed lines). As will become evident below, $\psi_b(g)$ distributions of seed populations do change, particularly in relation to seed dormancy. Thus, both environmental conditions (soil type, rainfall, irrigation, salinity) and physiological characteristics (the value of ψ_b and its variation among seeds) determine the pattern of seed germination and subsequent seedling emergence.

These features of seed germination responses to ψ have some important implications for stand establishment. First, the speed and particularly the uniformity of germination are quite sensitive to even relatively slight drops in ψ. Thus, to achieve the most rapid and uniform germination and emergence, the water status of the seeds must be kept as high and constant as possible, but not submerging the seeds for extended periods, which leads to oxygen deprivation and poor performance (see Section 4.4). This is well known in the transplant industry, which utilizes uniform temperatures near the optimum and frequent light applications of water following seeding to maintain high and constant ψ during germination and seedling emergence. In the field, growers aim to plant seeds into adequate soil moisture and irrigate after planting if possible. Without irrigation, Fig. 5b illustrates the impacts that drying soil will have on germination rates and uniformity. However, if rainfall occurs at any time, seeds will immediately respond to the increase in ψ and shift up to a higher germination curve. In practice, this can result in a split emergence pattern, with some seeds emerging early due to the initial soil water present, followed by a second flush of emergence if a subsequent rain event occurs. This presents a very difficult management situation for the grower, as plant uniformity in crop stands is almost entirely dependent upon the uniformity of seedling emergence. After emergence, each plant grows at an initially exponential rate that becomes linear with leaf canopy closure, so later emerging plants do not actually catch up with their peers until late in the development process (Loomis and Connor, 1992). If they are shaded by neighbors, their growth rates are even further reduced. This results in problems for virtually all crop management decisions that depend upon the developmental stage, including flowering, pollination and maturity as well as fertilization and pest management (Finch-Savage and Bassel, 2016). Similarly, seed performance in a given hydration environment is dependent on the physiological quality of the seed lot, particularly the mean of the threshold distribution and the variance around it (Fig. 5c and d). A seed lot with a more negative and a narrow $\psi_b(g)$ distribution will be much more robust to reductions in soil ψ during the germination period. Understanding the hydrotime behavior of seed populations instills an appreciation for the high sensitivity of seeds to their hydration environment and the consequences of failing to provide optimal conditions during the earliest step in crop production.

4.3 Water and temperature (hydrothermal time model)

Temperature and water are two inescapable and variable environmental factors that are critical for seed germination. Thus, combining the thermal and hydrotime models into a single HTT model is an important step that also was conceived by Gummerson (1986). This was done for suboptimal temperatures by combining Equations 5 and 11.

$$\theta_{HT} = \left(\psi - \psi_b(g)\right)\left(T - T_b\right)t_g, \qquad (13)$$

$$\text{or,}\quad GR_g = 1/t_g = \left[\left(\psi - \psi_b(g)\right)\left(T - T_b\right)\right]/\theta_{HT}. \tag{14}$$

The HTT constant (θ_{HT}, MPa °C h or MPa °C d) incorporates the effects of ψ on germination on a thermal time basis. Although it was noted that germination responses to temperature are often analyzed in relation to the logarithm of time (Fig. 1b), the HTT equation uses actual times to germination (t_g). This is because the variation in $\psi_b(g)$ thresholds in the hydrotime model accounts for the skewness present in the time courses (Fig. 1c). It is more appropriate to think of the HTT model as being based on the hydrotime model but expressing the time in thermal time units to normalize for the effect of temperature. This model has proved to be a broadly applicable and quite accurate model for describing germination rates and percentages in response to both T and ψ (Dürr et al., 2015), with to date nearly 50 publications using the term 'hydrothermal' in their titles and nearly 300 total publications addressing various aspects of application of this concept in seed physiology, crop production, weed management and ecology. In the suboptimal temperature range, these equations account well for the increasing germination rate as T increases above T_b and the opposite effect as ψ decreases toward the ψ_b thresholds of different seed fractions (Figs. 4b and 5a).

The HTT model can also provide an explanation for the reversal in germination rates and inhibition of germination at supraoptimal temperatures (Fig. 6a) (Alvarado and Bradford, 2002). This behavior at supraoptimal temperatures can be explained by an increase in $\psi_b(g)$ as T exceeds T_o. That is, the $\psi_b(g)$ distribution shifts to progressively higher values as T increases above T_o (Fig. 6c). When the value of $\psi_b(g)$ reaches 0 MPa for a given fraction g, that temperature is the T_c value for that fraction (Fig. 6b and c). This progressive inhibition of germination rates as the threshold distribution shifts upward toward 0 MPa explains why the slopes of these curves are parallel above T_o and why unlike T_b, which is often a single value for a seed lot, T_c values vary among seed fractions. This behavior can be added to the HTT model by including a constant (k_T, MPa °C^{-1}) that indicates the slope of the increase in ψ_b per degree above T_o.

$$\theta_{HT} = \left[\left(\psi + k_T\left(T - T_o\right)\right) - \psi_b(g)\right]\left(T_o - T_b\right)t_g, \tag{15}$$

$$\text{or,}\quad GR_g = 1/t_g = \left[\left(\psi + k_T\left(T - T_o\right)\right) - \psi_b(g)\right]\left(T_o - T_b\right)/\theta_{HT}. \tag{16}$$

These equations reduce the difference between $\psi - \psi_b(g)$ by the value of k_T for each degree that T exceeds T_o, which is equivalent to shifting the entire $\psi_b(g)$ distribution to more positive values (Fig. 6c; Alvarado and Bradford, 2002). The larger the value of k_T, the steeper the slope of this increase would be, and

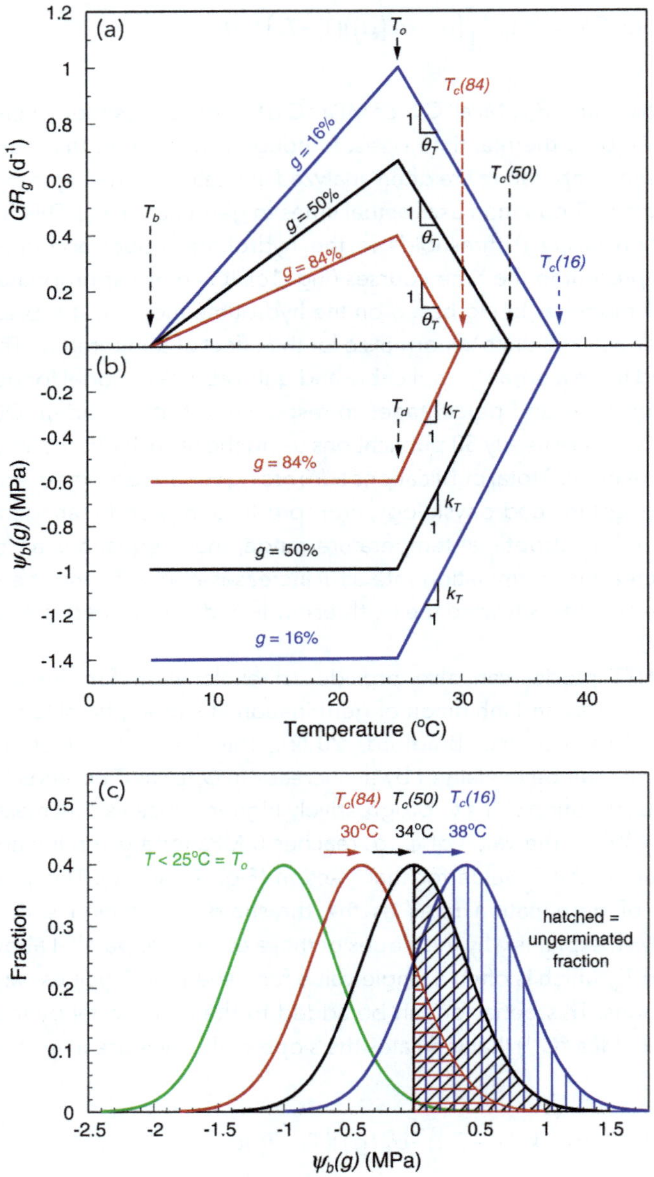

Figure 6 Relationships of germination rates to temperature and $\psi_b(g)$ distributions. (a) Illustration of the germination rates (GR_g) for g = 16%, 50% and 84% germination between T_b and $T_c(g)$ (as in Fig. 3b). (b) The $\psi_b(g)$ values of the seed population remain relatively constant below T_o, then increase with a slope of k_T per degree Celsius (Eqn. 15). The decreasing GR_g values at $T > T_o$ in panel (a) are due to this shift in the $\psi_b(g)$ distribution to higher (more positive) values. When the $\psi_b(g)$ values of a fraction of the seed population exceed 0 MPa, that fraction is unable to germinate at that temperature (GR_g = 0), which is the T_c value for that fraction. T_d indicates the T at which the $\psi_b(g)$

the sharper the drop in GR_g above T_o (Fig. 6a and b). In some cases, this can result in an inhibition of germination within a few degrees above the optimum temperature, also known as thermoinhibition (Bradford and Somasco, 1994).

Examples of how changes in the parameters of the HTT model influence germination speed are illustrated by plotting the percent germination after 5 days of imbibition across a matrix of temperatures and water potentials (Fig. 7). In an agricultural context, these examples reinforce the importance of maintaining near-optimal conditions in the seedbed or transplant tray during the initial imbibition and germination period. They also illustrate the expansion of the permissive hydrothermal space when T_b is lower or T_o is higher, or when the $\psi_b(g)$ distribution is more negative. In an ecological sense, these figures represent different possible germination niches for seeds having different HTT parameters (Liu et al., 2020). This will be discussed further subsequently in relation to seed dormancy (Section 5.1).

In Equation 15, and as discussed in relation to supraoptimal temperatures (Section 4.1), the thermal time component is accounted for by the $T_o - T_b$ term, which in essence caps the accumulation of thermal time at the amount associated with T_o. It would be expected that increasing T above T_o will continue to increase the speed of germination due to increasing thermal time accumulation. However, the germination rate is actually decreasing above T_o due to the upward shift in the ψ_b distribution. Capping thermal time accumulation at T_o solves this problem in some cases (Alvarado and Bradford, 2002). However, an alternative model has been proposed in which the increase in $\psi_b(g)$ begins at a temperature below T_o (termed T_d; Fig. 6b) and thermal time continues to accumulate with increasing T in the supraoptimal range even as the rise in $\psi_b(g)$ slows germination (Rowse and Finch-Savage, 2003; Finch-Savage et al., 2005). This results in a more gradual transition from increasing to decreasing GR rather than a sharp peak and reversal at T_o (Fig. 6). Both types of patterns have been observed (Bloomberg et al., 2009; Hardegree et al., 2015; Bakhshandeh et al., 2017), and both approaches give similar results with respect to the base and maximum temperatures and hydrotime parameters.

A key feature of thermal time models is that germination time courses at different temperatures (e.g., Fig. 1a) all collapse to a single time course when

Figure 6 (*Continued*)

distribution begins to increase, which is used in some models in which T_o is not a single T but can represent a more gradually changing transition (Rowse and Finch-Savage, 2003). (c) As the $\psi_b(g)$ distribution shifts toward 0 MPa in response to $T > T_o$, increasing fractions of the population will have ψ_b values equal to or exceeding 0 MPa; that fraction of the population is prevented from germinating (hatched areas of the $\psi_b(g)$ distributions). Thus, the slower germinating seeds (e.g., 84%) are the first to be inhibited at lower temperatures, while the faster-germinating seeds with lower ψ_b values can germinate to higher temperatures.

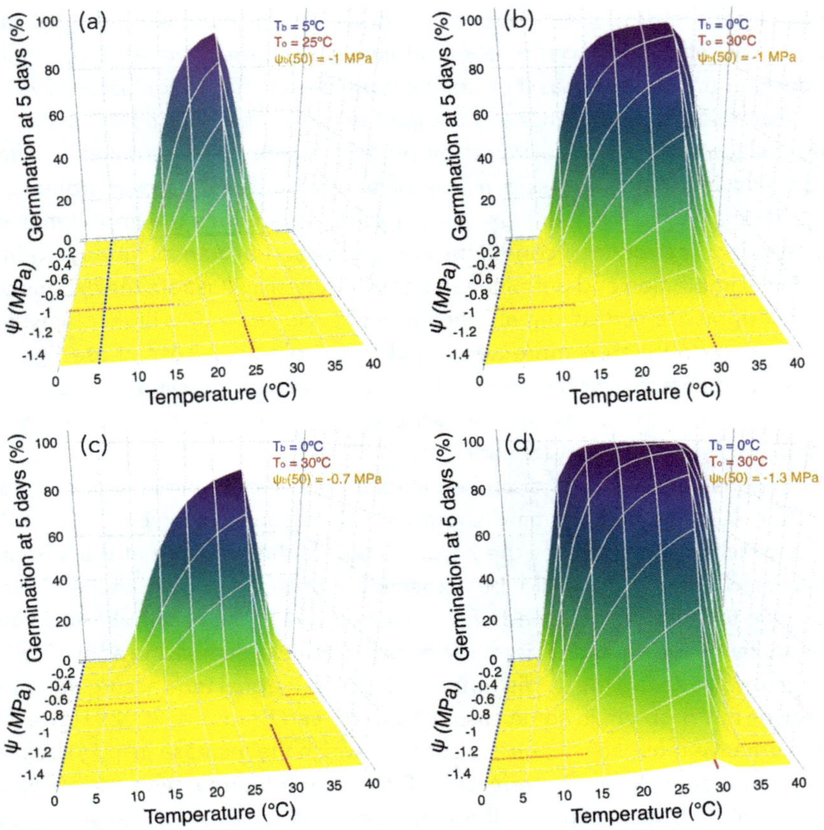

Figure 7 Examples of the effects of combinations of temperature (T) and water potential (ψ) on the percentages of seeds germinating within 5 days of planting. (a) Seeds with relatively higher T_b (5°C; blue dashed line on temperature axis) and low T_o (25°C; red dashed line on T axis). The hydrotime parameters are θ_{HT} = 50 MPa °C d; $\psi_b(50)$ = −1.0 MPa (dashed orange line on ψ axis); σ_{ψ_b} = 0.3 MPa; k_T = 0.1 MPa °C^{-1}. The narrow temperature range between T_b and T_o results in relatively slow germination except near the optimum temperature and germination decreases as ψ declines. (b) Using the same hydrotime parameters but reducing T_b to 0°C and increasing T_o to 30°C results in increased germination within 5 days at higher ψ and over a broader range of T. (c) Using the same cardinal temperatures as in panel (b), but raising the $\psi_b(50)$ to −0.7 MPa (equivalent to one standard deviation) slows germination despite a more favorable temperature range. (d) Using the same cardinal temperatures as in panel (b), but lowering the $\psi_b(50)$ to −1.3 MPa (equivalent to one standard deviation) speeds germination and expands the range of T and ψ over which high germination can occur within 5 days. Note the overall interchangeability of T and ψ in their effects on germination timing and their synergistic effects in expanding the hydrothermal germination space when both factors positively affect germination (low T_b, high T_o and low $\psi_b(50)$).

plotted on a thermal time scale (e.g., Dahal et al., 1990). This is expected from the concept of thermal time in which the time to the developmental event requires the same amount of thermal time accumulation at any suboptimal temperature (Fig. 4a). Similarly, germination time courses at different water potentials (e.g., Fig. 4b) can also be normalized using the simple function $[1 - (\psi/\psi_b(g))]\ t_g$, which will convert times to germination at any ψ to the corresponding time to germination at 0 MPa (Bradford, 1990; Dahal and Bradford, 1990), that is, on a normalized hydrotime scale. If effects of temperatures are accounted for by thermal time and those of water potentials by normalization of hydrotime, then germination time courses across all T and ψ can be plotted on a common normalized HTT scale (Dahal and Bradford, 1994; Alvarado and Bradford, 2002; Rowse and Finch-Savage, 2003; Alvarado and Bradford, 2005). Thus, the two main factors affecting germination and seedling emergence can be understood as acting on the 'biotime' scale at which germination occurs after imbibition (Bradford and Trewavas, 1994).

It is useful to think in terms of normalized HTTs, that is, the combination of T and ψ effects, in understanding and predicting seedling emergence patterns in both agricultural and ecological situations (Bradford, 1995; Finch-Savage et al., 2005; Meyer and Allen, 2009; Finch-Savage and Bassel, 2016). While the models have generally been developed using constant T and ψ, both of these vary daily in the field. Degree-day models generally use the average daily temperature, often also putting a cap on the maximum temperature, as the degree-days relationship for plant development also does not increase linearly above certain upper limits. Soil ψ also varies diurnally, daily and seasonally, with the magnitude varying with soil type and planting depth. Between irrigations or rainfall, soil ψ will decline, with a steeper gradient near the soil surface. On the other hand, soil ψ near the surface will increase quickly after irrigation or rainfall, and seeds previously prevented from germinating by low soil ψ will then germinate quickly (Finch-Savage et al., 2005). Lower water potentials are also experienced by seeds in saline soils, as the dissolved ions lower the osmotic potential of the soil solution. Halophytes (plants that can grow and reproduce in saline environments >200 mM NaCl, equivalent to ~−1 MPa) can germinate at low water potentials, and a halothermal time model based on the HTT model can describe their germination behavior in saline solutions (Seal et al., 2018). While the seed coat of the halophyte *Suaeda maritima* largely excludes the NaCl from the embryo (Szymansky et al., 2021), nonhalophytic species such as chicory (*Cichorium intybus* L.) may absorb significant quantities of salt ions, contributing to the osmotic adjustment of the embryos and enabling germination to somewhat lower water potentials (Bakhshandeh et al., 2020). The extent of this osmotic adjustment could be estimated by analyzing germination time courses using the halothermal time model. By lowering the osmotic potential of the soil solution, salinity will delay germination and can cause major problems for stand establishment in drying soils, as salts are concentrated near the surface.

4.4 Oxygen (oxygen-time model)

Another environmental factor influencing germination in the soil is the availability of oxygen (Corbineau and Côme, 1995). As the diffusion of oxygen through water is much slower than through air, seeds of some semiaquatic species, such as rice, are adapted to germinate at quite low oxygen concentrations (hypoxia), but germination of most seeds begins to be inhibited when oxygen percentage in the gaseous phase falls below 10% (Yentur and Leopold, 1976). Thus, saturation of soils by flooding can prevent germination due to restrictions of oxygen availability. Oxygen consumption rates of seeds are closely related to their germination rates and are affected by T and ψ in similar ways (Dahal et al., 1996; Bello and Bradford, 2016). It is therefore unsurprising that seed germination responses to oxygen are also described by the PBT model (Bradford et al., 2007). In this case, germination varies with the logarithm of the oxygen percentage in the gaseous phase according to the following equation:

$$\theta_{Ox} = \left(\log Ox - \log Ox_b(g)\right)t_g, \tag{17}$$

where θ_{Ox} is the oxygen-time constant [log (O_2%) d or log (O_2%) h], Ox is the percentage of O_2, and $Ox_b(g)$ is the base oxygen percentage that prevents germination of fraction g of the population. $Ox_b(50)$ values for different species varied from less than 1% to over 9%, with rice having the lowest value at 0.02%, or anaerobic. After-ripening of dormant barley grains lowered their median oxygen threshold from 36% (only 30% germination in ambient 21% O_2) to 0.3% by both increasing the permeability of the hull surrounding the grain and lowering the O_2 threshold of the embryo (Bradford et al., 2008). As hypoxia had been shown to increase the sensitivity of dormant grains to inhibition of germination by abscisic acid (ABA) (Benech-Arnold et al., 2006), PBT analyses of ABA and gibberellin (GA) effects on germination (see Section 5.5) found that sensitivity to these hormones also was influenced by oxygen availability (Bradford et al., 2008). The ability to quantify the effects of these factors and their interactions on germination revealed the complex relationships among internal (ABA and GA) and external (hull O_2 permeability) changes associated with after-ripening and dormancy loss (see also Section 5.1).

4.5 Priming and seed enhancement (hydrothermal priming time model)

Observations that seeds that had been hydrated to levels not allowing completion of germination could be dried without damage and subsequently would germinate more quickly when rehydrated led to the development of the practice of seed 'priming' (Heydecker et al., 1973; Hegarty, 1978). Seeds can be hydrated, incubated and dehydrated in a controlled manner, then distributed and planted in the normal way. Upon rehydration, these primed seeds will germinate more quickly and uniformly, resulting in improved stand establishment. Different approaches have been developed to control and limit seed water absorption during the priming treatment, using osmotica (osmopriming), clay particles (solid matrix priming) or simply limiting the amount of water provided (hydropriming) (Halmer, 2000; Halmer, 2004). Priming has also been used to pretreat seeds to break dormancy (Valdes and Bradford, 1987) or inoculate them with biological agents (biopriming) (Lutts et al., 2016). Seed priming has become a standard commercial practice in a number of higher value crops, including lettuce, tomato, pepper and celery.

A limitation for seed priming in practice is that the specific conditions for optimal treatment, primarily the T, ψ and duration of the treatments, vary among species and even seed lots within a species or variety. Thus, a range of trial treatments must be conducted empirically to identify the optimal conditions prior to treating a large commercial lot. The risk of failure is considerable, as if the ψ is too high and/or the duration is too long at the selected T, some fraction of the seeds can complete germination during the treatment and will be damaged upon dehydration. As priming is used primarily on high-value vegetable seeds, such mistakes can be quite costly. Thus a method to consistently pretest seed lots to determine their sensitivity characteristics and their responses to test conditions is valuable.

The germination kinetics of primed seeds can also be characterized by HTT models (Dahal and Bradford, 1990; Dahal et al., 1990; Bradford and Somasco, 1994; Cheng and Bradford, 1999; Patanè et al., 2016; Tatari et al., 2020). The improvements in germination rates and uniformity are primarily due to smaller values for θ_{HT} rather than to the lowering of $\psi_b(50)$. Effectively, the hydration period at reduced ψ contributes to the total HTT required for germination, so the time required after rehydration is reduced. As a lower θ_{HT} requirement shortens the time to germination for all fractions of the seed population, the variation in time of germination becomes smaller, or germination is more synchronous. Initial experiments indicated that up to a point, increasing durations of priming increased the subsequent GR_{50} (Tarquis and Bradford, 1992). In this case, since priming occurs at ψ levels below the lowest ψ_b values for the seed population (to prevent germination), but metabolic advancement can still occur, a new threshold term is required (ψ_{min}), which is the lowest ψ at which the priming effect will occur; that is, incubation at lower ψ values does not result in the advancement of GR. The same is the case for temperature, with priming time accumulating above a minimum threshold, or T_{min}. Thus, the

GR_{50} of the seed population after priming can be described on the basis of hydrothermal priming time as follows:

$$GR_{50} = GR_i + \left[\left(\psi - \psi_{min}\right)\left(T - T_{min}\right)t_p\right] / \theta_{HTP}, \tag{18}$$

where GR_i is the initial GR_{50} of the seed lot before priming (h^{-1} or d^{-1}), t_p is the duration of priming (h or d) and θ_{HTP} is a hydrothermal priming time (HTP) constant (MPa °C h or MPa °C d) (Bradford and Haigh, 1994). In practice, a matrix of test priming treatments at two or three T and ψ values and durations are conducted, the seeds are then dried, and the GR_{50} values for each test condition are determined in germination tests. With these data, regression methods can be used to estimate the effective values of ψ_{min}, T_{min} and θ_{HTP} for the seed lot. There can be complications in some cases. For example, some seeds can shift their ψ_b distribution to more negative values during extended incubation at reduced water potentials, potentially resulting in unexpected germination during the treatment (Dahal and Bradford, 1994). Some seeds are also induced into secondary dormancy by extended incubation at reduced water potentials (Gianinetti and Cohn, 2007; Hawkins et al., 2017; Malavert et al., 2021b).

While fitting PBT models requires additional preliminary pretests for each seed lot, an advantage of doing these in a systematic way and calculating the HTP parameters is that similar seed performance will be obtained from different combinations of T, ψ and t_p conditions if the total hydrothermal priming time, or $\theta_{HTP} = (\psi - \psi_{min})(T - T_{min})\,t_p$, is the same. That is, T, ψ and t_p are interchangeable as long as the total θ_{HTP} is the same. This allows optimization of materials, equipment and time for conducting priming treatments on a large scale. For example, adjusting ψ may require an osmoticant that is expensive or results in waste for disposal, so priming at higher ψ may be more economical. Similarly, raising T may shorten the actual priming time required and increase product throughput. As long as pregermination is prevented by keeping the ψ below the lowest ψ_b in the seed population, or keeping t_p shorter than the time needed for any seeds to complete germination under the chosen conditions, modification of the parameters to achieve a desired HTP time should increase GR by the predicted amount, and improved uniformity automatically accompanies increased GR. Thus, conducting the pretest systematically and using the results to calculate the HTP parameters would enable adjustment of the key parameters of the priming treatment to achieve optimal results for each seed lot with the greatest efficiency and least risk. Experience with this approach could perhaps identify consistent HTP parameters for a given species or variety of interest, allowing reduction in the number of pretest conditions required to characterize a specific lot, but as seed germination characteristics can change due to development, harvest and storage conditions, it is likely that such pretests will continue to be a component of successful priming operations.

5 Population-based models for physiological and other factors affecting seed behavior: dormancy, aging, seed vigor, respiration and other characteristics

5.1 *Dormancy (hydrothermal time and temperature limits models)*

A reduction or elimination of seed dormancy is one of the early hallmarks of the domestication of crop species. A plethora of dormancy mechanisms have been described that enable species to be successful and reproduce in their native environments (Baskin and Baskin, 2014; Willis et al., 2014). Once humans began to harvest seeds and purposefully replant them, dormancy was strongly selected against, as seeds that failed to germinate did not contribute progeny to the subsequent harvest, and it has been completely eliminated in some of our most important agricultural species. In wheat, for example, research on seed dormancy is now focused primarily on how to restore sufficient resistance to germination after development to prevent precocious sprouting in the event of untimely rain before harvest (Schramm et al., 2010; Nonogaki and Nonogaki, 2017). However, this domestication process is still in progress for many crop species, particularly those in which the seed is not the primary food product, such as the biennial vegetables. Many of these can exhibit seed dormancy and can require specific conditions to alleviate it, such as dry after-ripening, imbibed chilling or light. This may be as simple as holding seeds in a warehouse until they have after-ripened sufficiently, as for sunflowers or melons, or may involve priming treatments, such as those used to prevent the expression of thermoinhibition in lettuce seeds. It is evident that the presence of dormant seeds in a seed lot will reduce the percentage of usable transplants or affect the intended plant populations in direct-seeded crops.

Seed dormancy is a critical factor in ecological applications, where undomesticated species are being seeded into natural conditions. Native species are often exquisitely attuned to their environments, moving into and out of dormancy while in a seed bank that constitutes the future of the species (Donohue et al., 2010; Rubio de Casas et al., 2015; Finch-Savage and Footitt, 2017; ten Brink et al., 2020; Gianella et al., 2021). Successful reintroduction of species requires not only seeds from genotypes adapted to the environment of the location but also the timing of seed distribution to match seeds' expectations or requirements (Baskin and Baskin, 2020). In many cases, very low success rates are observed in such restoration efforts, while conversely, seed dormancy characteristics may play a critical role in the competitive advantages of invasive species (Meyer and Allen, 2009; Commander et al., 2020; Gioria et al., 2021).

A functional definition of dormancy is when a viable seed fails to germinate when provided conditions under which a nondormant seed of the same species would be able to complete germination. This self-referential definition is not entirely satisfactory, but it is the case that seed dormancy is generally a transient developmental state that eventually disappears to allow the seed to germinate. Between a fully dormant state in which a seed will not germinate under any conditions and a fully nondormant state in which the seed will germinate across a wide range of conditions, there are many intermediate states in which germination is conditional on the environment being within certain ranges (e.g., T or ψ) or providing certain signals (e.g., light). PBT models can therefore be useful for quantifying the status of a seed population in terms of the environmental thresholds permitting germination at a given time, and how these thresholds change in response to dormancy-breaking conditions.

The population-based HTT models have been effectively employed for both physiological and ecological analyses of seed dormancy. With respect to physiological aspects of dormancy, both the water potential and the thermal components can contribute. An early demonstration of this showed that after-ripening and dormancy loss in *Bromus tectorum* L. seeds were associated with decreases in the mean ψ_b of the seed population (Christensen et al., 1996; Bair et al., 2006), which resulted in more rapid germination and higher final percentages. Subsequent work confirmed that the loss of dormancy in seeds was associated with the progressive shift of the $\psi_b(g)$ distribution to more negative values, which enables more seeds to germinate and to do so more quickly (Alvarado and Bradford, 2005; Meyer and Allen, 2009). Manipulation of hormonal regulators of dormancy, such as ABA or fluridone (an ABA biosynthesis inhibitor), had the expected corresponding effects on the $\psi_b(g)$ threshold distribution. This is essentially the reverse of the upward shift in $\psi_b(g)$ when the temperature exceeds T_o (Fig. 6c); instead, as after-ripening (or other dormancy-breaking treatment) continues, $\psi_b(g)$ moves progressively more negative, allowing more seeds to germinate and speeding germination. Germination time courses as seed populations leave dormancy resemble those shown in Fig. 5b, but going in the direction from lower to higher ψ. As the $\psi_b(g)$ distribution shifts to more negative values, seeds with ψ_b thresholds more negative than the current ψ will start to germinate, with the time to germination inversely proportional to the difference between the soil ψ and the seed's ψ_b threshold. This gradually makes more seeds germinable over time, while still retaining sensitivity to current hydration conditions.

In addition to this mechanism based on shifts in $\psi_b(g)$, dormancy is also controlled by shifting of the temperature thresholds for germination (Batlla and Benech-Arnold, 2015). The cardinal temperatures discussed earlier (Fig. 4) assume that the seeds tested lack dormancy and that the T_b, T_o and $T_c(g)$ values are reasonably constant within a species. However, seeds of many species will not germinate across this entire temperature range when they are dormant. Instead, these cardinal temperatures represent end-point values that can

Published by Burleigh Dodds Science Publishing Limited, 2022.

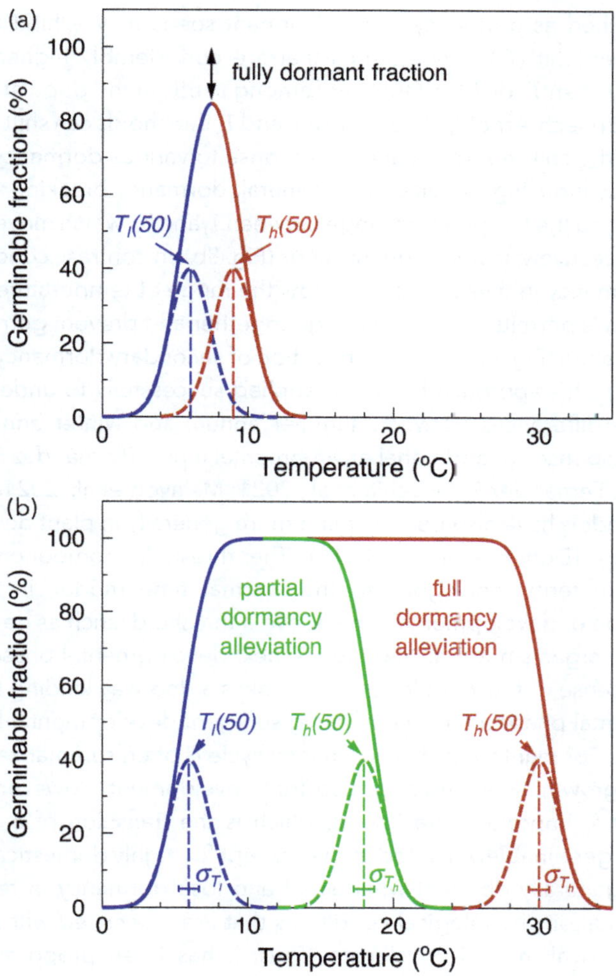

Figure 8 Dormancy alleviation by shifting of upper-temperature thresholds in a winter annual species. (a) A seed lot exhibiting dormancy due to a low high-temperature limit (T_h) for germination such that its distribution overlaps that of the lower temperature limit (T_l). This results in a narrow temperature range in which some seeds are above their T_l but still below their T_h, but a fraction of seeds are unable to germinate at any temperature. $T_l(50)$ and $T_h(50)$ indicate the medians of the lower and higher temperature limit distributions. (b) After being exposed to conditions that alleviate dormancy (e.g., chilling or after-ripening), the T_h distribution shifts to higher temperatures (green) and eventually to near the maximum temperature for germination (red curve, i.e., $T_c(g)$ for the species) when dormancy is fully alleviated. The opposite occurs in summer annual species, with inhibition of germination occurring initially at lower temperatures, and the T_l distribution shifts to lower temperatures in response to dormancy-alleviating conditions. Adapted from Batlla and Benech-Arnold (2015).

be approached as dormancy is lost. Dormant seeds may exhibit lower-limit (T_l) or higher-limit (T_h) temperatures that are considerably higher or lower, respectively, than T_b or T_c and that vary among seeds in the population (Fig. 8) (Batlla and Benech-Arnold, 2015). These T_l and T_h thresholds can shift seasonally in seeds in the soil seed bank and in response to various dormancy-breaking or dormancy-inducing conditions. In general, dormancy-breaking conditions tend to expand the temperature range between T_l and T_h, which move toward T_b or $T_c(g)$, respectively, in nondormant seeds (Fig. 8b). In contrast, conditions that induce dormancy in these species narrow the range of temperatures at which germination is permitted and can merge together and prevent germination at any temperature (Fig. 8a), as in the induction of secondary dormancy (Malavert et al., 2017). This approach has been applied successfully to understand the emergence differences between summer annual and winter annual weeds and their responses to additional environmental inputs (Fernández Farnocchia et al., 2019; Fernández Farnocchia et al., 2021; Malavert et al., 2021a).

PBT models have also been applied more generally in plant development and ecology (Donohue et al., 2015). The threshold component often is described in terms analogous to the thermal time model, in which the accumulation of developmental time above a threshold (such as degree-days) enables the organism to transition to the next developmental phase of its life cycle. This sense of the word 'threshold' evokes a doorway leading to the next developmental phase after accumulating sufficient developmental time in the prior phase. For plants, this developmental cycle is often summarized as seed, vegetative growth, flowering and seed/fruit development. However, this cycle omits another important checkpoint, which is the transition of a seed from dormant to germinable. As noted above, except for highly domesticated crops, seeds are generally dormant when shed and lose dormancy in response to environmental or physiological conditions that are associated with successful seedling survival in a given climate. Thus, it has been proposed that the transition of seeds from dormant to germinable is a legitimate developmental transition for plants, akin to vernalization that enables the vegetative to flowering transition (Huo et al., 2016). It is clearly the case that the initiation of germination is a singular once-in-a-lifetime event for an individual seed and that it triggers a massive cascade of gene expression and growth (Cadman et al., 2006; Finch-Savage et al., 2007). There are important follow-on consequences, as earlier germinators tend to have better access to light, while later emerging seedlings are likely to be shaded by neighbors and stunted. Environmental and seasonal germination niches have been identified for diverse species, which interact with dormancy thresholds to restrict germination to more favorable seasons or to take advantage of highly favorable conditions (Larson and Funk, 2016; Saatkamp et al., 2019). As temperature and water availability are the two most critical environmental factors for seedling success, it is no surprise that hydrothermal parameters characterizing seed dormancy loss and germination (e.g., as in Fig. 7) were highly correlated with species demographics in the field over decades in a desert plant community (Huang et al., 2016; Liu et al., 2020). Seed germination responses to light are also amenable to analysis by PBT models (Batlla and Benech-Arnold, 2005; Bradford, 2005; Malavert et al., 2021a). Similar approaches could be extended to apply to chilling treatments to break dormancy in both seeds and tree buds (Penfield and Springthorpe, 2012; Tarancón et al., 2017; Martínez-Berdeja et al., 2020), vernalization to promote flowering (Angel et al., 2015) and other aspects of plant growth and development (Bradford, 2018).

5.2 Aging and seed vigor (aging time model)

Another application of PBT models is to quantify seed aging and loss of vigor following harvest and during storage. The Ellis-Roberts seed viability equation is fundamental to modeling the deaths of increasing fractions of a seed lot over time when stored under specific conditions of T and ψ (Ellis and Roberts, 1981; Roberts and Ellis, 1989). Its population basis is integrated into the model, which is as follows:

$$v = K_i - p / \sigma, \tag{19}$$

where v is the probit of percent viability after storage period p, K_i is the initial viability (on a probit scale; see Section 6.1) and σ is the standard deviation of the normal distribution of seed deaths over time (days or years). As for germination, the most critical factors affecting seed aging rates are temperature and water content (Ellis and Roberts, 1981). These factors are accounted for by adjusting the value of σ depending upon the seed moisture content (m, fresh weight basis) and temperature (T) during the storage period according to the following equation:

$$\log\sigma = K_E - C_W \log m - C_H T - C_Q T^2, \tag{20}$$

where K_E is a species constant that accounts for the differences in inherent storability among species and C_W, C_H and C_Q are constants determined empirically in storage trials across a range of m and T conditions. It is important to note the logarithmic effect of m and the quadratic effect of T and that increases in either of these reduce storage life. These nonlinear effects of m and T on seed longevity are consistent with the earlier work of Harrington, whose 'thumb rules' for seed storage posited that every 1% increase in m or 10°F (6°C) increase in T reduced storage life by half, with compounding additive effects (Harrington, 1972).

It is common in the seed trade to measure seed moisture content (m) as a percentage of the initial fresh weight, while researchers generally express it as a percentage of dry weight. On either scale, the seed moisture content at a given ψ varies among species, primarily due to differences in seed oil content (Cromarty et al., 1982). Thus, Roberts and Ellis (1989) modified their longevity equation to be based on equilibrium relative humidity (*eRH*, %) rather than moisture content:

$$\log\sigma = K_E - C_W eRH - C_H T - C_Q T^2, \tag{21}$$

where the C_W constant has its meaning as earlier but with different values. This modification has not been widely adopted for the longevity equation, but there are considerable advantages to measuring seed water content as its equivalent *eRH* (see http://www.dryingbeads.org/tools for a conversion tool). While optimal seed moisture contents have to be adjusted among species, their storage characteristics are more similar on an *eRH* basis. That is, while a starchy seed (e.g., maize) might have a moisture content of 12% (fresh weight basis) at 50% eRH, an oily seed (e.g., sunflower) at the same *eRH* would have a moisture content of only 8%. However, their aging behaviors would be much more similar at the same *eRH* compared to being at the same moisture content (Bradford et al., 2018). Thus, it is standard practice in seed banks to pre-equilibrate seed lots at low RH (15%) prior to storage (see Section 7.3). In addition, while measuring seed moisture by heating in an oven requires at least several hours, electronic water potential meters can measure water activity (=*eRH*/100) in less than a minute (e.g., https://www.metergroup.com/), allowing real-time monitoring of seed moisture content during packaging, for example. Even simpler and cheaper are humidity indicator papers that can be used without any instrumentation to measure *eRH* of a seed batch for estimating harvest maturity or checking storage conditions (Bradford et al., 2016; Thompson et al., 2017).

The general pattern of seed aging begins with an initial plateau period of variable length during which the final viability remains essentially constant but the time to germination increases steadily (Fig. 9a and b). Eventually, the viability declines in a sigmoidal pattern and times to germination increase more rapidly (Fig. 9c). This pattern indicates that all seeds do not age at the same rate or die at the same time. Rather, seeds in a seed lot have a range of potential lifetimes that match well to a normal distribution, as demonstrated by the importance of σ in the Ellis–Roberts equation (Eqns. 20 and 21). However, as valuable as it is, the viability equation only predicts the fraction of seeds that are capable of germinating after a given storage period and does not address directly the variation among seed lots in storage times before viability begins to decline (see, e.g., Walters et al., 2005; Fleming et al., 2019).

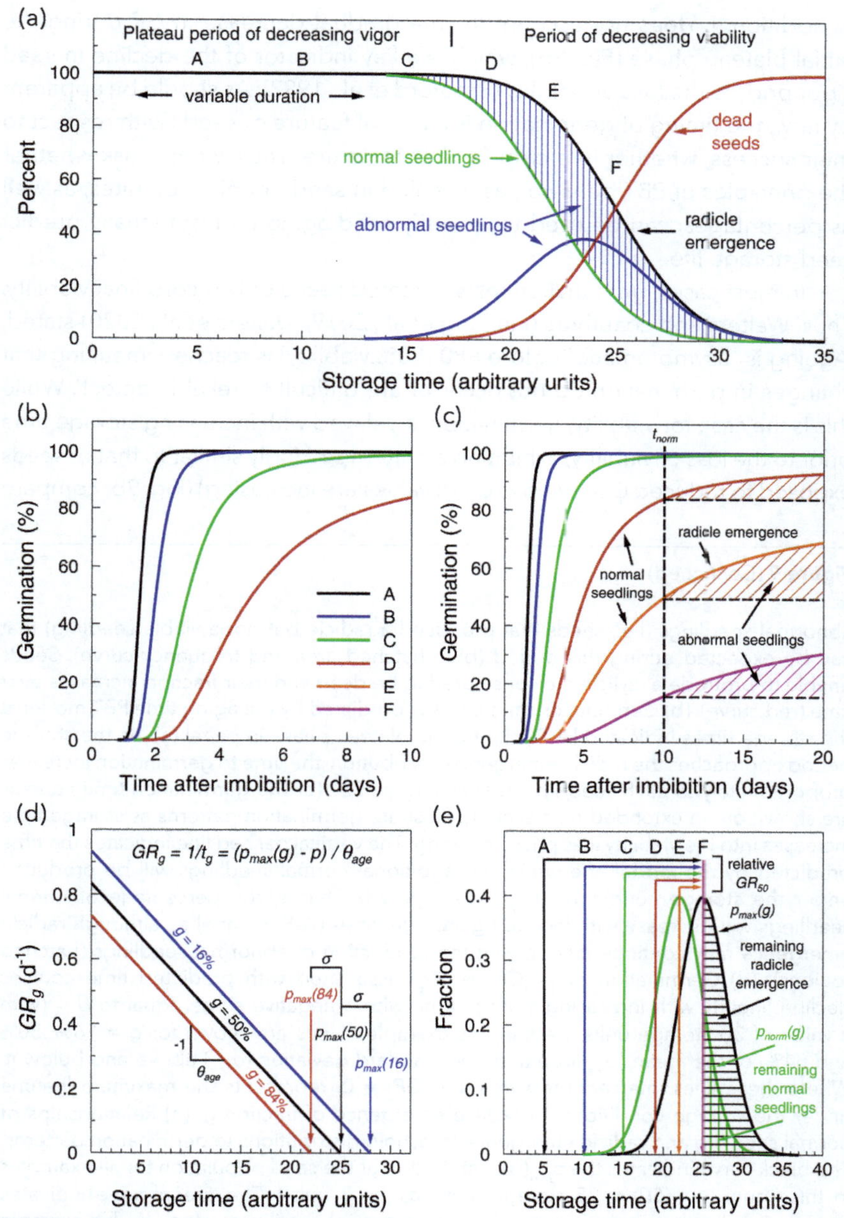

Figure 9 Threshold model for the effects of storage time on seed vigor and viability. (a) During seed storage, there is often an initial plateau period of variable length in which viability remains constant but vigor decreases followed by a period of relatively rapid loss of viability. The ability to produce normal or viable seedlings is lost first (green curve), followed by the loss of ability to even achieve radicle emergence or germination *sensu stricto* (black curve). The difference between the two curves indicates the percentage of

In addition, it does not account for the gradual decrease in *GR* during the initial plateau phase (Fig. 9d), which is a key indicator of the decline in seed vigor prior to the loss of viability (Bradford et al., 1993). As should be apparent by now, the timing of germination is a critical feature of seeds with respect to their success, whether in a crop field or in nature. Thus, we can ask whether the principles of PBT behavior, as revealed in seed germination rates, as well as percentages, can be used to quantify seed aging and potentially predict seed storage life.

In most cases, germination tests of stored seeds only record final viability. Thus, Walters and coauthors (Fleming et al., 2019; Walters et al., 2020) stated, 'Ageing is "asymptomatic" before P80 [80% viability] is reached, meaning that changes in germination do not occur or are difficult to reliably detect'. While this is the case for viability, germination is delayed with increasing storage time prior to the loss of viability, exhibiting a pattern strikingly similar to that of seeds exposed to reduced ψ (or whose $\psi_b(g)$ values are increasing) (Fig. 9b; compare

---

Figure 9 (*Continued*)

abnormal seedlings (i.e., seeds that produced a radicle but not a viable seedling) that can be expected during this period (blue hatched area and frequency curve). Seeds unable to protrude a radicle are considered to be dead and their fraction increases over time (red curve). (b) Germination time courses predicted by the aging time PBT model at the storage times indicated (A–D) during the plateau phase in panel (a). As the storage period approaches the radicle emergence distribution, the time to germination increases proportionately to each seed's $p_{max}(g)$ value (Eqn. 22). (c) The germination time courses are shown on an extended time scale to illustrate germination patterns as storage time increases into the viability loss phase of aging. The vertical dashed line indicates the time predicted by the model after which no additional normal seedlings will be produced under the storage conditions (t_{norm}; see Eqn. 23). That is, the percentage of normal seedlings will decrease with increasing storage times (A–F in panel a), although radicle emergence may continue longer, with the production of abnormal seedlings (hatched regions). (d) Germination rates ($GR_g = 1/t_g$) associated with predicted time courses decline linearly with increasing storage time with a negative slope equal to θ_{age} (with a value of 30 storage units, days in this example). Lines are shown for g = 16%, 50% and 84%, or the mean ($p_{max}(50)$) and one standard deviation ($\sigma_{p_{max}}$) above and below it. Where these lines intersect the *x*-axis (i.e., $GR_g = 0$) represents the maximum lifetime under the storage conditions for radicle emergence of fraction *g*. (e) Relationships of normal seedling and radicle emergence threshold distributions to germination patterns. The black curve indicates the $p_{max}(g)$ distribution of the seed population for all examples in this figure ($p_{max}(50)$ = 25 storage units, $\sigma_{p_{max}}$ = 3 units). The green curve indicates the $p_{norm}(g)$ distribution for the production of normal seedlings, which in this example occurs one standard deviation earlier in the storage period than the radicle emergence distribution. As the storage time (*p*) increases from A to F (vertical arrows), the difference between *p* and $p_{max}(g)$ decreases and t_g increases proportionately (Eqn. 22), resulting in the decreasing GR_g values shown in panel (d). At storage time F (= $p_{max}(50)$), only half of the seeds would be able to complete radicle emergence (eventually), and only 16% would be able to produce normal seedlings (c.f., panel c), as indicated by the hatched areas under the two distributions. Adapted from Bradford et al. (1993).

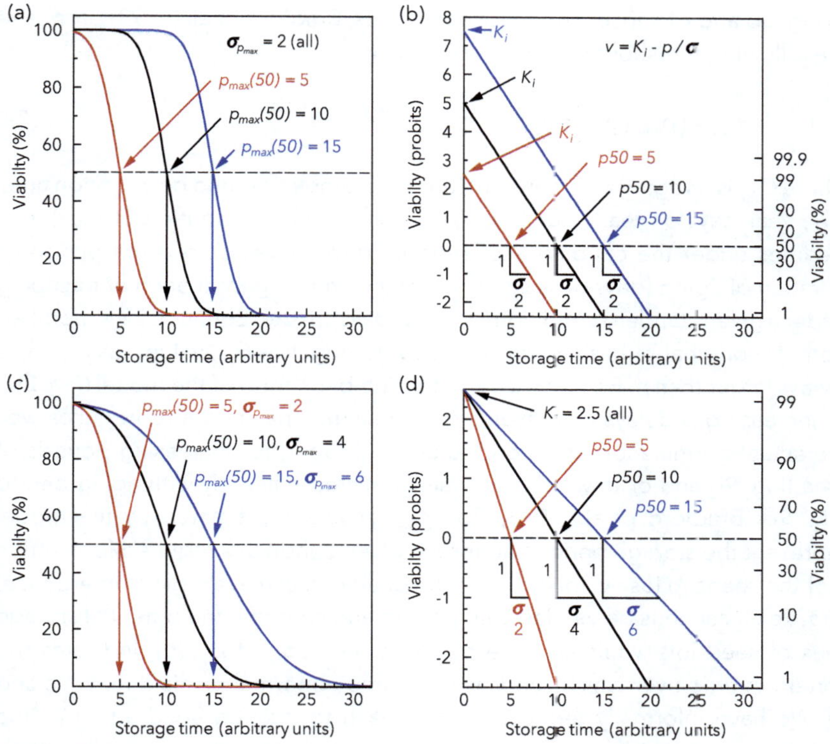

Figure 10 Comparison of the aging time model and the Ellis-Roberts longevity model parameters for the viability loss period of seed aging under different conditions. (a) For cases in which there is a plateau period prior to the beginning of viability loss (as in Fig. 9a), three examples are shown having different $p_{max}(50)$ values but the same $\sigma_{p_{max}}$ values. This could be the case for seed lots of a given species differing in initial quality but stored under the same storage conditions. Seeds with higher initial quality would have a longer plateau period (blue curve) than seeds with a lower initial quality (red curve). (b) The same viability loss time courses as in (a) analyzed by the Ellis-Roberts method, using only data after viability loss had begun. The data are plotted on a probit scale (left *y*-axis), which uses standard deviation units and transforms sigmoidal distributions into straight lines; the corresponding percentage values are shown on the right *y*-axis. The Ellis-Roberts *p50* values correspond to the $p_{max}(50)$ values in the aging model and the σ values are all equal, as the storage conditions are the same (Eqn. 20). Higher K_i values indicate the relative differences in initial seed quality among the lots (i.e., different lengths of their plateau phases), although the actual percentage values are not relevant, as the probit scale continues to approach 100% indefinitely. (c) Examples of the effect of different storage conditions on viability loss of subsamples of the same seed lot, in this case without a plateau period. Both the $p_{max}(50)$ and $\sigma_{p_{max}}$ values would increase under better storage conditions (lower *T* or *eRH*; blue curve) compared to poor storage conditions (higher *T* or *eRH*; red curve). (d) Ellis-Roberts analysis of the data in panel (c) indicates a common K_i value (it is the same seed lot in this case), which is equal to the initial viability percentage. The slopes of the lines increase (viability loss is more rapid) as the storage conditions become less favorable for seed longevity. In addition to the same information about the viability loss distribution as the Ellis-Roberts model, the aging time model also provides information on the germination kinetics during both the plateau and viability loss phases (Fig. 9b and c).

to Fig. 5b and d). Observing these similarities, Bradford et al. (1993) proposed the following threshold model of seed aging:

$$\theta_{age} = \left(p_{max}(g) - p\right)t_g, \tag{22}$$

where θ_{age} is an aging time constant (units of storage time and germination time, e.g., year days or year hours), $p_{max}(g)$ is the normal distribution of maximum lifetimes under the conditions of storage (days or years), p is the period or duration of aging (days or years) and t_g is the time to germination of fraction g under a given standard condition (hours or days). [Note: Equation 22 is modified from the original published version to avoid negative θ_{age} values, as $p_{max}(g)$ is always larger than p. Equation 22 is in the 'inhibitor' form of the model (Fig. 2b), as increasing p delays or prevents germination.] This model reproduces well the actual germination time courses of seeds aged for increasing periods of time (Fig. 9b and c), in which GR_{50} values decrease linearly with aging period (Fig. 9d; Bradford et al., 1993). The $p_{max}(g)$ values estimated by the model represent the storage periods under the given conditions required for fraction g of the seeds to lose viability (i.e., to be unable to achieve radicle emergence) (Fig. 9e). It can thus be used to characterize and compare the potential storage lives of seed lots, while $\sigma_{p_{max}}$ (the standard deviation of $p_{max}(g)$) indicates the spread in time over which seeds will become abnormal or die (Fig. 9a, c and e). We have informally termed this approach to quantify seed aging as the 'metronome' rule. It suggests that at maturity and harvest, a given seed has a fixed p_{max} period before death that is dependent upon the conditions during its maturation and in which it is stored subsequently. The seed can therefore be considered to have a fixed number of clicks of a metronome in its lifespan; however, like a metronome, those clicks can occur very slowly or very rapidly in time, depending upon the storage conditions (primarily T and eRH). Thus, the goal of the seed conservator is to enable the seed to play out its lifetime at an *adagio* (low T and eRH) rather than a *presto* (high T and eRH) tempo.

As the Ellis-Roberts model has been widely used and verified for a large number of species (https://data.kew.org/sid/viability/), it is relevant to compare it to the aging time model. With respect to the period of viability loss, the aging time model can identify the time until 50% viability loss ($p_{max}(50)$) and the standard deviation around it ($\sigma_{p_{max}}$; Fig. 10a and c). The Ellis-Roberts model uses probit analysis to identify the K_i (initial viability percentage on a probit scale), *p50* (the time to loss of 50% viability) and the slope of the probit viability loss line (the inverse of σ) (Fig. 10b and d). These values, except for K_i, are the same in both models. The K_i values initially assumed that they are the intercept of the probit lines with the y-axis, or initial viability, which works well when there is no plateau period of storage (Fig. 10d). However, when there is a plateau period of constant viability, only the data from the viability loss period should be used

Published by Burleigh Dodds Science Publishing Limited, 2022.

in the Ellis–Roberts analysis, and the K_i values for seed populations exhibiting a plateau phase will be unreasonably high when converted to percentages (Fig. 10b). This is not particularly a problem, as the differences in K_i values still indirectly indicate the relative differences in initial plateau periods among the seed lots. Thus, with respect to identifying useful parameters applicable to characterize the viability loss period of seed aging, both models provide similar information. However, the aging time model also characterizes the changes in germination time courses throughout the aging period, providing an aging phenotype for the period when viability alone is uninformative.

The aging time model also reveals other interesting features. For example, the parameters of this equation during the plateau period of the viability time course shifted to different values once viability started to decline (Bradford et al., 1993). That is, θ_{age} and $p_{max}(50)$ had larger values during the plateau period when viability was constant and had smaller values (more rapid loss with time) once viability began to decline. This is consistent with observations that once viability begins to decline, improving the storage conditions by lowering *T* is less effective than it is during the plateau period (Walters et al., 2004). The transition from the plateau to the viability loss period seems to be accompanied by increases in the rate of damage accumulation and/or decreases in the ability to repair such damage upon hydration, denoted by the change in aging time parameters. While it has been standard to use 50% viability (*p50*) as the germination percentage for characterizing longevity, it may in fact be more appropriate to use a higher value nearer to this transition point. In seed conservation, a primary objective is to preserve genetic diversity, so regeneration of accessions is recommended before viability falls to 85% of the initial value (De Vitis et al., 2020; Solberg et al., 2020). In the case of commercial vegetable seeds, even a fall in germination percentage below 95% can be too much for competitive markets. With either the viability equations or the aging time model, the time to any viability percentage *g* can be estimated, as the distribution in time is related to $p_{max}(50)$ and $\sigma_{p_{max}}$ or to *p50* and σ (Fig. 9a and e and 10). When σ is large, the time for viability to fall from 85% to 50% can be many years under optimal seed bank storage conditions. It may therefore be more appropriate to standardize longevity predictions at the threshold percentage for regeneration of accessions rather than the median of the population (Nagel and Börner, 2010). As it is the identification of when this phase of rapid deterioration and loss of viability begins to occur that is critical for both germplasm and agricultural seed storage, periodic measurements of germination rates under the storage conditions being used and fitting of the PBT aging model to the data could be used to predict approximate storage times when the rapid loss of viability would begin to occur on a lot by lot basis. This predictive ability for a given seed lot is not present in the Ellis–Roberts approach, which can only be determined retroactively after a significant fraction of viability has been lost, and the potential duration of the plateau period cannot be predicted. On the other hand, change in the parameters of the aging time model once viability begins to be lost suggests that its predictions during the plateau phase will somewhat overestimate the actual $p_{max}(50)$ and $\sigma_{p_{max}}$ values; that is, the rate of viability loss once it begins may be quicker than estimated from the decline in *GR* during the plateau phase. This again recommends putting emphasis on estimating the storage period at which viability begins to decline rather than when it has reached 50%.

The change in seed aging kinetics after an initial plateau period of storage suggests that during the plateau period, physiological or molecular damage (oxidation) is occurring but is largely repairable after imbibition (Zinsmeister et al., 2020). The delay in radicle emergence likely represents the additional time required for such repair before germination *per se* begins (Matthews and Khajeh-Hosseini, 2007). For example, molecular mechanisms assess DNA damage upon seed imbibition, and germination is delayed until such damage has been repaired (Waterworth et al., 2016). At some duration of aging, the accumulated damage is too great to repair fully upon imbibition, and seeds either produce abnormal seedlings or do not germinate (Fig. 9). Aged seeds that do produce normal seedlings, however, showed no reductions in root growth rates after radicle emergence (Tarquis and Bradford, 1992), indicating that the delay in germination allows repair to occur and that there are no persistent consequences for seedling growth. However, in a standard seedling growth vigor test, all seeds are planted at the same time and harvested at the same time a number of days later. In this case, the delayed germination will translate into reduced seedling size, not because the seedlings grow more slowly or are permanently handicapped but simply because they had a shorter time to grow after emergence due to their delayed germination. In fact, such root or seedling growth assays are actually indirect measures of the timing and variation in germination times. As discussed previously, delayed germination itself can result in reduced yield or management issues, but this is primarily due to the slower emergence rather than to the persistent effects of aging *per se* on the subsequent growth potential of the individual plant.

Another useful feature of the PBT aging model is that it allows the prediction of when abnormal seedlings will begin to appear in a germination time course. It is generally the case that the curves for loss of capacity to produce normal seedlings and loss of radicle emergence ability are parallel, and the difference between them represents the percentage of abnormal seedlings that will be observed after that aging time (Fig. 9a; Bradford et al., 1993). For species that produce more abnormal seedlings (e.g., soybean), this difference is large, while for other species (e.g., lettuce), relatively few abnormal seedlings are observed, and the difference in storage time between these curves is comparatively small. Once the aging properties of a seed lot have been determined using the aging time model, it is possible to calculate the time after which no additional normal seedlings are expected to appear (Fig. 9b and c) (Bradford et al., 1993),

$$t_{norm} < \theta_{age} / \left(p_{max}(g) - p_{norm}(g)\right) \tag{23}$$

where t_{norm} is the imbibition time after which normal seedlings will not appear (Fig. 9c) and $p_{norm}(g)$ is the distribution of aging times at which fraction g of normal seedlings occurs (green curves in Fig. 9a and e). This makes sense, as the normal distribution of seed deaths in time also represents the relative quality of seeds in the lot, with the higher quality seeds (having longer p_{max} values, lower ψ_b values, etc.) germinating earlier than the poorer quality seeds. Thus, extending the aging time model into seed testing could potentially reduce reliance on subjective scoring of seed abnormalities and instead use germination rates (i.e., percentage of seeds germinated within a specific time period) to rank seed lots for their vigor. This is essentially the same as the early count in seed testing (Ilbi et al., 2020), but the timing of such observations could be chosen based on the aging time model to approximate the actual normal seed percentage that would result from extending the test period and subjectively evaluating all of the seeds for normality. It is, of course, possible that some abnormal seedlings could appear prior to t_{norm}, which would likely be due to physical damage or some cause other than aging *per se*.

The similar patterns of germination time courses in response to aging, T and ψ (Figs. 1, 5 and 9) suggest that these could be combined into a 'hydrothermal aging time' model. Of course, the ψ values for optimal seed storage are in the range of −200 MPa (15% *eRH*), while few seeds of any species can germinate below −2 MPa (Walters et al., 2002; Bewley et al., 2013), so quite different mechanisms apply. However, we already understand fairly well how T and ψ (*eRH*) affect seed aging (Roberts and Ellis, 1989; Hay et al., 2003; Ellis and Hong, 2006; Ellis and Hong, 2007; De Vitis et al., 2020), so we can anticipate that conducting germination time course experiments after storage at multiple T and *eRH* conditions would enable incorporation of their effects on germination kinetics into the aging time model by adjusting $p_{max}(g)$ values as a function of T and *eRH*. Accelerated aging or controlled deterioration conditions (i.e., higher T and *eRH* than would be used for storage) have long been used to speed seed deterioration and assess potential longevity in much shorter times (Zhang and McDonald, 1997; Elias et al., 2012). However, the type of aging occurring at the very high eRH range (at or above equilibrium with saturated NaCl, or 75% *eRH*) is different from that occurring at lower values typical for seed storage (Schwember and Bradford, 2010). Controlled deterioration at more moderate ψ levels (e.g., in equilibrium with ~60% relative humidity) and elevated temperatures (up to 50°C) can result in aging time courses within a reasonable time (a month or so) that are likely to be transferrable to storage conditions using known relationships of aging rates to T and *eRH* (Walters et al.,

2020; Zinsmeister et al., 2020). That is, equations comparable to Equations 20 and 21 could be developed to adjust the aging model parameters (θ_{age}, $p_{max}(50)$ and $\sigma_{p_{max}}$) in response to the *T* and *eRH* conditions of the storage environment as in the Ellis–Roberts model. Further research is needed to confirm this, but the practical consequences for germplasm storage would be significant (see Section 7.3).

5.3 Seed respiration rates (all models)

While this chapter is more about modeling seed behavior than identifying the underlying causes, it is worth noting that the PBT models described here can also be applied to seed respiration rates. This requires measurements on a single-seed basis, as the models describe population behavior, not simply average behavior obtained from bulk seed measurements. However, equipment is available for measuring respiration (oxygen consumption) rates of individual seeds during imbibition and germination (Bradford et al., 2013). When this is done, the resulting time courses of oxygen consumption rates match very closely to those of germination across all of the factors described above, and the same PBT models can be applied to the respiration rate data (Bello and Bradford, 2016). This provides the opportunity for automating the collection of germination time course data, the major limitation to the application of PBT analyses.

These results also raise interesting questions for further study. While it is relatively straightforward to imagine how changes in *T* could affect metabolic rates through energy considerations, this is less clear for ψ effects at the quite high range at which effects on *GR* are evident. The close correspondence between germination rates and respiration rates in individual seeds suggests that seed metabolic rates are adjusted in response to environmental inputs and physiological states such as dormancy. For example, oxygen consumption rates of lettuce seeds imbibed at a high temperature that prevents germination (thermoinhibition) drop almost immediately to essentially zero (Ardura, 2017). It makes perfect sense that dormant seeds must conserve their resources even when hydrated, as in many climates, the soil seed bank will be hydrated essentially continuously, yet seeds can persevere in a dormant state for years and retain sufficient reserves for germination (Finch-Savage and Footitt, 2017). Despite early work on respiratory inhibitors and a possible role of the pentose phosphate pathway in dormancy (Roberts and Smith, 1977), we have relatively little understanding of metabolic rates in imbibed dormant seeds or how they are regulated, although some leads may be emerging from research on a metabolic 'power saving mode' in dormant plant vegetative buds (Martin-Fontecha et al., 2018). ABA is involved in dormancy in both buds and seeds, and the Sucrose-Non-Fermenting-1-Related Protein Kinase complex, which coordinates energy balance, anabolic versus catabolic metabolism and growth

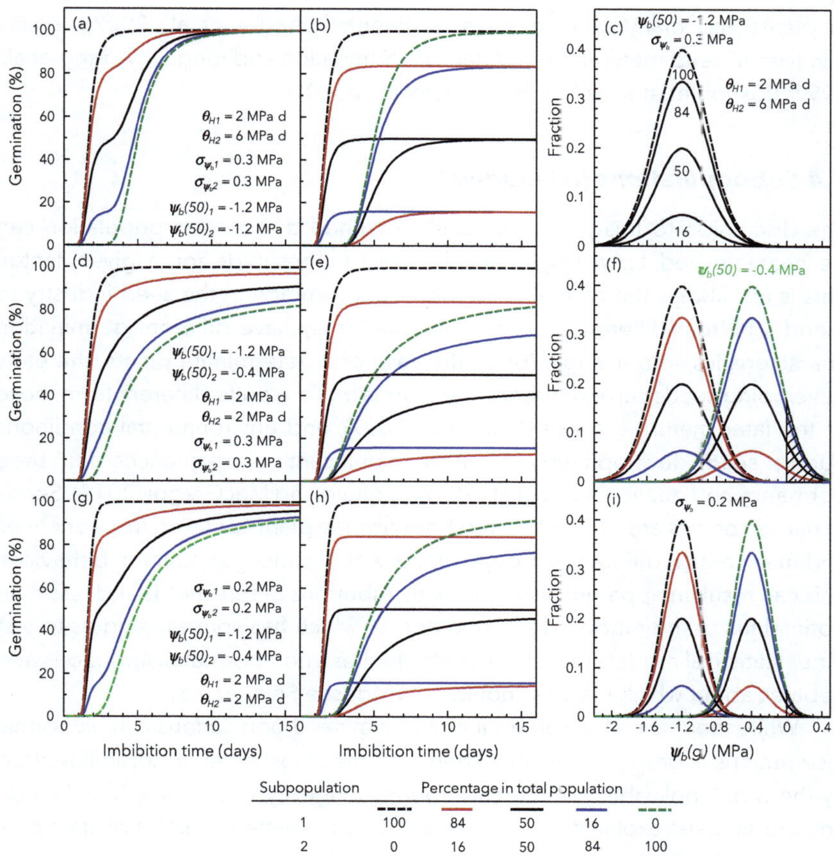

Figure 11 Illustration of the effects of subpopulations of seeds in a seed lot on germination time courses. Panels (a), (d) and (g) show the predicted germination time courses of the total populations, panels (b), (e) and (h) show the corresponding time courses of the component subpopulations composing the blended lots (same colors), and panels (c), (f) and (i) show the underlying $\psi_b(g)$ distributions of the two subpopulations. In panels (a) to (f), the dashed black line is for a single seed lot with the following hydrotime parameters: θ_H = 2 MPa d, $\psi_b(50)$ = −1.2 MPa and σ_{ψ_b} = 0.3 MPa (e.g., panel c), and all seeds were imbibed at 0 MPa. The solid curves represent seed lots in which 16% (red), 50% (black) or 84% (blue) of the seeds in population 1 are replaced by seeds from population 2 (a second seed lot; dashed green line). In (a) and (b), the two subpopulations have the same $\psi_b(g)$ distribution (c), but the θ_H value of subpopulation 2 seeds is increased to 6 MPa d. In (d), (e) and (f), the θ_H and σ_{ψ_b} values are the same in the two subpopulations, but the $\psi_b(50)$ value is changed to −0.4 MPa in subpopulation 2 (f). In (g), (h) and (i), an additional change of σ_{ψ_b} to 0.2 MPa is made in both populations (i). The hatched regions in (f) and (i) indicate that the $\psi_b(g)$ thresholds of seeds in this region are above 0 MPa, and thus that fraction of seeds would be unable to germinate on water. The examples in (a), (d) and (g) indicate the diversity of shapes of germination time courses that can be generated by changing the values in the hydrotime model and the fraction of seeds in each subpopulation. See Section 5.4 for further discussion.

in plants and many other organisms (Baena-González et al., 2007), is also involved in seed maturation, dormancy, germination and longevity (Yang et al., 1997; Rosnoblet et al., 2007; Bolingue et al., 2010).

5.4 Subpopulations (all models)

The discussion to this point has tacitly assumed that a seed population can be characterized by a single distribution of thresholds for a given factor. This is not always the case. For example, it is common in the seed industry to blend lots from different productions, which may have different germination characteristics, into a single lot. In the case of indeterminate plants, the early developing seeds can experience conditions that are quite different from those of the later maturing seeds (Donohue, 2009), and environmental conditions during seed development can have significant consequences for seed dormancy and quality (Auge et al., 2017; Penfield and MacGregor, 2017). Seeds of native species are often collected from diverse plants or fallen fruits or cones and may contain different subpopulations with distinct germination behaviors. This can result in apparent non-normal distributions of thresholds and less than optimal fits of the models described above, which has spurred studies to test other statistical models that might better fit these non-normal distributions and debates about which models should be used (see Section 6.3).

While there is no logical requirement that seed populations exhibit normal distributions of properties, in our experience, they most often do so, as illustrated by the wide applicability of the Ellis–Roberts longevity equation. When they do not, the simplest explanation is that it is due to the tested seed lot containing a mixture of different source populations. While these can be modeled by more complex functions and adjustment of additional empirical parameters, it makes more sense biologically to look for evidence of mixtures of subpopulations and test for their presence. For example, it is not uncommon in germination time course studies to observe what appear to be breaks in the time course pattern (e.g., Fig. 11a and g) or curves that are less sigmoidal (Fig. 11d). These will be evident only with closely spaced observations or, as in the case of single-seed respiration measurements or imaging, with data for each seed. It is very difficult to distinguish subpopulations with widely spaced observations during the germination time course, but if such subpopulations are known to be present, predictions based on them can fit such data better (Bello and Bradford, 2016). If we assume that there are two populations of seeds mixed together, we can fit two PBT models simultaneously, using the following general equation for a factor X:

$$D(g) = f_1\left[\left[X - (\theta_{X1}/t_{g1}) - X_b(50)_1\right]/\sigma_{X1}\right] + (1-f_1)\left[\left[X - (\theta_{X2}/t_{g2}) - X_b(50)_2\right]/\sigma_{X2}\right], \quad (24)$$

Published by Burleigh Dodds Science Publishing Limited, 2022.

where $D(g)$ is the sensitivity distribution of the total population, f_1 is the fraction of the total population in subpopulation 1, X is the level of factor X, θ_{X1} and θ_{X2} are the time constants for subpopulations 1 and 2, t_{g1} and t_{g2} are the germination times of individual seeds in subpopulations 1 and 2. $X_b(50)_1$ and $X_b(50)_2$ are the medians of the distributions of X_b values for subpopulations 1 and 2, and σ_{X1} and σ_{X2} are the standard deviations of X_b values among seeds in subpopulations 1 and 2. The values of f_1 and the parameters of both subpopulation models can be optimized by adjusting the parameters to minimize the deviation of the total population model from the actual data (Bello and Bradford, 2016) (see Sections 6.1 and 6.2).

Even relatively small differences in the hydrotime parameters of two populations can result in changes in the time courses, dependent also on the fraction of each population present (examples in Fig. 11). The predicted time courses of each subpopulation (Fig. 11b, e, and h) illustrate how these can overlap and meld together into relatively normal-looking time courses (Fig. 11d). For temperature (Bello, 2020) and aging (Bello and Bradford, 2016), this approach is capable of matching the actual total germination time courses of the mixed populations, while also accurately determining the model parameters for each subpopulation. It is also the case that seed populations can change their PBT parameters during a germination test, such as during extended incubation at lower ψ values (Dahal and Bradford, 1994). Testing data for the presence of additional subpopulations could identify such shifts and quantify their parameters.

This approach to deconvolute non-normal threshold distributions into multiple subpopulations with normal distributions is analogous to Fourier transformation of complex frequency patterns into their underlying sinusoidal components (Rahman, 2011). Theoretically, additional subpopulation terms could be added to the equation above (i.e., $f_3 = 1 - f_1 - f_2$, etc.) if justified by the data. Of course, larger seed populations and frequent data collection would be required to identify more subtle modifications of germination time course patterns. By varying the fraction and the parameters of each subpopulation, a wide diversity of total germination time course patterns can be generated, some of which will better match more complex or skewed distributions. For example, priming primarily decreases the value of the time constant, while aging increases the time constant; if subpopulations are present that did not respond similarly to priming or were aged differently, the patterns shown in Figure 11a can result (Bello and Bradford, 2016). If the median threshold value is different between the subpopulations, more normal-looking curves can result, but in fact, two distinct but overlapping subpopulations of seeds generate the time courses in Figure 11d and g. Changing the standard deviations of the threshold distributions can also result in distinct time courses (Fig. 11g and i). Mixed populations of seeds with different temperature thresholds can also be resolved from their germination time courses (Bello, 2020), a situation that likely occurs in species in which dormancy is regulated by shifts in such thresholds (Batlla and Benech-Arnold, 2015).

Thus, for understanding the biological meaning of such patterns, it is more appropriate to assume that they are caused by multiple normally distributed subpopulations than to seek a specific function to empirically fit each novel pattern. The latter may fit the observed data well, but the biological meaning of the additional parameters required for such models can be obscure (Watt et al., 2010; Mesgaran et al., 2013; Chen et al., 2021). In contrast, the subpopulation approach is directly amenable to practical use, for example, if there is a less vigorous minor subpopulation in the seed lot (e.g., Fig. 11a, population 2). Characterizing it in terms of the fraction that it represents in the total population and the time at which germination of the seeds in each population is most distinct (Fig. 11b) could guide sorting equipment to remove those seeds from the lot. This is particularly feasible if imaging approaches can also be applied to identify unique characteristics of the two populations by scanning them prior to germination testing, so that the timing of each seed's germination is known and can be associated with detectable physical characteristics for sorting (Bello and Bradford, 2021; Nehoshtan et al., 2021).

5.5 Hormone time or biotime models

The focus of this chapter is to address factors in seed behavior that impact seed quality and influence seedling emergence. However, seed dormancy and germination are largely controlled by plant hormones, particularly by GA and ethylene acting to promote germination and ABA to inhibit germination (Bewley et al., 2013; Cao et al., 2019; Nonogaki, 2019). Thus, we will briefly point out that PBT models can also be used to quantify the sensitivities to hormones that influence seed germination (Ni and Bradford, 1992; Ni and Bradford, 1993; Dutta and Bradford, 1994; Bradford et al., 2008). For ABA, a germination inhibitor, the relevant equation is as follows:

$$\theta_{ABA} = \left(\log\left[ABA_b(g)\right] - \log\left[ABA\right]\right)t_g, \tag{25}$$

where θ_{ABA} is the time constant (log (M) h or log (M) d), $ABA_b(g)$ is the distribution of base ABA concentrations that prevent germination of fraction g of the population (M), ABA is the concentration of ABA (M) applied in the test. For the germination promoter GA, the equation is expressed as follows:

$$\theta_{GA} = \left(\log\left[GA\right] - \log\left[GA_b(g)\right]\right)t_g \tag{26}$$

where θ_{GA} is the time constant (log (M) h or log (M) d), GA is the concentration of GA (M) applied in the test and $GA_b(g)$ is the distribution of base GA concentrations that just permit germination of fraction *g* of the population. For ABA, increasing the concentration applied to seeds will result in a series of time courses similar to those in Fig. 5b, with delayed and inhibited germination as the ABA concentration increases. For example, the ABA-time PBT model was used to quantify the ABA sensitivity of embryo growth and germination of celery seeds (Walker et al., 2021). When GA was applied to tomato seeds that were mutated to prevent GA biosynthesis, the GA-time PBT model closely matched the more rapid germination and higher final percentages as the GA concentration increased (Ni and Bradford, 1993).

PBT models specifically quantify the sensitivity of seed germination to these factors and the speed of response to them once their sensitivity thresholds have been exceeded. They therefore incorporate the two key aspects of hormonal regulation, the level of the hormone and the sensitivity of the cells or tissue to it, into a 'biotime' model that addresses both the recruitment of additional individuals (proteins, cells, seeds) and rate effects or turnover or growth/development as hormone levels increase (Bradford and Trewavas, 1994; Trewavas, 2012; Bradford, 2018). In addition, it was possible to model the effects and interactions of multiple different factors simultaneously, such as ψ and ABA or GA, by combining the t_g predictions of their PBT models for a given fraction and subtracting the t_g for that fraction in water (Ni and Bradford, 1992, 1993). As tools are now available to investigate hormonal regulation of germination on a gene-by-gene and cell-by-cell basis (Topham et al., 2017; Xu et al., 2020; Abley et al., 2021), it is important to understand the inherent variation among seeds in their sensitivity thresholds. For example, it is often the case that a given hormone dosage (e.g., of GA) might alleviate dormancy only for a specific fraction of the seed population It is not uncommon to see this referred to as being 'partially effective' in breaking dormancy (e.g., Xu et al., 2020). However, as germination is an all-or-none response for an individual seed, this actually indicates that the hormone was 100% effective in alleviating dormancy in a certain fraction of the population; higher dosages could potentially recruit more seeds to germinate, or if not, the remaining seeds require a different signal to trigger germination.

It may well be that within seed populations, particularly of undomesticated species or those in the process of meeting their dormancy requirements, multiple signals might be effective in stimulating germination, but only in certain subfractions of the seed population. Rather than being considered as stochastic noise (Johnston and Bassel, 2018), we could view this sensitivity variation as an integral feature of seed behavior and use it to partition the seed population into subfractions that exhibit specific responses to regulatory factors. For example, a 'partially dormant' seed lot might be imbibed to allow all of the nondormant seeds to germinate. Exposure of the remaining seeds to a hormone or signal (e.g., inducing expression of a specific gene) might then stimulate an additional cohort of seeds to germinate, which could be collected and analyzed. Exposing a dormant seed population to increasing dosages of a dormancy-breaking factor, allowing all responding seeds to complete germination and removing them before applying the next increment of the factor dosage would be an experimental approach to achieve fractionation of the population according to seeds' threshold sensitivities. Our usual procedure of exposing the entire population to a treatment and then looking for an increase in the total germination mixes the dormant and nondormant seeds together in the germination time course and obscures the signal from those seeds that are specifically responding to the applied signal. The overall response of the seed population to dormancy-breaking factors is likely the sum of the behaviors of subpopulations that are waiting for different combinations of signals to trigger germination, enabling the variation in dormancy that is the basis of bet-hedging behavior (Willis et al., 2014; Gianella et al., 2021).

6 Applying population-based threshold models

6.1 A brief history of fitting population-based models to germination data

As noted in Section 3.2, Eric Roberts initiated the application of population approaches to seed biology (Roberts, 1960, 1961). He and Richard Ellis pioneered the application of probit analysis to quantify the distribution of seed deaths in time that underlies their seed longevity equation (Ellis and Roberts, 1980, 1981). The basis of probit analysis is to transform germination percentage (or fraction) into 'probability units', which are equal to the standard deviation of the distribution of the parameter values in the population. The median (50%) is converted to 0 probits, and increases and decreases in frequencies from the median are plotted in standard deviation units. This crowds together percentages near the median and expands the scale at both extremes (see scales at top of Fig. 3a or 10a and d). When plotted on a cumulative scale, this converts a symmetric sigmoidal curve into a straight line. Thus, for seed viability or thermal time data, converting germination fractions into probit units enabled the use of linear regression to identify the median (when probit = 0; equal also to the mean in a normal distribution) and standard deviation (the inverse of the slope) of the population. While this approach has its critics (see below), it is worth remembering that at the time in the 1960s and 1970s, electronic calculators and personal computers did not exist to do regression analyses or fit complex statistical models. The ability to visually or graphically use linear relations to solve such computational problems is underappreciated now but represented a major advance at the time (see, e.g., the nomograms developed to solve the longevity equation in Ellis, 1988).

Published by Burleigh Dodds Science Publishing Limited, 2022.

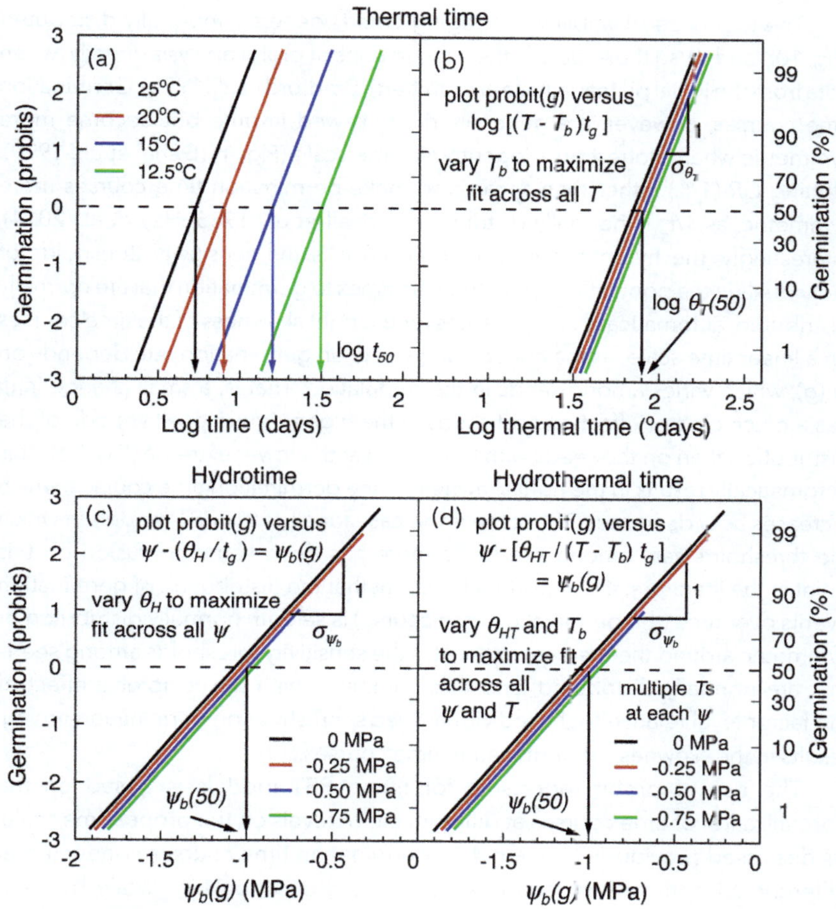

Figure 12 Application of repeated probit analysis to identify thermal time, hydrotime and hydrothermal time parameters of seed germination. (a) The germination time courses at different suboptimal temperatures shown in Fig. 1(b) are plotted here after transforming the percentage values into probits (y-axis). The units on a probit scale are standard deviations of a normal distribution (comparable percentages shown on right y-axis). The probit transformation converts normal sigmoidal cumulative distributions into straight lines. (b) Plotting the data of panel (a) on a log thermal time basis, or log $[(T - T_b)\, t_g]$, results in a single common regression line when the optimal value of T_b is used. These are shown as parallel lines for illustration; with actual data, a data cloud will coalesce around the optimal regression line (least residual error) when the best T_b value is used. (c) The same approach can be used to estimate hydrotime parameters by plotting probit (g) (y-axis) as a function of $\psi - (\theta_H/t_g)$ for a series of time courses at different water potentials (Fig. 1c) and varying θ_H until the optimal regression is obtained. The value of $\psi_b(50)$ is where the regression crosses probit (50) = 0, and σ_{ψ_b} is the inverse of the slope. Note that unlike suboptimal temperatures, in which most viable seeds will eventually germinate, germination is inhibited for an increasing fraction of the population as ψ decreases, so the data available for more negative ψ values will be limited to lower values of g. This basic approach can be applied for all of the PBT models in which there is a distribution of

The loss of seed viability over aging time is generally normally distributed (Fig. 10b and d), so those data fit the expectations of probit analysis directly (when data from the initial plateau phase are omitted) (Bradford et al., 1993). Germination time courses, however, are generally right skewed in time but become more symmetric when plotted on a logarithmic time scale (Fig. 1) (Dahal et al., 1990). Plotting *GR* ($1/t_g$) rather than t_g can also make germination time courses more symmetric, as $1/t_g$ is normally distributed (Covell et al., 1986; Hay et al., 2014). Interestingly, the hydrotime model, in which it is the sensitivity thresholds of individuals in the population rather than the times to germination that are normally distributed, automatically compensates for the right skewness of the time courses on a linear time scale, as the effect of a given ψ on germination rate depends on $\psi_b(g)$, which varies among seeds in the population. That is, a small change in ψ has a much greater effect on t_g of seeds at the higher (more positive) side of the distribution than on the seeds with thresholds at the lower extreme (Fig. 5b). This automatically results in the right skewness in the germination time courses, which increases as ψ decreases. The same is the case for all of the PBT models in which the thresholds vary among seeds. As there has been some confusion on this point in the literature, it is important to repeat that the distribution of germination events over time (the germination time course) is seldom normally distributed or symmetric around the median; rather, it is the sensitivity thresholds among seeds that are normally distributed, and this, combined with the reciprocal effect of the factor level relative to the individual seed's threshold on germination timing, results in the skewness of germination time courses.

The use of probit regression for fitting PBT models is based on the normalization of time courses at different factor levels on the proper time scale. As discussed previously (Section 4.1), germination time courses conducted at different suboptimal temperatures and plotted on a log time scale become parallel sigmoidal curves (Fig. 1b). If the germination scale is transformed to probit values, which are in units of standard deviations, these sigmoidal curves are converted to straight lines (Fig. 12a). The concept of thermal time indicates that when expressed on a thermal time scale, all of these lines should coalesce onto a common line on a thermal time scale based upon a common value of T_b. Thus, if we plot probit (g) versus the log thermal time ($\log(T - T_b)$) to germination for each observation time (t_g) (Fig. 12b), the result should be a single common regression according to the following equation:

Figure 12 *(Continued)*

thresholds among seeds in the population. (d) The hydrothermal time parameters can be determined as in (c), except that there can be multiple temperatures for each ψ value. The temperature is included as thermal time, by plotting probit (g) versus $\psi - \theta_{HT}/[(T - T_b)\, t_g]$. In this case, both θ_{HT} and T_b must be varied to find the optimal fit. A separate experiment conducted in water to determine T_b can give a good starting estimate of this value.

$$\text{probit}(g) = \left[\left(\log(T - T_b)t_g\right) - \log\theta_T(50)\right] / \sigma_{\theta_T}. \quad (27)$$

By varying the value of T_b and conducting repeated regressions on the complete data set conducted at multiple temperatures, the value resulting in the minimum residual error, or maximum R^2 value, can be identified. The value of log $\theta_T(50)$ is the thermal time at which the line crosses probit (50%) = 0, and the inverse of the slope of the line is the value of σ_{θ_T}, indicating the variation in θ_T across germination fractions (Fig. 12b; Covell et al., 1986). Similarly, time courses at different constant ψ values can be normalized on a probit regression representing the sensitivity thresholds among seeds in the population (Fig. 12c):

$$\text{probit}(g) = \left[\psi - (\theta_H / t_g) - \psi_b(50)\right] / \sigma_{\psi_b}. \quad (28)$$

To determine the parameters of this equation, the term $\psi - (\theta_H/t_g)$ (which is equivalent to $\psi_b(g)$) is plotted on the x-axis versus the probit of the percentage g corresponding to t_g for each observed data point across ψ values. The value of θ_H is varied until the optimal regression across all data is attained, and the x value where the regression crosses probit = 0 is equal to $\psi_b(50)$ and σ_{ψ_b} is the inverse of the slope (Bradford, 1990). These can be combined into the HTT model

$$\text{probit}(g) = \left[\psi - \left(\theta_{HT} / (T - T_b)t_g\right) - \psi_b(50)\right] / \sigma_{\psi_b}, \quad (29)$$

which can be solved in the same manner by plotting the probit of g versus $\psi - [\theta_{HT}/(T - T_b)\, t_g]$ and varying both θ_{HT} and T_b until the regression across all tested ψ and T values is optimized (Fig. 12d). A good estimate of T_b values can be obtained by first analyzing only the data in water (ψ = 0 MPa) by the thermal time approach. This approach can be applied similarly to the other PBT models in which the threshold values vary among seeds in the population.

An important point can be made here regarding which data points are relevant for inclusion in fitting PBT models. The fundamental experimental data are observations of the additional seeds that have germinated since the last observation, resulting in a cumulative germination time course when summed over time. It is fundamental to the PBT model that the term t_g refers to the time at which cumulative fraction (or percentage) g was achieved. There is always some error in this, as one seldom observes the exact moment of radicle protrusion for a given seed. However, in the most sensitive period for data collection when germination percentage is increasing rapidly, this error is minimal when referring to the highest germination percentage achieved at that observation time. On the other hand, later in the germination time course, there are often observation times at which no additional germinated seeds are recorded. In terms of the PBT model, these are not valid times for incorporation in fitting the model, as no additional germination occurred, so g has not increased and would therefore be the same as at the last observation. In terms of the model, each value of g is uniquely associated with a given time after imbibition, or t_g. We argue, therefore, that only observation times at which g has increased should be included in the data set for fitting PBT models. If this is not done, many germination time courses will end with a string of such null data points as germination does not increase as observation time is extended. This occurs often in hydrotime studies, as reducing the ψ will result in time courses that plateau at lower and lower percentages (e.g., Fig. 5). If null data points for the same percentage are included in probit analyses, for example, those points will skew the regressions to the right, resulting in inaccurately large values for σ_{ψ_b}. This can confound attempts to compare different distribution models that may handle these null points differently. The same is the case for not including viability data during an initial plateau phase in the calculation of parameter values for the Ellis–Roberts model (Fig. 10b). In addition to skewing fits to the data, the inclusion of these null data points can also inflate the number of data points in error estimations. Some statistical models automatically remove such null points by only considering the increase since the last observation time, and if this is zero, the observation effectively does not contribute to the estimation. It is a simple step to test whether a data point at time t is greater than at $t - 1$ and only including valid t_g values in fitting PBT models.

This approach to fitting data to these models involves identifying the parameter values of the sensitivity distribution and the time constant that result in all of the actual data points coalescing in a common linear regression on a probit scale. This has been done by iteration (i.e., 'repeated probit analyses') to identify the values that account for the greatest percentage of the variance or that minimize the residual error (Covell et al., 1986; Bradford, 1990; Bello and Bradford, 2016). Weighting should also be used in probit regressions, which gives greater weight to values nearer the median than to those at the extremes, and can be done automatically with statistical software. In fact, fitting of these models can now be done using nonlinear least-squares curve-fitting methods that do not require probit transformation of the germination fraction data (Hardegree, 2006; Hardegree and Winstral, 2006; Chantre et al., 2009). These routines compare the predicted values using an initial set of parameters to the actual values across all factor levels and iteratively adjust the parameter values to minimize the residual error across all data points. Thus, while still convenient (e.g., can be done easily in a spreadsheet) and of assistance in understanding the underlying basis of the models, the use of probit transformations and repeated regressions is no longer necessary to fit PBT models.

6.2 R programs to fit PBT models

We have developed R scripts to fit germination data for all of the PBT models discussed above. The *'pbtm'* package combines existing PBT model calculations, parameter outputs and respective plots (www.pbtmodels.org). The package uses a nonlinear least-squares function (Bates and Watts, 1988; Bates and Chambers, 1992) to fit each model by comparing the raw data and respective treatments directly with the curve predicted by the selected model.

The R pnorm function computes the cumulative density function (cdf) which expresses the probability of a variable *X* to take a similar or lower value than *x*, respecting a normal distribution with a certain mean and standard deviation (sd). The pnorm function only requires three parameters (*X*, mean and sd) to output the probability but also provides options such as the direction of the evaluation (lower.tail) where the standard value (lower.tail=TRUE) assumes the probability of *x* to take similar or lower value (cdf for the lower or left area of the distribution) or this can be inverted to compute the probability of *x* to take a similar or higher value (cdf for higher or right side of the distribution). In the context of PBT models, we utilize the equations described above for each model but isolating the varying factor, such as $\theta_T(g)$ for the thermal time model or $X_b(g)$ for all other models. These equations are then used within the pnorm function with a varying mean, standard deviation and the direction of the distribution (lower.tail) depending on whether it is a promotive or inhibitory factor. The model can be optimized using a nonlinear least-squares algorithm (nls function in the R software) that refines all parameters involved by successive iterations, minimizing the sum of squares estimates of errors (SSE) or least squares when compared to the actual germination or respiration data. As discussed above, a tool in the R package also provides the option of removing null data observations when no additional germination occurred prior to model calculation.

6.3 Alternative approaches

There has been considerable debate regarding whether probit regression analysis is appropriate for analyzing seed germination data, and if not, what statistical approach should be used and what the underlying population distribution model should be (Watt et al., 2010, 2011; Mesgaran et al., 2013; Hay et al., 2014; Onofri et al., 2014; Hardegree et al., 2015; Mesgaran et al., 2017; Onofri et al., 2018; Gianinetti, 2020; Moltchanova et al., 2020). We have focused on the normal distribution here, but other models such as the Weibull, logistic and others have been proposed and tested (Watt et al., 2010; Mesgaran et al., 2013). There is no doubt that across diverse species and seed lots within those species, variation is present among the germination patterns observed. We have argued above (Section 5.4) that in such cases, the differences could also be due to mixtures of seeds from multiple populations having different distributions, and in some cases, making that assumption and fitting multiple subpopulations can resolve the differences in goodness of fit. It has also been proposed that the experimental protocols used in seed germination violate assumptions of regression analysis, and other methods such as time-to-event or 'survival' analyses are more appropriate (Hay et al., 2014; Moltchanova et al., 2020; Romano and Stevanato, 2020). Gianinetti (2020) provides an in-depth analysis of these statistical issues and discusses the assumptions and complications inherent in seed germination data as well as advice on experimental design and execution in relation to the statistical approach utilized.

We will not go into the details of these discussions here, as there are valid arguments for different viewpoints, depending upon the statistical assumptions that are made and the parameters that are of most interest to the investigator, as well as the inherent variations in seed characteristics when comparing among species and seed lots. However, we are in agreement with Roberts (1999): "It is not often very useful to fit entirely arbitrary models to data: it is usually much more productive to seek solutions in which the coefficients have some biological meaning." Our focus has been on understanding seed behavior through the PBT modeling approach, and in most cases, all of the statistical approaches give quite similar values for the model parameters, with the differences among them being in the fraction of variation accounted for or in calculating the error terms to be used when comparing to other seed lots. Thus, for the purposes of estimating PBT model parameters and using them to inform us about the underlying seed behavior, the choice of method is less critical than when specific statistical comparisons (such as whether a given treatment significantly differs from another treatment) are the objective.

6.4 Experimental recommendations for fitting PBT models

In many cases in the literature, the quality of the original data is the limiting factor and makes the debate about different statistical models moot. As illustrated above in discussing subpopulations, discrimination among models describing germination time courses on the basis of their shape requires closely spaced observations. Often, the frequency of observations is simply too low or poorly spaced in time relative to germination frequency to have confidence in the ability to discriminate between alternative models (Bello and Bradford, 2016; Gianinetti, 2020). For this reason, and because we are primarily concerned with seed performance, we will discuss some experimental and technical factors that should be considered when collecting data for use in applying PBT models.

Fitting PBT models requires data for germination time courses under different conditions that must be known as accurately as possible. It is relatively easy to maintain constant temperatures using growth chambers or temperature gradient tables, but it is critical to actually measure temperatures in the vicinity of the seeds rather than relying solely on the growth chamber's thermostat. Most researchers do not have the facilities to replicate multiple chambers at each temperature (Hardegree, 2006), so actual temperatures should be measured during experiments. It is not essential that a specific absolute temperature was maintained, but the actual temperature(s) during the experiment should be used when fitting the models. Similarly, using published recipes to prepare osmotic solutions to control ψ during germination experiments can result in significant errors, particularly with polyethylene glycol (PEG) solutions. Instead, the solution water potentials should be measured using an osmometer or similar instrument, and the seeds should be transferred to fresh plates every few days in order to assure that the ψ remains constant. For the most precise work, filter discs can be placed on the germination blotters with the seeds and moved directly into the osmometer for measurement. Again, it is less critical to maintain a specific target ψ value than it is to know what the actual value is during the experiment. When fitting the models, the factor level is essential information that is associated specifically with the seed lot's germination behavior. If that value is incorrect, it will cause error that is impossible for any statistical method to correct. Thus, if one wanted temperatures of 10°C, 15°C and 20°C or water potentials of −0.2, −0.4, and −0.6 MPa, and the actual values were 10.5°C, 14.8°C and 19°C or −0.24 MPa, −0.5 MPa and −0.7 MPa, that is no problem if those actual values are associated with the observed germination time courses, as the fitting routines will use those values in estimating the model parameters. However, using the expected values rather than the actual values will result in poor fits, incorrect parameter values and limited reproducibility. In our experience, seeds have a remarkable ability to sense their conditions, and when we have found anomalous behavior in a given experiment, we generally can trace it to human error or inaccurate control of the conditions; the seeds' responses across temperatures or dosages were virtually always accurate for the environment that they actually were experiencing.

The second important feature is to schedule observations to obtain at least 3–5 data points between the initiation of germination and ~80% of the final percentage for each condition. For example, lettuce seeds in optimal conditions can begin germinating within 12 h and be completed within 20 h; thus, an experiment planned for daily observations will have missed all of the relevant data before the first scheduled observation. Rather, experiments must be designed to obtain several observations during the time frame in which germination is occurring, such as by imbibing the seeds the previous evening and making observations frequently during the next day. This makes it clear why automated imaging, robotic scoring or correlated phenotypes such as respiration rates that can be frequently monitored are highly desirable. On the other hand, at low T or ψ, or for more slowly germinating species, germination events will be delayed and more widely spaced in time and daily observations may be sufficient. It is often worth running a small-scale preliminary test to identify the key times for observations when planning a large experiment to be sure to schedule observations at the proper times for all of the treatment conditions. We note that some researchers fit observed germination data to arbitrary multiparameter models and then predict values for specific germination fractions (e.g., every 10%) for use in fitting PBT models. We are not in favor of this approach, as it can artificially increase the number of data points that are not associated with actual experimental observations. As statistical procedures are sensitive to the number of observations, this practice can artificially reduce estimates of residual error, for example.

On the other hand, in a seed testing lab that is familiar with a given species and the characteristics of its germination, it may be sufficient to collect only a few observations to place a given seed lot in a vigor ranking system based on its *GR*. Observation times could be identified and standardized such that germination percentages at only one or two time points could be converted to parameter values. If one is familiar with a species' characteristics, tests at only two temperatures and two water potentials are sufficient to estimate the HTT model parameters for a seed lot (Bradford and Still, 2004). It is not necessary or even desirable to estimate T_b or ψ_b by testing under conditions close to these values. This greatly extends the time required and allows extraneous factors to intervene (e.g., fungal contamination, evaporation from plates, etc.). Instead, temperatures near to T_o and 5°C below it or at 0 and −0.3 MPa may be sufficient since the response of *GR* is linear and the values of T_b or $\psi_b(50)$ can be determined by extrapolation to the intercept on the *x*-axis (Figs. 4 and 5). Similarly, if initial tests using controlled deterioration are used fit the aging time model, subsequent tests to check the vigor of stored seed lots would require only a single test under the same test conditions to estimate the seeds' aging rate by changes in *GR*.

7 Future trends in research

The demonstrated and potential utility of PBT models of seed germination have been discussed above. In this section, we mention some specific cases where the models, or the viewpoint on seed biology that they have revealed, can have beneficial applications. In addition, we note where additional research is needed to realize this potential.

7.1 Seed testing

There is a large literature on sampling, statistical comparisons, etc. for the seed testing industry (Elias et al., 2012). For universality, consistency and labeling, these are valid and needed. However, despite much work on vigor tests (Baalbaki et al., 2009), and their widespread use in the seed industry, few have been incorporated into seed laws or adopted across species. Most vigor tests put seeds under some type of stress (e.g., cold, osmotic or salt stress), rapidly age the seeds (e.g., artificial aging, controlled deterioration) or measure seedling growth after some period of time. The case has been made here that all of these either directly impact the speed of germination (stress and age) or are indirect results of variation in germination timing (growth tests). Thus, the analysis of complete germination time courses would be the most direct approach to quantifying and comparing seed vigor. In fact, many companies use internal vigor tests that may utilize smaller sample sizes, focus on rates and uniformity in addition to total viability, and explore limits of environmental responses rather than optimal conditions, to select and create high-performing seed lots.

Nonetheless, germination time courses are seldom measured in standard seed testing due to the labor required to collect frequent observations to characterize such time courses in detail. However, this situation is rapidly changing. Methods for imaging of test plates and automated extraction of germination timing information have been available for a decade (Joosen et al., 2010), and improvements in both imaging and analysis methods are occurring rapidly (Silva et al., 2019; Colmer et al., 2020; Bello and Bradford, 2021; https://www.seed-x.com; https://www.vibeia.com/; https://videometer.com/). In most cases, however, it still has been necessary for a person to move the test plates individually under the imaging system to collect the data, again limiting observations to the times that workers are available. That is changing now with inexpensive electronic cameras that can be placed over each sample to collect images as frequently as desired and transmit the extracted data to a computer (Colmer et al., 2020). This fully automates the collection of germination time courses. Indirect approaches such as robotic single-seed respiration measurements can also be converted into time courses closely paralleling germination timing (Bello and Bradford, 2016). Thus, the impediments to using germination rates as a universal vigor test are being overcome, and the application of PBT models to such data allows precise quantification of the critical features of the seed population, including uniformity and stress sensitivity. Just as the data acquisition can be automated, the application of the PBT models could be incorporated into the software, reducing the labor requirements to setting up the experiments and subsequently receiving and interpreting the model parameters and graphs quantifying seed performance.

7.2 Seed production, conditioning and enhancement

A key factor in the production of high-quality seeds is their maturity at harvest. In general, seeds should be allowed to complete their full maturation program and dry below the level of metabolic activity before harvest (Ellis, 2019). On the other hand, delaying harvest risks field weathering, preharvest sprouting and aging that will reduce seed quality. Postharvest technologies cannot reverse such damage but are able to remove immature, aged and damaged seeds, thereby upgrading seed lots. Methods to detect seed maturity specifically using chlorophyll fluorescence (high levels are associated with less mature seeds) can be used to upgrade seed quality in a number of species (Jalink et al., 1998; Kenanoglu et al., 2015; Bello and Bradford, 2021). Variation in seed maturity of indeterminate crops is a clear example of seed lots that contain multiple subpopulations (Still and Bradford, 1998), which is the foundation of standard seed conditioning equipment using seed size, shape, density, color or other physical characteristics to fractionate seed lots. More advanced imaging and association of spectral traits or internal structures (visible by X-ray) with quality features now enable sorting of seed lots on parameters directly associated with seed performance (Vanderburg et al., 1994; Bruggink and van Duijn, 2017; https://www.seed-x.com/product/sorter/). PBT models can be used to quantify seed lot performance and potentially to identify specific subpopulations of under-performing seeds for removal. These methods have also been applied to weed seed samples to estimate their germination potential and even herbicide tolerance (Matzrafi et al., 2017). Considerable research is needed to test and adapt these new technologies to diverse species and to identify the physical and/or physiological features that are closely associated with seed quality. The rapid deployment of artificial intelligence methods to extract such relationships from large data sets makes this a promising avenue for future research and commercialization. Structuring such algorithms to look for population-based relationships and patterns rather than simply independent stochastic correlations could make them even more powerful for identifying features for classifying seeds according to their performance potential.

7.3 Seed storage

Seed longevity models have been quite useful in identifying the storage characteristics of many species and improving seed storage conditions for commercial seed storage and germplasm conservation (Ellis, 1988). A major issue for seed bank managers is the variation in times at which seed lots, even of the same species, can lose viability (Fleming et al., 2019; Solberg et al., 2020). This requires periodic viability tests, which eventually consume seeds in the stored sample even if they are still viable. However, reducing the frequency of testing risks allowing viability to fall below the recommended 85% and potential loss of genetic diversity (Ellis et al., 2019). As stated by Hay and Whitehouse (2017), 'What a genebank manager ideally needs to manage collections better is a test at the time the seeds are placed into storage that either predicts when the seed lots will reach the viability threshold [. . .] or that ranks the seed lots for relative longevity'. Despite the importance of estimating the length of the initial high viability period for individual seed lots, seed banks in general have not adopted routine procedures to estimate this when the sample has entered into storage. Instead, studies of seed longevity in seed banks are typically retrospective analyses of samples after extended storage periods (Walters et al., 2005; Nagel and Börner, 2010; Fleming et al., 2019; Solberg et al., 2020).

The PBT aging model shows that the reduction in *GR* during storage before any loss of viability can be used to estimate the maximum viability period under the conditions of the test (Section 5.2; Fig. 9). Pretests using controlled deterioration (CD) conditions and measuring germination rates under standardized conditions could be used to estimate the ρ_{max} values for a desired viability percentage. Conducting the same *GR* test after a storage period would reveal where a lot is in its progression toward loss of vigor and viability under the current storage conditions, calibrating the aging model to the actual storage conditions. Even if exact calibration is not possible, thresholds for reductions in *GR* in response to aging could be used to rank seed lots into potential longevity categories. Using the automated methods mentioned above, this could actually reduce the labor required for seed testing in seed banks, as it would allow prediction of when sampling during storage would be the most informative regarding potential loss of viability and reduce the number of uninformative tests during the plateau phase based only on total viability. In commercial seeds, the time frame is much shorter, but in both cases, it is 95–85% viability that is critical, not 50%, as the seed lot either needs to be salable or retain the desired genetic diversity. Thus, measuring *GR* during the initial phase of the aging period (which can be done on as few as 25 or 50 seeds per sample when using single-seed methods, saving the seed resources) can predict when viability will start to decline. The critical need is therefore to calibrate specific CD test conditions with subsequent longevity under the seed bank or commercial warehouse conditions. Seeds in the latter already are being tested periodically under current protocols, so the key step is to initiate the CD pretesting protocols to enable storage life prediction using the aging PBT model to be calibrated to the actual conditions. As noted previously (Section 5.2), these CD tests should be conducted at an *eRH* of 60% or less and the temperature can be increased up to 50°C, but tests at higher *eRH* values may not be good predictors of longevity under drier conditions. In addition, adding *GR* measurements to routine inventory tests (usually every 6 months in commercial seeds) would quickly build up a database relating changes in *GR* to both vigor and viability for different species and varieties and over time would provide useful information regarding production, harvest and storage effects on seed quality.

7.4 Breeding for improved seed quality and performance

Seed quality and performance can be improved by selection, as is evident in the seed traits that are characteristic of domestication, such as loss of dormancy. In reality, however, seed quality and performance generally are low on the list of breeder priorities, unless it is a limiting factor for sales. For example, seed shattering is still a problem for seed production in many crops, particularly vegetables (Still and Bradford, 1998), and only recently have commercial canola varieties become available with an engineered trait reducing shattering (Lambert et al., 2015). Priming is widely used to remove dormancy and increase uniformity in high-value seeds, as it is easier than breeding out inherited dormancy behaviors in these less-domesticated species. In addition, phenotyping for *GR* has been labor intensive and subject to variation from nongenetic factors (Morris et al., 2016). The increasing automation of seed testing should be helpful here, and the quantitative parameters characterizing seed quality from the PBT models could be useful as phenotypes for selection (Saux et al., 2020). Until improved seed quality traits become genetically fixed, the hydropriming time model can be useful for developing effective and efficient seed enhancement protocols. As physiological and molecular research is uncovering in increasing depth the mechanisms regulating seed dormancy and germination, we can anticipate that targets for gene modification can be identified more quickly and utilized to improve seed performance (Nonogaki and Nonogaki, 2017; Xu et al., 2020; Bizouerne et al., 2021).

7.5 Crop and weed emergence models for integrated crop management

Between T_b and T_o, all of the viable seeds should germinate eventually, but it can take a long time at lower T (Fig. 1). Thus, it is common practice to plant when soil T exceeds practical minimums so that germination and emergence can proceed without delay and avoid subjecting the seeds to risk from pathogens or pests. Planting at T above T_o also can occur, as in lettuce plantings in late summer for fall and winter production, leading to thermoinhibition and poor stand establishment that can be overcome by seed priming (Valdes et al., 1985). While crop management planning on a thermal time basis is routine in agriculture, we should think of lower ψ as being similar to low T; even relatively small reductions in ψ will delay and reduce uniformity. Priming is often claimed to improve tolerance to salinity, for example, but in most cases, it primarily affects the speed of germination under water or salt stress, which is valuable, but it has less effect on the minimum ψ for germination (Dahal and Bradford, 1990; Bradford and Somasco, 1994; Cheng and Bradford, 1999; Tatari et al., 2020). However, improvements in speed and uniformity of emergence due to priming can be of significant value in the production of greenhouse crops or transplants.

Seed dormancy models are often associated with weed seeds, and emergence models are used to assist in controlling weeds (Allen et al., 2007; Boddy et al., 2012; Fernández Farnocchia et al., 2021). Management of weed seed production in crop fields and the associated seed banks is critical to extend the lifetime of herbicides and reduce reliance on them (Ramesh, 2015). The stand establishment period for the crop is the most critical time for weed control, so understanding the seed germination behavior of associated weeds may enable the implementation of management approaches. PBT models are being applied effectively for this purpose (Boddy et al., 2012; Batlla and Agostinelli, 2017; Matzrafi et al., 2020), and extension of this approach should be encouraged.

7.6 Ecological applications of PBT models

Ecologists studying germination, bet hedging, species demographics and climate change are also using PBT concepts to describe and model these processes (Huang et al., 2016; Larson and Funk, 2016; Gremer et al., 2020; Liu et al., 2020; Zhang et al., 2020). For example, the PBT models may be particularly useful in constructing ecological germination niche diagrams (Fig. 7). Conceptualizing seeds as populations, potentially with distinct functional subpopulations requiring different environmental signals to promote germination, could be particularly useful in studies of diverse dormancy and bet-hedging behaviors (Gianella et al., 2021). With ecological research now sharply focused on the implications of climate change, modeling approaches to identify 'climate clones' or locations around the world that have (or will have in the future) similar climates could be of considerable benefit for seed companies to extend the lifetimes of existing varieties adapted to specific climates and to conservationists seeking to identify locations to preserve species currently under threat (https://climate.ai/solutions/#seed).

7.7 Research on determinants of seed quality attributes

While remarkable progress has been made in understanding the genetic, molecular and physiological bases of seed vigor and dormancy, much research remains to be done to understand the specific signaling pathways translating environmental signals into plant metabolic, growth and developmental responses. We believe that progress in this area will require the incorporation of population-based thinking into experimental designs. In the early stages of such work, pooled samples of seeds at different stages of dormancy or germination can be adequate to identify major genes or the consequences of their mutation or loss of function. However, the closer we get to understanding specific mechanisms and pathways at the cellular or biochemical levels, the more important it is to move to single-seed approaches and fractionation of seed populations into similar cohorts on a developmental time scale. For example, single-seed assays indicated that some germination-related enzymes increased in activity 1000-fold as the seeds initiated germination, while no activity was expressed in slower germinating or dormant seeds at the same time (Still and Bradford, 1997). Pooling all seeds and extracting them together obscures this fact and leads to the assumption that all seeds are gradually increasing expression of this enzyme. Similarly, if a seed lot is imbibed and the entire population is harvested and pooled together periodically for extraction and mRNA analysis, there appears to be a smooth progression from development/dormancy-related transcripts to the expression of germination/growth-related genes associated with the germination time course. However, if the seeds are separated into germinated and ungerminated fractions before extraction, the germinated fraction will contain almost exclusively transcripts from germination/growth-related genes, while the transcript profile of the ungerminated seeds is still almost identical to that of dry seeds (Yoong, 2015, and authors' unpublished results). This indicates that imbibed but ungerminated seeds maintain a physiological state similar to that of mature seeds until they commit to germination, followed by a massive loss of maturation-related transcripts and correspondingly increased expression of genes associated with germination/growth in a relatively short time on a seed-by-seed basis. This makes sense for dormant or ungerminated seeds, as they are constantly at risk for soil drying and need to retain the dehydration tolerance mechanisms developed during the latter stages of seed development. However, once a seed commits to germination, there is no going back, so the seed can fully initiate reserve mobilization and growth-related gene expression. Pooling seeds in different developmental time stages obscures such timing and makes it difficult to decipher or reconstruct causal connections among specific signaling and response pathways occurring in a given seed.

The same is true for studies of changes occurring in seeds in association with aging and loss of viability. Under genebank storage conditions, it can be years between the time that the first 10% of seeds lost viability and when 50% of seeds had died. Thus, measuring chemical changes in a seed lot at its p_{50} will mix together seeds that are still viable and have years to go before dying with other seeds that have suffered extensive damage and have been dead for years already (Fleming et al., 2017; Zinsmeister et al., 2020). As oxidation is the primary chemical process associated with aging (Bailly et al., 2008; Groot et al., 2015), there will be an increase in oxidized products in such samples compared to unaged seeds (Nagel et al., 2019). However, it is impossible in such a pooled sample to determine whether such oxidation is a cause or effect of the loss of viability. That is, protective mechanisms may largely prevent such changes in viable seeds, and when these are exhausted, large-scale oxidation will occur; that is, some oxidation may be associated with the gradual loss of seed quality, but the large increase in such chemical signals over an aging time course may be primarily contributed by the increasing fraction of dead seeds in the sampled population. Thus, as we pursue the discovery of refined and specific physiological and biochemical mechanisms, pooling samples of disparate seed populations will prevent conclusively establishing causal relationships. Instead, it will be subject to the seed biology equivalent of what economists have called 'the flaw of averages' (Savage, 2012). One approach would be to cut seeds in half and save one half for extraction and analysis and assay the other half for viability, such as by the tetrazolium test. Once the viability status is known, the saved half-seeds of the known viability category could be extracted individually or pooled. In some species, such as soybean, one cotyledon can be removed and imbibed, and its subsequent color development is correlated with the viability of the embryonic seedling; this can be used to sort the individual seeds into viability categories for the analysis of the dry embryonic axis (Fleming et al., 2021). A more convenient solution could be advances in multispectral imaging that can presort seeds into viability categories and enable sampling of subpopulations of seeds with similar deterioration or viability characteristics (Olesen et al., 2015; ElMasry et al., 2019).

As seed and plant scientists adopt single-seed and single-cell methods to identify the actual causal factors of seed behavior and their underlying genes or regulatory pathways (Topham et al., 2017; Abley et al., 2021; Alamos et al., 2021), it will be important to also refine the experimental designs associated with the application of such methods. Those methods are very powerful, but practitioners are currently struggling to understand how to deal with the 'noise' in such studies. The PBT models described here imply that while seeds are highly variable, they form populations that are not randomly chaotic but rather are ordered based on their individual sensitivities to environmental and molecular signals. Such individual variation may be stochastic, but it is neither noise nor error, but rather is an integral component of how seeds fulfill their evolutionary function of introducing seedlings into environments conducive to their survival. Thus, although we have emphasized population-level analyses, the same considerations highlight 'the importance of individuality', which has been identified across levels of biological organization (Gilroy and Trewavas, 2001; Trewavas, 2014). We should therefore consider variation among individuals as a feature, not a bug, in the software of biology. Seed biology has been and can be a leader in illustrating how to utilize individual-based data to understand the emergent properties of cellular and molecular populations (Bradford, 2018; Alamos et al., 2021).

7.8 Broader implications

While we have focused here primarily on the population characteristics of seeds as revealed through PBT models, another interesting component is the ability to transform developmental processes occurring at different speeds onto a normalized time scale. This is quite familiar to agriculturalists through the use of thermal time or degree-days to normalize for changes in plant or insect development across variations in temperature. It seems natural for us to accept that warmer temperatures increase molecular motion and that physical or chemical processes speed up as temperatures rise. However, the actual basis of temperature sensing and response in plants remains largely obscure, and plants are able to compensate temperature differences if needed, such as to maintain an accurate circadian clock or accumulate chilling time for vernalization (Penfield, 2008; Penfield and MacGregor, 2014; Zhao et al., 2021). There is strong evidence that phytochrome B (PHYB) is involved in sensing temperature as well as light, but the underlying mechanisms are complex and still under study (Hernando et al., 2021). Nonetheless, seeds germinate almost perfectly in accordance with thermal time over most of the range between T_b and T_o. Similarly, their germination rates are also precisely adjusted by reductions in ψ, and in combination with temperature on the HTT scale, although the specific cellular mechanism responsible for sensing water potential also remains obscure (Dorone et al., 2021). Perhaps less expected is that the primary hormones regulating dormancy and germination, GA and ABA, also act in the same way, speeding or slowing the time to germination as their concentrations exceed thresholds that vary among seeds and in response to conditions such as chilling or after-ripening. Thus, through seeds, we can catch a glimpse of 'biotime', or biological time, that regulates metabolic and physiological processes and which may or may not coincide with clock time (Bradford and Trewavas, 1994). The relationships between dosages, thresholds and rates underlying PBT models (Fig. 2) may apply across biological scales and processes in a much more fundamental way than we currently appreciate.

In addition, the PBT models specifically incorporate the concept of sensitivity thresholds as being both a critical foundation for such biotime behavior and a source of variation among individuals that provide a wide range of flexibility in signal/response systems. As has been insightfully and entertainingly presented by Braitenberg (1984) in relation to neurological function, variation in the threshold sensitivities of simple autonomous units can be combined into surprisingly complex sensing and control systems. In essence, a normal distribution of sensitivities provides a reservoir of response units, some of which are highly sensitive and may be able to take care of the situation, but if not, and signal levels rise, more units (e.g., cells, enzymes, transcription factors, etc.) can be recruited from the population (Powers et al., 2019). This provides a simple, automatic mechanism for monitoring and maintaining a status quo state in a changing environment or the ability to recruit help and respond quickly to large changes in inputs or developmental phase changes. Examples of apparent PBT behavior across all scales of biology have been described (Bradford, 2018). Thinking in PBT terms changes the way that experiments are designed and their outcomes are interpreted. We believe that a PBT-based viewpoint is essential for seed biologists and will be instructive for all biologists.

8 Where to look for further information

We have organized this chapter to provide a general introduction to PBT models and then to describe their application to an array of topics related to seed quality and performance. Thus, readers interested in a specific topic or application can refer to those specific sections and follow up with the citations therein. A detailed introduction to the initial development and application of the thermal, hydrotime and HTT models can be found in Bradford (1995). Further development of the models and illustration of the importance of the original data quality can be found in Bello and Bradford (2016). Approaches and applications for seed dormancy and ecology can be found in Allen et al. (2007), Batlla and Benech-Arnold (2015) and Liu et al. (2020). Different viewpoints on the statistical analysis of seed germination patterns and the fitting of PBT models can be found in several publications (Hay et al., 2014; Gianinetti, 2020; Moltchanova et al., 2020; Romano and Stevanato, 2020). R scripts and a website for fitting PBT models to germination data can be accessed at www.pbtmodels.org.

9 Acknowledgements

The authors acknowledge the funding for research on this topic from their own lab over the past 30+ years that has come from the United States Department of Agriculture, the National Science Foundation and other sources. We particularly acknowledge the support of the Western Regional Seed Physiology Research Group, a voluntary consortium of seed companies that has supported research on PBT models and in particular recent efforts to develop R scripts and development of a website to facilitate access to tools for fitting PBT models. The authors also acknowledge the thorough and insightful feedback of Dr. Ger Lenssen on an early draft of this chapter.

10 Abbreviations

Term	Definition
ABA	Abscisic acid (concentration)
$ABA_b(g)$	Distribution of base ABA concentrations for germination fractions *g*
CD	Controlled deterioration
cdf	Cumulative density function
C_W, C_H and C_Q	Storability constants determined empirically in Ellis-Roberts equation
D(*g*)	Sensitivity distribution of the total population
eRH	Equilibrium relative humidity = water activity × 100
f_1	Fraction of the total population in subpopulation 1
g	Germination percentage
GA	Gibberellins (concentration)
$GA_b(g)$	Distribution of base GA concentrations for germination fractions *g*
GR	Germination rate
GR_g	GR for specific germination fraction *g* in the population
HT	Hydrotime
HTP	Hydrothermal priming time
HTT	Hydrothermal time
K_E	Species constant for storability in Ellis-Roberts model
K_i	Initial viability on probit scale in Ellis-Roberts model
k_T	Constant slope of the increase in $\psi_b(g)$ per degree above T_o
m	Seed moisture content fresh weight basis
MGT	Mean germination time
n	Germinated seeds or population
O_2	Oxygen
°C d	Degree-day
°C h	Degree-hour
Ox	Percentage of O_2
Ox_b	Base oxygen percentage that prevents germination
$Ox_b(50)$	Median Ox_b for population
$Ox_b(g)$	Distribution of Ox_b percentages of germination fractions *g*
p	Storage period
p50	Storage period to a loss of 50% viability
PBT	Population-based threshold
PEG	Polyethylene glycol
p_{max}	Maximum lifetimes under the conditions of storage
$p_{max}(50)$	Median p_{max} for population
$p_{max}(g)$	Distribution of p_{max} for germination fractions *g*

Published by Burleigh Dodds Science Publishing Limited, 2022.

Term	Definition
$p_{norm}(g)$	Distribution of aging times at which fraction g of normal seedlings occurs
probit	Probability unit = units of one standard deviation
ψ	Water potential, MPa
ψ_b	Minimum or base threshold ψ that prevents germination
$\psi_b(50)$	Median ψ_b for population
$\psi_b(g)$	Distribution of ψ_b for germination fractions g in the population
ψ_{min}	Lowest ψ at which the priming effect will occur
RH	Relative humidity
σ	Standard deviation
$\sigma_{p_{max}}$	Standard deviation for p_{max} distribution in population
σ_{ψ_b}	Standard deviation of ψ_b distribution in population
σ_{θ_T}	Standard deviation of θ_T distribution in population
σ_{X1} and σ_{X2}	Standard deviations of X_b values among seeds in subpopulations 1 and 2
T	Temperature
t	Time to germination
T_b	Minimum or base temperature
T_c	Maximum or ceiling temperature
$T_c(g)$	Distribution of T_c for germination fractions g in the population
t_g	Time to germination of percentage or fraction g
t_{g1} and t_{g2}	Germination times of fractions g in subpopulations 1 and 2
θ_{ABA}	ABA time constant
θ_{GA}	GA time constant
θ_H	Hydrotime constant
θ_{HT}	Hydrothermal time constant
θ_{HTP}	Hydrothermal priming time constant
θ_{Ox}	Oxygen-time constant
θ_T	Thermal time constant
$\theta_T(g)$	Distribution of θ_T for specific germination fractions g in the population
θ_X	Time constant for generic factor X
θ_{X1} and θ_{X2}	Time constants for subpopulations 1 and 2 of generic factor X
θ_{age}	Aging time constant
T_h	Dormancy higher limit T
T_l	Dormancy lower limit T
T_{min}	Lowest T at which the priming effect will occur
t_{norm}	Imbibition time after which normal seedlings will not appear
T_o	Optimum temperature

Term	Definition
t_p	Duration of priming
v	Percent viability on probit scale
X	Level or concentration of generic factor X
X_b	Base or threshold level for a response to X
$X_b(g)$	Distribution of X_b for germination fractions g in the population
$X_b(50)_1$ and $X_b(50)_2$	Medians of the distributions of X_b values for subpopulations 1 and 2

11 References

Abley, K., Formosa-Jordan, P., Tavares, H., Chan, E. Y. T., Afsharinafar, M., Leyser, O. and Locke, J. C. W. (2021). An ABA-GA bistable switch can account for natural variation in the variability of Arabidopsis seed germination time. *eLife* 10: e59485. 10.7554/eLife.59485.

Alamos, S., Reimer, A., Niyogi, K. K. and Garcia, H. G. (2021). Quantitative imaging of RNA polymerase II activity in plants reveals the single-cell basis of tissue-wide transcriptional dynamics. *Nat. Plants* 7(8): 1037–1049. 10.1038/s41477-021-00976-0.

Allen, P. S., Benech-Arnold, R. L., Batlla, D. and Bradford, K. J. (2007). Modeling of seed dormancy. In: Bradford, K. J. and Nonogaki, H. (Eds), *Seed Development, Dormancy and Germination*. Blackwell Publishing, Oxford, UK, pp. 72–112.

Alvarado, V. and Bradford, K. J. (2002). A hydrothermal time model explains the cardinal temperatures for seed germination. *Plant Cell Environ.* 25(8): 1061–1069.

Alvarado, V. and Bradford, K. J. (2005). Hydrothermal time analysis of seed dormancy in true (botanical) potato seeds. *Seed Sci. Res.* 15(2): 77–88.

Angel, A., Song, J., Yang, H., Questa, J. I., Dean, C. and Howard, M. (2015). Vernalizing cold is registered digitally at *FLC*. *Proc. Natl Acad. Sci. U. S. A.* 112(13): 4146–4151. 10.1073/pnas.1503100112.

Ardura, D. (2017). Mass spectrometry-based metabolomics applied to plant chemotaxonomy and to thermoinhibition of germination in lettuce (*Lactuca sativa*) seeds. PhD Dissertation. University of California, Davis, CA.

Auge, G. A., Leverett, L. D., Edwards, B. R. and Donohue, K. (2017). Adjusting phenotypes via within- and across-generational plasticity. *New Phytol.* 216(2): 343–349. 10.1111/nph.14495.

Baalbaki, R., Elias, S., Marcos-Filho, J. and McDonald, M. B. (2009). *Seed Vigor Testing Handbook*. Association of Official Seed Analysts, Ithaca, NY.

Baena-González, E., Rolland, F., Thevelein, J. M. and Sheen, J. (2007). A central integrator of transcription networks in plant stress and energy signalling. *Nature* 448(7156): 938–942. 10.1038/nature06069.

Bailly, C., El-Maarouf-Bouteau, H. and Corbineau, F. (2008). From intracellular signaling networks to cell death: the dual role of reactive oxygen species in seed physiology. *C. R. Biol.* 331(10): 806–814.

Bair, N. B., Meyer, S. E. and Allen, P. S. (2006). A hydrothermal after-ripening time model for seed dormancy loss in *Bromus tectorum* L. *Seed Sci. Res.* 16(1): 17–28.

Bakhshandeh, E., Bradford, K. J., Pirdashti, H., Vahabinia, F. and Abdellaoui, R. (2020). A new halothermal time model describes seed germination responses to salinity across both sub- and supra-optimal temperatures. *Acta Physiol. Plant.* 42(8): 137. 10.1007/s11738-020-03126-9.

Bakhshandeh, E., Jamali, M., Afshoon, E. and Gholamhossieni, M. (2017). Using hydrothermal time concept to describe sesame (*Sesamum indicum* L.) seed germination response to temperature and water potential. *Acta Physiol. Plant.* 39(11): 250. 10.1007/s11738-017-2549-8.

Baskin, C. C. and Baskin, J. M. (2014). *Seeds: Ecology, Biogeography, and Evolution of Dormancy and Germination* (2nd edn.). Elsevier Science, Burlington.

Baskin, C. C. and Baskin, J. M. (2020). Breaking seed dormancy during dry storage: a useful tool or major problem for successful restoration via direct seeding? *Plants (Basel)* 9(5): 17. 10.3390/plants9050636.

Baskin, J. M. and Baskin, C. C. (2021). The great diversity in kinds of seed dormancy: a revision of the Nikolaeva–Baskin classification system for primary seed dormancy *Seed Sci. Res.* 31: 249–277. 10.1017/S096025852100026X.

Bassel, G. W. (2016). To grow or not to grow? *Trends Plant Sci.* 21(6): 498–505. 10.1016/j.tplants.2016.02.001.

Bates, D. M. and Chambers, J. M. (1992). Nonlinear models. In: Chambers, J. M. and Hastie, T. J. (Eds), *Statistical Models*. New York, S. Chapman & Hall/CRC, pp. 421–454.

Bates, D. M. and Watts, D. G. (1988). *Nonlinear Regression Analysis and Its Applications*. New York, Wiley.

Batlla, D. and Agostinelli, A. M. (2017). Thermal regulation of secondary dormancy induction in *Polygonum aviculare* seeds: a quantitative analysis using the hydrotime model. *Seed Sci. Res.* 27(3): 231–242. 10.1017/S0960258517000198.

Batlla, D. and Benech-Arnold, R. L. (2005). Changes in the light sensitivity of buried *Polygonum aviculare* seeds in relation to cold-induced dormancy loss: development of a predictive model. *New Phytol.* 165(2): 445–452.

Batlla, D. and Benech-Arnold, R. L. (2015). A framework for the interpretation of temperature effects on dormancy and germination in seed populations showing dormancy. *Seed Sci. Res.* 25(2): 147–158. 10.1017/S0960258514000452.

Batlla, D., Verges, V. and Benech-Arnold, R. L. (2003). A quantitative analysis of seed responses to cycle-doses of fluctuating temperatures in relation to dormancy: development of a thermal time model for *Polygonum aviculare* L. seeds. *Seed Sci. Res.* 13(3): 197–207.

Bello, P. and Bradford, K. J. (2016). Single-seed oxygen consumption measurements and population-based threshold models link respiration and germination rates under diverse conditions. *Seed Sci. Res.* 26(3): 199–221.

Bello, P. and Bradford, K. J. (2021). Relationships of *Brassica* seed physical characteristics with germination performance and plant blindness. *Agriculture* 11(3): 220. 10.3390/agriculture11030220.

Bello, P. H. N. (2020). *Analytical and Computational Tools to Assess Seed Quality and Model Germination Rates*. MS. University of California, Davis, Davis, CA.

Benech-Arnold, R. L., Gualano, N., Leymarie, J., Côme, D. and Corbineau, F. (2006). Hypoxia interferes with ABA metabolism and increases ABA sensitivity in embryos of dormant barley grains. *J. Exp. Bot.* 57(6): 1423–1430.

Bewley, J. D., Bradford, K. J., Hilhorst, H. W. M. and Nonogaki, H. (2013). *Seeds: Physiology of Development, Germination and Dormancy* (3rd edn.). Springer, New York.

Bierhuizen, J. F. and Wagenvoort, W. A. (1974). Some aspects of seed germination in vegetables. I. The determination and application of heat sums and minimum temperature for germination. *Sci. Hortic.* 2(3): 213–219.

Bizouerne, E., Buitink, J., Vu, B. L., Vu, J. L., Esteban, E., Pasha, A., Provart, N., Verdier, J. and Leprince, O. (2021). Gene co-expression analysis of tomato seed maturation reveals tissue-specific regulatory networks and hubs associated with the acquisition of desiccation tolerance and seed vigour. *BMC Plant Biol.* 21(1): 124. 10.1186/s12870-021-02889-8.

Bliss, C. I. (1934). The method of probits. *Science* 79(2037): 38–39. 10.1126/science.79.2037.38.

Bloomberg, M., Sedcole, J. R., Mason, E. G. and Buchan, G. (2009). Hydrothermal time germination models for radiata pine (*Pinus radiata* D. Don). *Seed Sci. Res.* 19(3): 171–182.

Boddy, L. G., Bradford, K. J. and Fischer, A. J. (2012). Population-based threshold models describe weed germination and emergence patterns across varying temperature, moisture and oxygen conditions. *J. Appl. Ecol.* 49(6): 1225–1236. 10.1111/j.1365-2664.2012.02206.x.

Bolingue, W., Rosnoblet, C., Leprince, O., Vu, B. L., Aubry, C. and Buitink, J. (2010). The MtSNF4b subunit of the sucrose non-fermenting-related kinase complex connects after-ripening and constitutive defense responses in seeds of *Medicago truncatula*. *Plant J.* 61(5): 792–803. 10.1111/j.1365-313X.2009.04106.x.

Bradford, K. J. (1990). A water relations analysis of seed germination rates. *Plant Physiol.* 94(2): 840–849.

Bradford, K. J. (1995). Water relations in seed germination. In: Kigel, J. and Galili, G. (Eds), *Seed Development and Germination*. Marcel Dekker, Inc., New York, pp. 351–396.

Bradford, K. J. (2005). Threshold models applied to seed germination ecology. *New Phytol.* 165(2): 338–341.

Bradford, K. J. (2018). Interpreting biological variation: seeds, populations and sensitivity thresholds. *Seed Sci. Res.* 28(3): 158–167. 10.1017/S0960258518000156.

Bradford, K. J., Bello, P., Fu, J.-C. and Barros, M. (2013). Single-seed respiration: a new method to assess seed quality. *Seed Sci. Technol.* 41(3): 420–438.

Bradford, K. J., Benech-Arnold, R. L., Côme, D. and Corbineau, F. (2008). Quantifying the sensitivity of barley seed germination to oxygen, abscisic acid and gibberellin using a population-based threshold model. *J. Exp. Bot.* 59(2): 335–347.

Bradford, K. J., Côme, D. and Corbineau, F. (2007). Quantifying the oxygen sensitivity of seed germination using a population-based threshold model. *Seed Sci. Res.* 17(1): 33–43.

Bradford, K. J., Dahal, P. and Bello, P. (2016). Using relative humidity indicator paper to measure seed and commodity moisture contents. *Agric. Environ. Lett.* 1(1). 10.2134/ael2016.04.0018: 160018.

Bradford, K. J., Dahal, P., Van Asbrouck, J., Kunusoth, K., Bello, P., Thompson, J. and Wu, F. (2018). The dry chain: reducing postharvest losses and improving food safety in humid climates. *Trends Food Sci. Technol.* 71: 84–93. 10.1016/j.tifs.2017.11.002.

Bradford, K. J. and Haigh, A. M. (1994). Relationship between accumulated hydrothermal time during seed priming and subsequent seed germination rates. *Seed Sci. Res.* 4: 1–10.

Bradford, K. J. and Somasco, O. A. (1994). Water relations of lettuce seed thermoinhibition. I. Priming and endosperm effects on base water potential. *Seed Sci. Res.* 4: 1–10.

Bradford, K. J. and Still, D. W. (2004). Applications of hydrotime analysis in seed testing. *Seed Technol.* 26: 75–85.

Bradford, K. J., Tarquis, A. M. and Duran, J. M. (1993). A population-based threshold model describing the relationship between germination rates and seed deterioration. *J. Exp. Bot.* 44(7): 1225–1234.

Bradford, K. J. and Trewavas, A. J. (1994). Sensitivity thresholds and variable time scales in plant hormone action. *Plant Physiol.* 105(4): 1029–1036.

Braitenberg, V. (1984). *Vehicles. Experiments in Synthetic Psychology*. MIT Press, Cambridge, MA.

Bruggink, H. and van Duijn, B. (2017). X-ray based seed analysis. *Seed Test. Int.* 153: 45–50.

Burghardt, L. T., Metcalf, C. J. E., Wilczek, A. M., Schmitt, J. and Donohue, K. (2015). Modeling the influence of genetic and environmental variation on the expression of plant life cycles across landscapes. *Am. Nat.* 185(2): 212–227. 10.1086/679439.

Cadman, C. S. C., Toorop, P. E., Hilhorst, H. W. M. and Finch-Savage, W. E. (2006). Gene expression profiles of Arabidopsis Cvi seeds during dormancy cycling indicate a common underlying dormancy control mechanism. *Plant J.* 46(5): 805–822. 10.1111/j.1365-313X.2006.02738.x.

Cao, H., Han, Y., Li, J., Ding, M., Li, Y., Li, X., Chen, F., Soppe, W. J. and Liu, Y. (2019). *Arabidopsis thaliana* SEED DORMANCY 4-LIKE regulates dormancy and germination by mediating the gibberellin pathway. *J. Exp. Bot.* 71(3): 919–933. 10.1093/jxb/erz471.

Chantre, G. R., Batlla, D., Sabbatini, M. R. and Orioli, G. (2009). Germination parameterization and development of an after-ripening thermal-time model for primary dormancy release of *Lithospermum arvense* seeds. *Ann. Bot.* 103(8): 1291–1301. 10.1093/aob/mcp070.

Chen, D., Chen, X., Wang, J., Zhang, Z., Wang, Y., Jia, C. and Hu, X. (2021). Estimation of thermal time model parameters for seed germination in 15 species: the importance of distribution function. *Seed Sci. Res.* 31(2): 83–90. 10.1017/S0960258521000040.

Cheng, Z. Y. and Bradford, K. J. (1999). Hydrothermal time analysis of tomato seed germination responses to priming treatments. *J. Exp. Bot.* 50(330): 89–99.

Christensen, M., Meyer, S. E. and Allen, P. S. (1996). A hydrothermal time model of seed after-ripening in *Bromus tectorum* L. *Seed Sci. Res.* 6: 1–7.

Colmer, J., O'Neill, C. M., Wells, R., Bostrom, A., Reynolds, D., Websdale, D., Shiralagi, G., Lu, W., Lou, Q., Le Cornu, T., Ball, J., Renema, J., Flores Andaluz, G., Benjamins, R., Penfield, S. and Zhou, J. (2020). SeedGerm: a cost-effective phenotyping platform for automated seed imaging and machine-learning based phenotypic analysis of crop seed germination. *New Phytol.* 228(2): 778–793. 10.1111/nph.16736.

Commander, L. E., Merino-Martín, L., Elliott, C. P., Miller, B. P., Dixon, K. and Stevens, J. (2020). Demographic, seed and microsite limitations to seedling recruitment in semi-arid mine site restoration. *Plant Soil* 457(1–2): 113–129. 10.1007/s11104-019-04081-2.

Corbineau, F. and Côme, D. (1995). Control of seed germination and dormancy by the gaseous environment. In: Kigel, J. and Galili, G. (Eds), *Seed Development and Germination*. Marcel Dekker, Inc., New York, pp. 397–424.

Covell, S., Ellis, R. H., Roberts, E. H. and Summerfield, R. J. (1986). The influence of temperature on seed germination rate in grain legumes. 1. A comparison of chickpea, lentil, soybean and cowpea at constant temperatures. *J. Exp. Bot.* 37(5): 705–715.

Cromarty, A. S., Ellis, R. H. and Roberts, E. H. (1982). *The Design of Seed Storage Facilities for Genetic Conservation*. International Board for Plant Genetic Resources, Rome.

Dahal, P. and Bradford, K. J. (1990). Effects of priming and endosperm integrity on seed germination rates of tomato genotypes. II. Germination at reduced water potential. *J. Exp. Bot.* 41(11): 1441-1453.

Dahal, P. and Bradford, K. J. (1994). Hydrothermal time analysis of tomato seed germination at suboptimal temperature and reduced water potential. *Seed Sci. Res.* 4(2): 71–80.

Dahal, P., Bradford, K. J. and Jones, R. A. (1990). Effects of priming and endosperm integrity on seed germination rates of tomato genotypes. I. Germination at suboptimal temperature. *J. Exp. Bot.* 41(11): 1431–1439.

Dahal, P., Kim, N. S. and Bradford, K. J. (1996). Respiration and germination rates of tomato seeds at suboptimal temperatures and reduced water potentials. *J. Exp. Bot.* 47(7): 941–947.

De Vitis, M., Hay, F. R., Dickie, J. B., Trivedi, C., Choi, J. and Fiegener, R. (2020). Seed storage: maintaining seed viability and vigor for restoration use. *Restor. Ecol.* 28(S3): S249–S255. 10.1111/rec.13174.

Donohue, K. (2009). Completing the cycle: maternal effects as the missing link in plant life histories. *Philos. Trans. R. Soc. Lond. B Biol. Sci.* 364(1520): 1059–1074. 10.1098/rstb.2008.0291.

Donohue, K., Burghardt, L. T., Runcie, D., Bradford, K. J. and Schmitt, J. (2015). Applying developmental threshold models to evolutionary ecology. *Trends Ecol. Evol.* 30(2): 66–77. 10.1016/j.tree.2014.11.008.

Donohue, K., Rubio de Casas, R., Burghardt, L., Kovach, K. and Willis, C. G. (2010). Germination, postgermination adaptation, and species ecological ranges. *Annu. Rev. Ecol. Evol. Syst.* 41(1): 293–319. 10.1146/annurev-ecolsys-102209-144715.

Dorone, Y., Boeynaems, S., Flores, E., Jin, B., Hateley, S., Bossi, F., Lazarus, E., Pennington, J. G., Michiels, E., De Decker, M., Vints, K., Baatsen, P., Bassel, G. W., Otegui, M. S., Holehouse, A. S., Exposito-Alonso, M., Sukenik, S., Gitler, A. D. and Rhee, S. Y. (2021). A prion-like protein regulator of seed germination undergoes hydration-dependent phase separation. *Cell*. 10.1016/j.cell.2021.06.009.

Ducournau, S., Feutry, A., Plainchault, P., Revollon, P., Vigouroux, B. and Wagner, M. H. (2005). Using computer vision to monitor germination time course of sunflower (*Helianthus annuus* L.) seeds. *Seed Sci. Technol.* 33(2): 329–340. 10.15258/sst.2005.33.2.06.

Dürr, C., Dickie, J. B., Yang, X. -Y. and Pritchard, H. W. (2015). Ranges of critical temperature and water potential values for the germination of species worldwide: contribution to a seed trait database. *Agric. Forest Meteorol.* 200: 222–232. 10.1016/j.agrformet.2014.09.024.

Dutta, S. and Bradford, K. J. (1994). Water relations of lettuce seed thermoinhibition. II. Ethylene and endosperm effects on base water potential. *Seed Sci. Res.* 4(1): 11–18.

Egli, D. B. (2017). *Seed Biology and the Yield of Grain Crops* (2nd edn.). CAB International, New York.

Elias, S. G., Copeland, L. O., McDonald, M. B. and Baalbaki, R. Z. (2012). *Seed Testing. Principles and Practices*. Michigan State University Press, East Lansing, MI.

Ellis, R. H. (1988). The viability equation, seed viability nomographs, and practical advice on seed storage. *Seed Sci. Technol.* 16: 29–50.

Ellis, R. H. (2019). Temporal patterns of seed quality development, decline, and timing of maximum quality during seed development and maturation. *Seed Sci. Res.* 29(2): 135–142. 10.1017/S0960258519000102.

Ellis, R. H., Covell, S., Roberts, E. H. and Summerfield, R. J. (1986). The influence of temperature on seed germination rate in grain legumes. 2. Intraspecific variation in chickpea (*Cicer arietinum* L.) at constant temperatures. *J. Exp. Bot.* 37(10): 1503–1515.

Ellis, R. H. and Hong, T. D. (2006). Temperature sensitivity of the low-moisture-content limit to negative seed longevity-moisture content relationships in hermetic storage. *Ann. Bot.* 97(5): 785–791. 10.1093/aob/mcl035.

Ellis, R. H. and Hong, T. D. (2007). Quantitative response of the longevity of seed of twelve crops to temperature and moisture in hermetic storage. *Seed Sci. Technol.* 35(2): 432–444.

Ellis, R. H., Nasehzadeh, M., Hanson, J., Ndiwa, N. and Woldemariam, Y. (2019). Medium-term seed storage of diverse genera of forage grasses, evidence-based GeneBank monitoring intervals, and regeneration standards. *Genet. Resour. Crop Evol.* 66(3): 723–734. 10.1007/s10722-019-00748-y.

Ellis, R. H. and Roberts, E. H. (1980). Improved equations for the prediction of seed longevity. *Ann. Bot.* 45(1): 13–30.

Ellis, R. H. and Roberts, E. H. (1981). The quantification of aging and survival in orthodox seeds. *Seed Sci. Technol.* 9: 373–409.

ElMasry, G., Mandour, N., Wagner, M. H., Demilly, D., Verd er, J., Belin, E. and Rousseau, D. (2019). Utilization of computer vision and multispectral imaging techniques for classification of cowpea (*Vigna unguiculata*) seeds. *Plant Methods* 15: 24. 10.1186/s13007-019-0411-2.

Fernández Farnocchia, R. B., Benech-Arnold, R. L. and Batlla, D. (2019). Regulation of seed dormancy by the maternal environment is instrumental for maximizing plant fitness in *Polygonum aviculare*. *J. Exp. Bot.* 70(18): 4793–4806. 10.1093/jxb/erz269.

Fernández Farnocchia, R. B., Benech-Arnold, R. L., Mantese, A. and Batlla, D. (2021). Optimization of timing of next-generation emergence in *Amaranthus hybridus* is determined via modulation of seed dormancy by the maternal environment. *J. Exp. Bot.* 72(12): 4283–4297. 10.1093/jxb/erab141.

Fernández-Pascual, E., Carta, A., Mondoni, A., Cavieres, L. A., Rosbakh, S., Venn, S., Satyanti, A., Guja, L., Briceño, V. F., Vandelook, F., Mattana, E., Saatkamp, A., Bu, H., Sommerville, K., Poschlod, P., Liu, K., Nicotra, A. and Jiménez-Alfaro, B. (2021). The seed germination spectrum of alpine plants: a global meta-analysis. *New Phytol.* 229(6): 3573–3586. 10.1111/nph.17086.

Finch-Savage, W. E. and Bassel, G. W. (2016). Seed vigour and crop establishment: extending performance beyond adaptation. *J. Exp. Bot.* 67(3): 567–591. 10.1093/jxb/erv490.

Finch-Savage, W. E., Cadman, C. S. C., Toorop, P. E., Lynn, J. R. and Hilhorst, H. W. M. (2007). Seed dormancy release in Arabidopsis Cvi by dry after-ripening, low temperature, nitrate and light shows common quantitative patterns of gene expression directed by environmentally specific sensing. *Plant J.* 51(1): 60–78. 10.1111/j.1365-313X.2007.03118.x.

Finch-Savage, W. E. and Footitt, S. (2017). Seed dormancy cycling and the regulation of dormancy mechanisms to time germination in variable field environments. *J. Exp. Bot.* 68(4): 843–856. 10.1093/jxb/erw477.

Finch-Savage, W. E., Rowse, H. R. and Dent, K. C. (2005). Development of combined imbibition and hydrothermal threshold models to simulate maize (*Zea mays*) and chickpea (*Cicer arietinum*) seed germination in variable environments. *New Phytol.* 165(3): 825–838.

Fleming, M. B., Hill, L. M. and Walters, C. (2019). The kinetics of ageing in dry-stored seeds: a comparison of viability loss and RNA degradation in unique legacy seed collections. *Ann. Bot.* 123(7): 1133–1146. 10.1093/aob/mcy217.

Fleming, M. B., Patterson, E. L. and Walters, C. (2021). Too little, too late: transcription during imbibition of lethally aged soybean seeds is weak and delayed, but not aberrant. *bioRxiv*. 10.1101/2021.03.25.437023.

Fleming, M. B., Richards, C. M. and Walters, C. (2017). Decline in RNA integrity of dry-stored soybean seeds correlates with loss of germination potential. *J. Exp. Bot.* 68(9): 2219–2230. 10.1093/jxb/erx100.

Folberth, C., Khabarov, N., Balkovič, J., Skalský, R., Visconti, P., Ciais, P., Janssens, I. A., Peñuelas, J. and Obersteiner, M. (2020). The global cropland-sparing potential of high-yield farming. *Nat. Sustainability* 3(4): 281–289. 10.1038/s41893-020-0505-x.

Galletti, P. A., Carvalho, M. E. A., Hirai, W. Y., Brancaglioni, V. A., Arthur, V. and Barboza da Silva, C. (2020). Integrating optical imaging tools for rapid and non-invasive characterization of seed quality: tomato (*Solanum lycopersicum* L.) and carrot (*Daucus carota* L.) as study cases. *Front. Plant Sci.* 11: 577851. 10.3389/fpls.2020.577851.

Gianella, M., Bradford, K. J. and Guzzon, F. (2021). Ecological, (epi)genetic and physiological aspects of bet-hedging in angiosperms. *Plant Reprod.* 34(1): 21–36. 10.1007/s00497-020-00402-z.

Gianinetti, A. (2020). Basic features of the analysis of germination data with generalized linear mixed models. *Data* 5(1): 6. 10.3390/data5010006.

Gianinetti, A. and Cohn, M. A. (2007). Seed dormancy in red rice. XII: population-based analysis of dry-afterripening with a hydrotime model. *Seed Sci. Res.* 17(4): 253–271. 10.1017/S0960258507841523.

Gilroy, S. and Trewavas, A. (2001). Signal processing and transduction in plant cells: the end of the beginning? *Nat. Rev. Mol. Cell Biol.* 2(4): 307–314.

Gioria, M., Carta, A., Baskin, C. C., Dawson, W., Essl, F., Kreft, H., Pergl, J., van Kleunen, M., Weigelt, P., Winter, M. and Pyšek, P. (2021). Persistent soil seed banks promote naturalisation and invasiveness in flowering plants. *Ecol. Lett.* 24(8): 1655–1667. 10.1111/ele.13783.

Gremer, J. R., Chiono, A., Suglia, E., Bontrager, M., Okafor, L. and Schmitt, J. (2020). Variation in the seasonal germination niche across an elevational gradient: the role of germination cueing in current and future climates. *Am. J. Bot.* 107(2): 350–363. 10.1002/ajb2.1425.

Gremer, J. R., Kimball, S. and Venable, D. L. (2016). Within-and among-year germination in Sonoran Desert winter annuals: bet hedging and predictive germination in a variable environment. *Ecol. Lett.* 19(10): 1209–1218. 10.1111/ele.12655.

Groot, S. P. C., de Groot, L., Kodde, J. and van Treuren, R. (2015). Prolonging the longevity of *ex situ* conserved seeds by storage under anoxia. *Plant Genet. Resour.* 13(1): 18–26. 10.1017/S1479262114000586.

Grundy, A. C., Phelps, K., Reader, R. J. and Burston, S. (2000). Modelling the germination of *Stellaria media* using the concept of hydrothermal time. *New Phytol.* 148(3): 433–444.

Gummerson, R. J. (1986). The effect of constant temperatures and osmotic potential on the germination of sugar beet. *J. Exp. Bot.* 37(6): 729–741.

Halmer, P. (2000). Commercial seed treatment technology. In: Black, M. and Bewley, J. D. (Eds), *Seed Technology and Its Biological Basis*. Sheffield Academic Press, Sheffield, UK, pp. 257–286.

Halmer, P. (2004). Methods to improve seed performance in the field. In: Bench-Arnold, R. L. and Sanchez, R. A. (Eds), *Handbook of Seed Physiology. Applications to Agriculture*. Food Products Press, New York, pp. 125–166.

Hardegree, S. P. (2006). Predicting germination response to temperature. I. Cardinal-temperature models and subpopulation-specific regression. *Ann. Bot.* 97(6): 1115–1125.

Hardegree, S. P., Walters, C. T., Boehm, A. R., Olsoy, P. J., Clark, P. E. and Pierson, F. B. (2015). Hydrothermal germination models: comparison of two data-fitting approaches with probit optimization. *Crop Sci.* 55(5): 2276–2290. 10.2135/cropsci2014.10.0703.

Hardegree, S. P. and Winstral, A. H. (2006). Predicting germination response to temperature. II. Three-dimensional regression, statistical gridding and iterative-probit optimization using measured and interpolated-subpopulation data. *Ann. Bot.* 98(2): 403–410.

Harrington, J. F. (1972). Seed storage and longevity. In: Kozlowski, T. T. (Ed.), *Seed Biology. Volume III. Insects, and Seed Collection, Storage, Testing, and Certification*. Academic Press, New York, pp. 145–245.

Hawkins, K. K., Allen, P. S. and Meyer, S. E. (2017). Secondary dormancy induction and release in *Bromus tectorum* seeds: the role of temperature, water potential and hydrothermal time. *Seed Sci. Res.* 27(1): 12–25. 10.1017/S0960258516000258.

Hay, F. R., Mead, A. and Bloomberg, M. (2014). Modelling seed germination in response to continuous variables: use and limitations of probit analysis and alternative approaches. *Seed Sci. Res.* 24(3): 165–186. 10.1017/S096025851400021X.

Hay, F. R., Mead, A., Manger, K. and Wilson, F. J. (2003). One-step analysis of seed storage data and the longevity of *Arabidopsis thaliana* seeds. *J. Exp. Bot.* 54(384): 993–1011. 10.1093/jxb/erg103.

Hay, F. R. and Whitehouse, K. J. (2017). Rethinking the approach to viability monitoring in seed genebanks. *Conserv. Physiol.* 5(1): cox009. 10.1093/conphys/cox009.

Hegarty, T. W. (1978). The physiology of seed hydration and dehydration, and the relation between water stress and the control of germination: a review. *Plant Cell Environ.* 1(2): 101–119.

Hernando, C. E., Murcia, M. G., Pereyra, M. E., Sellaro, R. and Casal, J. J. (2021). Phytochrome B links the environment to transcription. *J. Exp. Bot.* 72(11): 4068–4084. 10.1093/jxb/erab037.

Heydecker, W., Higgins, J. and Gulliver, R. L. (1973). Accelerated germination by osmotic seed treatment. *Nature* 246(5427): 42–44.

Huang, Z., Liu, S., Bradford, K. J., Huxman, T. E. and Venable, D. L. (2016). The contribution of germination functional traits to population dynamics of a desert plant community. *Ecology* 97(1): 250–261. 10.1890/15-0744.1.

Huo, H., Wei, S. and Bradford, K. J. (2016). DELAY OF GERMINATION1 (DOG1) regulates both seed dormancy and flowering time through microRNA pathways. *Proc. Natl Acad. Sci. U. S. A.* 113(15): E2199–E2206. 10.1073/pnas.1600558113.

Ilbi, H., Powell, A. A. and Alan, O. (2020). Single radicle emergence count for predicting vigour of marigold (*Tagetes* spp.) seed lots. *Seed Sci. Technol.* 48(3): 381–389. 10.15258/sst.2020.48.3.06.

ISTA (2020). International Seed Testing Association. International rules for seed testing. *Rules* 2020. Available at: https://www.seedtest.org/upload/cms/user/OGM19-06RulesProposalsforISTA2020.pdf.

Jalink, H., Van der Schoor, R., Frandas, A., Van Pijlen, J. G. and Bino, R. J. (1998). Chlorophyll fluorescence of *Brassica oleracea* seeds as a non-destructive marker for seed maturity and seed performance. *Seed Sci. Res.* 8(4): 437–443.

Johnston, I. G. and Bassel, G. W. (2018). Identification of a bet-hedging network motif generating noise in hormone concentrations and germination propensity in Arabidopsis. *J. R. Soc. Interface* 15(141) 10.1098/rsif.2018.0042.

Joosen, R. V. L., Kodde, J., Willems, L. A. J., Ligterink, W., van der Plas, L. H. W. and Hilhorst, H. W. M. (2010). GERMINATOR: a software package for high-throughput scoring and curve fitting of Arabidopsis seed germination. *Plant J.* 62(1): 148–159.

Kenanoglu, B. B., Demir, I. and Jalink, H. (2015). Improvement of seed germination performance of stored commercial pepper seed lots with chlorophyll fluorescence sorting method. *Am. J. Exp. Agric.* 10(4): 1–6. 10.9734/AJEA/2016/20467.

Klupczyńska, E. A. and Pawłowski, T. A. (2021). Regulation of seed dormancy and germination mechanisms in a changing environment. *Int. J. Mol. Sci.* 22(3): 1357. 10.3390/ijms22031357.

Lambert, B., Denolf, P., Engelen, S., Golds, T., Haesendonckx, B., Ruiter, R., Robbens, S., Bots, M. and Laga, B. (2015). Omics-directed reverse genetics enables the creation of new productivity traits for the vegetable oil crop canola. *Procedia Environ. Sci.* 29: 77–78. 10.1016/j.proenv.2015.07.167.

Larson, J. E. and Funk, J. L. (2016). Regeneration: an overlooked aspect of trait-based plant community assembly models. *J. Ecol.* 104(5): 1284–1298. 10.1111/1365-2745.12613.

Leprince, O., Pellizzaro, A., Berriri, S. and Buitink, J. (2017). Late seed maturation: drying without dying. *J. Exp. Bot.* 68(4): 827–841. 10.1093/jxb/erw363.

Liu, S., Bradford, K. J., Huang, Z. and Venable, D. L. (2020). Hydrothermal sensitivities of seed populations underlie fluctuations of dormancy states in an annual plant community. *Ecology* 101(3): e02958. 10.1002/ecy.2958.

Liu, S., Streich, J., Borevitz, J. O., Rice, K. J., Li, T., Li, B. and Bradford, K. J. (2019). Environmental resource deficit may drive the evolution of intraspecific trait variation in invasive plant populations. *Oikos* 128(2): 171–184. 10.1111/oik.05548.

Long, R. L., Gorecki, M. J., Renton, M., Scott, J. K., Colville, L., Goggin, D. E., Commander, L. E., Westcott, D. A., Cherry, H. and Finch-Savage, W. E. (2015). The ecophysiology of seed persistence: a mechanistic view of the journey to germination or demise. *Biol. Rev. Camb. Philos. Soc.* 90(1): 31–59. 10.1111/brv.12095.

Loomis, R. S. and Connor, D. J. (1992). *Crop Ecology: Productivity and Management in Agricultural Systems*. Cambridge University Press, New York, NY.

Lutts, S., Benincasa, P., Wojtyla, L., Kubala, S., Pace, R., Lechowska, K., Quinet, M., Garnczarska, M. (2016). Seed priming: new comprehensive approaches for an old empirical technique. In: Araujo, S. and Balestrazzi, A. (Eds), *New Challenges in Seed*

Biology - Basic and Translational Research Driving Seed Technology. InTech, Rijeka. 10.5772/64420.

Mahmoud, A. M. A. (2020). Tomato rootstock breeding: evaluation of tomato interspecific hybrid rootstocks under greenhouse conditions. *Hort. J.* 89(5): 575–585. 10.2503/hortj.UTD-199.

Maity, A., Lamichaney, A., Joshi, D. C., Bajwa, A., Subramanian, N., Walsh, M. and Bagavathiannan, M. (2021). Seed shattering: a trait of evolutionary importance in plants. *Front. Plant Sci.* 12: 657773. 10.3389/fpls.2021.657773.

Malavert, C., Batlla, D. and Benech-Arnold, R. L. (2017). Temperature-dependent regulation of induction into secondary dormancy of *Polygonum aviculare* L. seeds: a quantitative analysis. *Ecol. Modell.* 352: 128-138. 10.1016/j.ecolmodel.2017.03.008.

Malavert, C., Batlla, D. and Benech-Arnold, R. L. (2021a). Light sensitivity changes during dormancy induction in *Polygonum aviculare* L. seeds: development of a predictive model of annual changes in seed-bank light sensitivity in relation to soil temperature. *Weed Res.* 61(2): 115–125. 10.1111/wre.12463.

Malavert, C., Batlla, D. and Benech-Arnold, R. L. (2021b). The role of seed water content for the perception of temperature signals that drive dormancy changes in *Polygonum aviculare* buried seeds. *Funct. Plant Biol.* 48(1): 28–39. 10.1071/FP20011.

Martin-Fontecha, E. S., Tarancon, C. and Cubas, P. (2018). To grow or not to grow, a power-saving program induced in dormant buds. *Curr. Opin. Plant Biol.* 41: 102-109. 10.1016/j.pbi.2017.10.001.

Martínez-Berdeja, A., Stitzer, M. C., Taylor, M. A., Okada M., Ezcurra, E., Runcie, D. E. and Schmitt, J. (2020). Functional variants of *DOG1* control seed chilling responses and variation in seasonal life-history strategies in *Arabidopsis thaliana*. *Proc. Natl Acad. Sci. U. S. A.* 117(5): 2526–2534. 10.1073/pnas.1912451117.

Matthews, S. and Khajeh-Hosseini, M. (2007). Length of the lag period of germination and metabolic repair explain vigour differences in seed lots of maize (*Zea mays*). *Seed Sci. Technol.* 35(1): 200–212. 10.15258/sst.2007.35.1.18.

Matzrafi, M., Herrmann, I., Nansen, C., Kliper, T., Zait, Y., Ignat, T., Siso, D., Rubin, B., Karnieli, A. and Eizenberg, H. (2017). Hyperspectral technologies for assessing seed germination and trifloxysulfuron-methyl response in *Amaranthus palmeri* (Palmer amaranth). *Front. Plant Sci.* 8: 474.

Matzrafi, M., Osipitan, O. A., Ohadi, S. and Mesgaran, M. B. (2020). Under pressure: maternal effects promote drought tolerance in progeny seed of Palmer amaranth (*Amaranthus palmeri*). *Weed Sci.* 69(1): 31–38. 10.1017/wsc.2020.75.

Mesgaran, M. B., Mashhadi, H. R., Alizadeh, H., Hunt, J., Young, K. R. and Cousens, R. D. (2013). Importance of distribution function selection for hydrothermal time models of seed germination. *Weed Res.* 53(2): 89–101. 10.1111/wre.12008.

Mesgaran, M. B., Onofri, A., Mashhadi, H. R. and Cousens, R. D. (2017). Water availability shifts the optimal temperatures for seed germination: a modelling approach. *Ecol. Modell.* 351: 87–95. 10.1016/j.ecolmodel.2017.02.020.

Meyer, S. E. and Allen, P. S. (2009). Predicting seed dormancy loss and germination timing for *Bromus tectorum* in a semi-arid environment using hydrothermal time models. *Seed Sci. Res.* 19(4): 225–239. 10.1017/S0960258509990122.

Mitchell, J., Johnston, I. G. and Bassel, G. W. (2017). Variability in seeds: biological, ecological, and agricultural implications. *J. Exp. Bot.* 68(4): 809–817. 10.1093/jxb/erw397.

Moltchanova, E., Sharifiamina, S., Moot, D. J., Shayanfar, A. and Bloomberg, M. (2020). Comparison of three different statistical approaches (non-linear least-squares regression, survival analysis and Bayesian inference) in their usefulness for estimating hydrothermal time models of seed germination. *Seed Sci. Res.* 30(1): 64–72. 10.1017/S0960258520000082.

Morris, K., Barker, G. C., Walley, P. G., Lynn, J. R. and Finch-Savage, W. E. (2016). Trait to gene analysis reveals that allelic variation in three genes determines seed vigour. *New Phytol.* 212(4): 964–976. 10.1111/nph.14102.

Mortensen, A. K., Gislum, R., Jørgensen, J. R. and Boelt, B. (2021). The use of multispectral imaging and single seed and bulk near-infrared spectroscopy to characterize seed covering structures: methods and applications in seed testing and research. *Agriculture* 11(4): 301. Available at: https://www.mdpi.com/2077-0472/11/4/301.

Nagel, M. and Börner, A. (2010). The longevity of crop seeds stored under ambient conditions. *Seed Sci. Res.* 20(1): 1–12. 10.1017/S0960258509990213.

Nagel, M., Seal, C. E., Colville, L., Rodenstein, A., Un, S., Richter, J., Pritchard, H. W., Brner, A. and Kranner, I. (2019). Wheat seed ageing viewed through the cellular redox environment and changes in pH. *Free Radic. Res.* 53(6): 641–654. 10.1080/10715762.2019.1620226.

Nehoshtan, Y., Carmon, E., Yaniv, O., Ayal, S. and Rotem, O. (2021). Robust seed germination prediction using deep learning and RGB image data. *Scientific Reports* 11: 22030. 10.1038/s41598-021-01712-6.

Ni, B. R. and Bradford, K. J. (1992). Quantitative models characterizing seed germination responses to abscisic acid and osmoticum. *Plant Physiol.* 98(3): 1057–1068.

Ni, B. R. and Bradford, K. J. (1993). Germination and dormancy of abscisic acid-deficient and gibberellin-deficient mutant tomato seeds: sensitivity of germination to abscisic acid, gibberellin, and water potential. *Plant Physiol.* 101(2): 607–617.

Nonogaki, H. (2019). Seed germination and dormancy: the classic story, new puzzles, and evolution. *J. Integr. Plant Biol.* 61(5): 541–563. 10.1111/jipb.12762.

Nonogaki, H., Chen, F. and Bradford, K. J. (2007). Mechanisms and genes involved in germination *sensu stricto*. In: Bradford, K. J. and Nonogaki, H. (Eds), *Seed Development*. Dormancy and Germination. Blackwell Publishing, Oxford, UK, pp. 264–304.

Nonogaki, M. and Nonogaki, H. (2017). Prevention of preharvest sprouting through hormone engineering and germination recovery by chemical biology. *Front. Plant Sci.* 8: 90.

Olesen, M. H., Nikneshan, P., Shrestha, S., Tadayyon, A., Deleuran, L. C., Boelt, B. and Gislum, R. (2015). Viability prediction of *Ricinus cummunis* L. seeds using multispectral imaging. *Sensors (Basel)* 15(2): 4592–4604.

Onofri, A., Benincasa, P., Mesgaran, M. B. and Ritz, C. (2018). Hydrothermal-time-to-event models for seed germination. *Eur. J. Agron.* 101: 129–139. 10.1016/j.eja.2018.08.011.

Onofri, A., Mesgaran, M. B., Neve, P. and Cousens, R. D. (2014). Experimental design and parameter estimation for threshold models in seed germination. *Weed Res.* 54(5): 425–435. 10.1111/wre.12095.

Patanè, C., Saita, A., Tubeileh, A., Cosentino, S. L. and Cavallaro, V. (2016). Modeling seed germination of unprimed and primed seeds of sweet sorghum under PEG-induced water stress through the hydrotime analysis. *Acta Physiol. Plant.* 38(5): 115. 10.1007/s11738-016-2135-5.

Penfield, S. (2008). Temperature perception and signal transduction in plants. *New Phytol.* 179(3): 615–628.

Penfield, S. and MacGregor, D. (2014). Temperature sensing in plants. In: Franklin, K. A. and Wigge, P. A. (Eds), *Temperature and Plant Development*. Wiley Blackwell, Oxford, UK, pp. 1–18.

Penfield, S. and MacGregor, D. R. (2017). Effects of environmental variation during seed production on seed dormancy and germination. *J. Exp. Bot.* 68(4): 819–825. 10.1093/jxb/erw436.

Penfield, S. and Springthorpe, V. (2012). Understanding chilling responses in Arabidopsis seeds and their contribution to life history. *Philos. Trans. R. Soc. Lond. B Biol. Sci.* 367(1586): 291–297. 10.1098/rstb.2011.0186.

Perotti, V. E., Larran, A. S., Palmieri, V. E., Martinatto, A. K. and Permingeat, H. R. (2020). Herbicide resistant weeds: a call to integrate conventional agricultural practices, molecular biology knowledge and new technologies. *Plant Sci.* 290: 110255. 10.1016/j.plantsci.2019.110255.

Powers, S. K., Holehouse, A. S., Korasick, D. A., Schreiber, K. H., Clark, N. M., Jing, H., Emenecker, R., Han, S., Tycksen, E., Hwang, I., Sozzani, R., Jez, J. M., Pappu, R. V. and Strader, L. C. (2019). Nucleo-cytoplasmic partitioning of ARF proteins controls auxin responses in *Arabidopsis thaliana*. *Mol. Cell* 76(1): 177–190.e5. 10.1016/j.molcel.2019.06.044.

Rahman, M. (2011). *Applications of Fourier Transforms to Generalized Functions*. WIT Press, Boston, MA.

Ramesh, K. (2015). Weed problems, ecology, and management options in conservation agriculture: issues and perspectives. *Adv. Agron.* 131: 251–303, 10.1016/bs.agron.2014.12.003.

Roberts, E. H. (1960). The viability of cereal seed in relation to temperature and moisture: with eight figures in the text. *Ann. Bot.* 24(1): 12–31. 10.1093/oxfordjournals.aob.a083684.

Roberts, E. H. (1961). Dormancy of rice seed. I. The distribution of dormancy periods. *J. Exp. Bot.* 12(2): 319–329. 10.1093/jxb/12.2.319.

Roberts, E. H. (1973). Predicting the storage life of seeds. *Seed Sci. Technol.* 1: 499–514.

Roberts, E. H. (1999). A search for pattern and form. *Seed Sci. Res.* 9(3): 181–208. 10.1017/S0960258599000203.

Roberts, E. H. and Ellis, R. H. (1989). Water and seed survival. *Ann. Bot.* 63(1): 39–52. 10.1093/oxfordjournals.aob.a087727.

Roberts, E. H. and Smith, R. D. (1977). Dormancy and the pentose phosphate pathway. In: Khan, A. A. (Ed.), *The Physiology and Biochemistry of Seed Dormancy and Germination*. Elsevier/North-Holland Biomedical Press, Amsterdam, pp. 385–411.

Romano, A. and Stevanato, P. (2020). Germination data analysis by time-to-event approaches. *Plants (Basel)* 9(5): 617. 10.3390/plants9050617.

Rosnoblet, C., Aubry, C., Leprince, O., Vu, B. L., Rogniaux, H. and Buitink, J. (2007). The regulatory gamma subunit SNF4b of the sucrose non-fermenting-related kinase complex is involved in longevity and stachyose accumulation during maturation of *Medicago truncatula* seeds. *Plant J.* 51(1): 47–59. 10.1111/j.1365-313X.2007.03116.x.

Rowse, H. R. and Finch-Savage, W. E. (2003). Hydrothermal threshold models can describe the germination response of carrot (*Daucus carota* L.) and onion (*Allium cepa* L.) seed populations across both sub- and supra-optimal temperatures. *New Phytol.* 158: 101–108.

Royal Botanic Gardens (2021). Kew Seed Information Database (SID). Version 7.1. Available at: http://data.kew.org/sid/.

Rubio de Casas, R., Donohue, K., Venable, D. L. and Cheptou, P.-O. (2015). Gene-flow through space and time: dispersal, dormancy and adaptation to changing environments. *Evol. Ecol.* 29(6): 813–831. 10.1007/s10682-015-9791-6.

Saatkamp, A., Cochrane, A., Commander, L., Guja, L. K., Jimenez-Alfaro, B., Larson, J., Nicotra, A., Poschlod, P., Silveira, F. A. O., Cross, A. T., Dalziell, E. L., Dickie, J., Erickson, T. E., Fidelis, A., Fuchs, A., Golos, P. J., Hope, M., Lewandrowski, W., Merritt, D. J., Miller, B. P., Miller, R. G., Offord, C. A., Ooi, M. K. J., Satyanti, A., Sommerville, K. D., Tangney, R., Tomlinson, S., Turner, S. and Walck, J. L. (2019). A research agenda for seed-trait functional ecology. *New Phytol.* 221(4): 1764–1775. 10.1111/nph.15502.

Saux, M., Bleys, B., André, T., Bailly, C. and El-Maarouf-Bouteau, H. (2020). A correlative study of sunflower seed vigor components as related to genetic background. *Plants (Basel)* 9(3): 386. 10.3390/plants9030386.

Savage, S. L. (2012). *The Flaw of Averages: Why We Underestimate Risk in the Face of Uncertainty*. John Wiley &Sons, Hoboken, NJ.

Schramm, E. C., Abellera, J. C., Strader, L. C., Campbell, K. G. and Steber, C. M. (2010). Isolation of ABA-responsive mutants in allohexaploid bread wheat (*Triticum aestivum* L.): drawing connections to grain dormancy, preharvest sprouting, and drought tolerance. *Plant Sci.* 179(6): 620–629.

Schwartz-Lazaro, L. M. and Copes, J. T. (2019). A review of the soil seedbank from a seed scientists perspective. *Agronomy* 9(7): 369. 10.3390/agronomy9070369.

Schwember, A. R. and Bradford, K. J. (2010). Quantitative trait loci associated with longevity of lettuce seeds under conventional and controlled deterioration storage conditions. *J. Exp. Bot.* 61(15): 4423–4436. 10.1093/jxb/erq248.

Seal, C. E., Barwell, L. J., Flowers, T. J., Merrett Wade, E. and Pritchard, H. W. (2018). Seed germination niche of the halophyte *Suaeda maritima* to combined salinity and temperature is characterised by a halothermal time model. *Environ. Exp. Bot.* 155: 177–184. 10.1016/j.envexpbot.2018.06.035.

Silva, L. Jd, Medeiros, A. Dd and Oliveira, A. M. S. (2019). SeedCalc, a new automated R software tool for germination and seedling length data processing. *J. Seed Sci.* 41(2): 250–257.

Solberg, S. Ø., Yndgaard, F., Andreasen, C., von Bothmer, R., Loskutov, I. G. and Asdal, Å. (2020). Long-term storage and longevity of orthodox seeds: a systematic review. *Front. Plant Sci.* 11: 1007.

Soltani, E., Baskin, J. M. and Baskin, C. C. (2019). A review of the relationship between primary and secondary dormancy, with reference to the volunteer crop weed oilseed rape (*Brassica napus*). *Weed Res.* 59(1): 5–14. 10.1111/wre.12342.

Steiner, J. J. and Akintobi, D. C. (1986). Effect of harvest maturity on viability of onion seed. *HortScience* 21: 1220–1221.

Still, D. W. and Bradford, K. J. (1997). Endo-beta-mannanase activity from individual tomato endosperm caps and radicle tips in relation to germination rates. *Plant Physiol.* 113(1): 21–29. 10.1104/pp.113.1.21.

Still, D. W. and Bradford, K. J. (1998). Using hydrotime and ABA-time models to quantify seed quality of Brassicas during development. *JASHS* 123(4): 692–699.

Szymansky, C.-M., Muscolo, A., Yeo, M., Colville, L., Clatworthy, I., Salge, T. and Seal, C. E. (2021). Elemental localisation and a reduced glutathione redox state protect seeds of the halophyte *Suaeda maritima* from salinity during over-wintering and germination. *Environ. Exp. Bot.* 190: 104569. 10.1016/j.envexpbot.2021.104569.

Tarancón, C., González-Grandío, E., Oliveros, J. C., Nicolas, M. and Cubas, P. (2017). A conserved carbon starvation response underlies bud dormancy in woody and herbaceous species. *Front. Plant Sci.* 8: 788.

Tarquis, A. M. and Bradford, K. J. (1992). Prehydration and priming treatments that advance germination also increase the rate of deterioration of lettuce seeds. *J. Exp. Bot.* 43(3): 307–317.

Tatari, S., Ghaderi-Far, F., Yamchi, A., Siahmarguee, A., Shayanfar, A. and Baskin, C. C. (2020). Application of the hydrotime model to assess seed priming effects on the germination of rapeseed (*Brassica napus* L.) in response to water stress. *Botany* 98(5): 283–291. 10.1139/cjb-2019-0192.

ten Brink, H., Gremer, J. R. and Kokko, H. (2020). Optimal germination timing in unpredictable environments: the importance of dormancy for both among- and within-season variation. *Ecol. Lett.* 23(4): 620–630. 10.1111/ele.13461.

Thompson, J. F., Reid, M. S., Felix, L., Donis-Gonzalez, I., Mjawa, B. and Ambuko, J. (2017). Dry Card™ -- A low-cost dryness indicator for dried products. *AIMS Agric. Food* 2: 339–344. 10.3934/agrfood.2017.4.339.

Topham, A. T., Taylor, R. E., Yan, D., Nambara, E., Johnston, I. G. and Bassel, G. W. (2017). Temperature variability is integrated by a spatially embedded decision-making center to break dormancy in Arabidopsis seeds. *Proc. Natl Acad. Sci. U. S. A.* 114(25): 6629–6634. 10.1073/pnas.1704745114.

Trewavas, A. J. (2012). Information, noise and communication: thresholds as controlling elements in development. In: Witzany, G. and Ba uska, F. (Eds), *Biocommunication of Plants. Signaling and Communication in Plants vol. 14*. Springer-Verlag, Berlin, pp. 11–35.

Trewavas, A. J. (2014). *Plant Behaviour and Intelligence*. Oxford University Press, Oxford, UK.

Trudgill, D. L., Honek, A., Li, D. and Van Straalen, N. M. (2005). Thermal time – concepts and utility. *Ann. Appl. Biol.* 146: 1–14. 10.1111/j.1744-7348.2005.04088.x.

Valdes, V. M. and Bradford, K. J. (1987). Effects of seed coating and osmotic priming on the germination of lettuce seeds. *J. Amer. Soc. Hort. Sci.* 112: 153–156.

Valdes, V. M., Bradford, K. J. and Mayberry, K. S. (1985). Alleviation of thermodormancy in coated lettuce seeds by seed priming. *HortScience* 20: 1112–1114.

Vanderburg, W. J., Aartse, J. W., Vanzwol, R. A., Jalink, H. and Bino, R. J. (1994). Predicting tomato seedling morphology by X-ray-analysis of seeds. *J. Amer. Soc. Hort. Sci.* 119: 258–263.

Walker, M., Pérez, M., Steinbrecher, T., Gawthrop, F., Pavlović, I., Novák, O., Tarkowská, D., Strnad, M., Marone, F., Nakabayashi, K. and Leubner-Metzger, G. (2021). Molecular mechanisms and hormonal regulation underpinning morphological dormancy: a case study using *Apium graveolens* (Apiaceae). *Plant J.* 108(4): 1020–1036 10.1111/tpj.15489.

Walters, C., Farrant, J. M., Pammenter, N. W. and Berjak, P. (2002). Desiccation stress and damage. In: Black, M. and Pritchard, H. W. (Eds) *Desiccation and Survival in Plants. Drying without Dying*. CABI Publishing, Wallingford, Oxon, UK, pp 263–291.

Walters, C., Fleming, M. B., Hill, L. M., Dorr, E. J. and Richards, C. M. (2020). Stress-response relationships related to ageing and death of orthodox seeds: a study comparing viability and RNA integrity in soya bean (*Glycine max*) cv. Williams 82. *Seed Sci. Res.* 30(2): 161–172. 10.1017/S0960258520000197.

Walters, C., Wheeler, L. and Stanwood, P. C. (2004). Longevity of cryogenically stored seeds. *Cryobiology* 48(3): 229–244. 10.1016/j.cryobiol.2004.01.007.

Walters, C., Wheeler, L. M. and Grotenhuis, J. M. (2005). Longevity of seeds stored in a genebank: species characteristics. *Seed Sci. Res.* 15(1): 1–20.

Waterworth, W. M., Footitt, S., Bray, C. M., Finch-Savage, W. E. and West, C. E. (2016). DNA damage checkpoint kinase ATM regulates germination and maintains genome stability in seeds. *Proc. Natl Acad. Sci. U. S. A.* 113(34): 9647–9652.

Watt, M. S. and Bloomberg, M. (2012). Key features of the seed germination response to high temperatures. *New Phytol.* 196(2): 332–336.

Watt, M. S., Bloomberg, M. and Finch-Savage, W. E. (2011). Development of a hydrothermal time model that accurately characterises how thermoinhibition regulates seed germination. *Plant Cell Environ.* 34(5): 870–876. 10.1111/j.1365-3040.2011.02292.x.

Watt, M. S., Xu, V. and Bloomberg, M. (2010). Development of a hydrothermal time seed germination model which uses the Weibull distribution to describe base water potential. *Ecol. Modell.* 221(9): 1267–1272. 10.1016/j.ecolmodel.2010.01.017.

Willis, C. G., Baskin, C. C., Baskin, J. M., Auld, J. R., Venable, D. L., Cavender-Bares, J., Donohue, K., Rubio de Casas, R. and NESCent Germination Working Group (2014). The evolution of seed dormancy: environmental cues, evolutionary hubs, and diversification of the seed plants. *New Phytol.* 203(1): 300–309. 10.1111/nph.12782.

Xu, H., Lantzouni, O., Bruggink, T., Benjamins, R., Lanfermeijer, F., Denby, K., Schwechheimer, C. and Bassel, G. W. (2020). A molecular signal integration network underpinning Arabidopsis seed germination. *Curr. Biol.* 30(19): 3703–3712.e4. 10.1016/j.cub.2020.07.012.

Yang, H., Dahal, P., Cheng, Z. Y., Abe, H. and Bradford, K. J. (1997). Expression of a SNF4-like protein kinase-related gene in tomato seeds. *Plant Physiol.* 114: 1398–1398.

Yentur, S. and Leopold, A. C. (1976). Respiratory transition during seed germination. *Plant Physiol.* 57(2): 274–276. 10.1104/pp.57.2.274.

Yoong, F. Y. (2015). Natural variation and genetic analyses of high temperature germination in lettuce: from QTL mapping to candidate gene identification. PhD. University of California, Davis.

Zhang, J. H. and McDonald, M. B. (1997). The saturated salt accelerated aging test for small-seeded crops. *Seed Sci. Technol.* 25: 123–131.

Zhang, R., Luo, K., Chen, D., Baskin, J., Baskin, C., Wang, Y. and Hu, X. (2020). Comparison of thermal and hydrotime requirements for seed germination of seven *Stipa* species from cool and warm habitats. *Front. Plant Sci.* 11: 560714. 10.3389/fpls.2020.560714.

Zhao, Y. S., Zhu, P., Hepworth, J., Bloomer, R., Antoniou-Kourounioti, R. L., Doughty, J., Heckmann, A., Xu, C. Y., Yang, H. C. and Dean, C. (2021). Natural temperature fluctuations promote *COOLAIR* regulation of *FLC*. *Genes Dev.* 35(11–12): 888–898. 10.1101/gad.348362.121.

Zinsmeister, J., Leprince, O. and Buitink, J. (2020). Molecular and environmental factors regulating seed longevity. *Biochem. J.* 477(2): 305–323. 10.1042/BCJ20190165.

Chapter 4

Biotic sources of seed losses influencing germination and emergence success in crop plants and agricultural weeds

James W. Dalling and Adam S. Davis, University of Illinois Urbana-Champaign, USA

1 Introduction

It is estimated that overall, between 17% and 30% of the yield of the world's major crops is lost to pests, pathogens, and competition from weeds (Oerke, 2006, Savary et al., 2019). The magnitude of these losses varies considerably among major crop species, ranging from on average ~26% in soybeans to 40% in potatoes. The biotic sources of these losses are also variable. Most yield reduction can be attributed to competition from weeds, for example 38–80% of yield loss in soybean has been attributable to competition from *Amaranthus* species (Bensch et al., 2003), but insect pests and diseases are also important. Furthermore, despite increases in pesticide use, crop losses have not significantly decreased over the last half century (Oerke, 2006, Savary et al., 2019). While increases in the use of herbicides and pesticides have allowed the development of more intensive agricultural practices that can offset higher crop losses, economic impacts resulting from eliminating the natural enemies of plant pests, and the development of pesticide or herbicide resistance, are most strongly felt in areas with low food security and fast-growing populations including sub-Saharan Africa and the Indian sub-continent (Savary et al., 2019). Economic losses from pests and pathogens often also extend beyond direct declines in yield to include aesthetics, harvestability, nutrient content, storage characteristics, and contamination by pests or the toxins that pests and pathogens produce (reviewed by Oerke, 2006).

http://dx.doi.org/10.19103/AS.2022.0105.06

In this chapter, we highlight pests and pathogens that impact seed development on the maternal plant, persistence in the soil, and that influence seedling emergence success (Fig. 1). Our focus extends beyond focal crop species to also consider the plant species that compete with crop plants for light and nutrients. Many agrestal weed species evolved in environments that selected for traits that have enabled them to exploit the conditions intended to support vigorous crop growth in agricultural production fields, including repeated annual disturbance, high soil nutrient levels, and ample soil moisture (Baker, 1974, Holm et al., 1997). However, unlike their crop counterparts, which are the product of intensive selection for traits that confer uniformity and tractability, such as loss of dormancy, genetic homogeneity, and low plasticity, weeds remain wild, with enormous genetic diversity and phenotypic plasticity at their disposal to help them survive and thrive without human caretakers.

Improved understanding of how pests and pathogens locate susceptible seeds and breach their defenses can therefore be used to develop management practices that reduce crop seed losses, increase weed seed losses, or both. Much of the literature on the interaction of seeds with pests and pathogens comes from weed scientists, who have recognized the importance of depleting seed banks as a management approach, particularly for species with annual

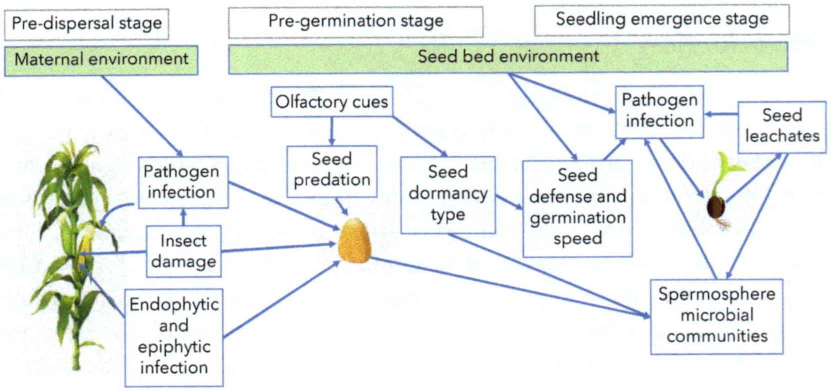

Figure 1 Critical biotic interactions that influence seed losses for crops and agrestal weed species. While at the pre-dispersal stage seeds can pick up microbial infections from the maternal plant that enter the seed or can remain on external tissues after dispersal. In addition to direct seed losses, damage from insect pests that breach protective tissues may provide opportunities for subsequent pathogen infection. Once seeds are dispersed, predators may detect seeds based on visual or olfactory cues, which in turn depend on whether seeds are permeable and therefore also on their dormancy class. Dormancy can also influence investment in chemical and physical defenses and germination speed, which in turn affect susceptibility to pathogen infection. At the emergence stage, release of seed leachates can affect spermosphere microbial communities that in turn mediate seed-pathogen interactions. In this review we highlight differences in susceptibility to, or protection against, seed losses at each of these stages in crop plants and weeds and highlight management approaches that can be developed from these biotic interactions.

life histories (Davis, 2006). However recent research in natural systems has also begun to reveal the range of interactions between seeds and microbes, as well as defense traits that are yet to be fully explored in agricultural settings, or potentially that have previously been bred out of crop plants. We therefore include these studies when considering potential targets for more sustainable crop production.

2 Major biotic sources of seed losses

Sources of seed loss can be categorized into two groups. Those that occur during the seed maturation process (pre-dispersal), and those that occur once seeds are on or in the soil (post-dispersal). For crop plants, where seeds are harvested before dispersal, and then spend relatively short periods when sewn into the soil prior to germination, pre-dispersal losses are likely to be most important. In contrast, post-dispersal losses are most important for most weed species, many of which have the capacity for prolonged seed dormancy and therefore can spend years or decades in the soil (Roberts and Feast, 1973, Kivilaan and Bandurski, 1981, Wilson and Lawson, 1992). Losses can be further categorized according to whether they are attributable to pathogens or predators. The relative importance of these sources of natural enemies varies among crop and weed species and is described in the sections below. It is important, however, to recognize that the activities of these enemies are not independent. For example, damage to seeds from probing insects may create openings in seed protective structures or release seed exudates that provide opportunities for subsequent fungal or bacterial infection (Mills, 1983, Kremer and Spencer, 1989). Alternatively, the breach of impermeable seed or fruit walls may release olfactory cues that potentially attract vertebrate seed predators (Paulsen et al., 2013). Seeds that are damaged by predators and pathogens may also pass standard viability tests using vital stains yet may also exhibit increased rates of lethal germination. Lethal germination can occur when seeds germinate at inappropriate times, either because damage to the seed results in the loss of physical dormancy, or potentially because microbial infection results in the release of plant hormones that stimulate germination (Goggin et al., 2015).

2.1 Vertebrate and invertebrate sources of pre-dispersal seed predation

The predation of weed seeds has largely been studied in the context of developing the ecological knowledge necessary to support effective weed biocontrol strategies based on increasing weed seed predation rates. Sarabi (2019) argued that pre-dispersal predation of weed seeds should be the most important target for biocontrol of weeds based on granivory, since the period between ovule formation and seed maturation is the most vulnerable stage of seed development, accounting for the majority of seed mortality (Fenner and Thompson, 2005). Perhaps because of the relative difficulty of quantifying pre-dispersal seed predation levels, the scientific literature is less well represented in this area (18 studies published between 1994 and 2021) than are studies of post-dispersal predation of weed seeds (70 studies published between 1999 and 2021).

Invertebrates can contribute substantially to weed seed losses due to pre-dispersal seed predation. Seed feeding activity by the granivorous weevil *Rynocyllus conicus* reduced Musk thistle (*Carduus nutans*) seed production from 24% to 78%, compared to a weevil-free population, under field conditions in central Nebraska, USA, over a two-year period. Pre-dispersal seed feeding by the scentless plant bug (*Niesthrea louisianica*) on field populations of velvetleaf (*Abutilon theophrasti*) in central Missouri, USA, reduced seed viability by 16–18% compared to insect-free conditions (Kremer and Spencer, 1989). Seed losses of field-grown redroot pigweed (*Amaranthus retroflexus*) to pre-dispersal predation by *Cleophora lineapulvella* (Lepidoptera: Coleophoridae) in Guelph, Ontario, ranged from 42% to 93% over a two-year period. Interactions between weeds and their invertebrate pre-dispersal seed predators appear to support maintenance of genetic diversity in some weed species. Seed losses to granivorous weevils in jimson weed (*Datura stramonium*) were greater in inbred *D. stramonium* populations than in outcrossing populations, despite increased preference by the weevils for outcrossing populations (Bello-Bedoy et al., 2011).

The comprehensive review of weed seed predation by Sarabi (2019) does not include any studies of pre-dispersal weed seed predation by vertebrates. However, there is evidence that some waterfowl species ingest weed seeds prior to dispersal. A study of contents of digestive tracts of migratory waterfowl (Farmer et al., 2017) indicated the importance of pre-dispersal predation of weed seeds by non-resident ducks and geese in central Missouri, USA. The team found seeds of 59 species in the crop, oesophagus, and stomachs of 529 ducks and geese, with more diversity of seeds recovered from ducks than from geese. Of the 59 plant species represented, only 2 were crop species (corn and soybean) and accounted for only 0.06% and 0.02% of the 35 020 recovered seeds, respectively. The remainder of the plant species were primarily weed species, with 13 weeds of high economic importance disproportionately represented among recovered seeds. At least for the migratory waterfowl in this study, most of the granivory recorded focused on pre-dispersal predation of weed seeds rather than crop species.

In contrast, many other avian species can be primary drivers of pre-dispersal seed predation in crops (Elphick, 2015, Gebhardt et al., 2011). For example, pre-dispersal seed predation by the rose-ringed parakeet (*Psittacula krameri*) in Pakistan can cause yield losses of 40–80% in maize and sunflower (Ahmad et al., 2012). However, there is also a tendency for producers to overestimate the seed depredation damage caused by avian pests (Can-Hernandez et al., 2019), and to overlook the net impacts of birds in agroecosystems (Pejchar et al., 2018). The impact of genetic variation in seed traits on avian pre-dispersal seed predation may provide clues to reducing crop seed losses without direct suppression of avian populations. A suite of reproductive morphological traits, including long involucral bracts, horizontally oriented heads, concave heads, and long head-to-stem distances, confer some degree of resistance to pre-dispersal seed predation by birds in sunflower (Gross and Hanzel, 1991, Khaleghizadeh, 2011). Lower seed predation among weedy sunflower genotypes suggests that there may be valuable genetic diversity to explore in weedy crop relatives to breed for resistance to predation. Preferential pre-dispersal predation of crop-wild hybrids of sunflowers, compared to wild sunflowers, appears to limit gene flow from cultivated sunflower populations to wild populations (Cummings et al., 1999). On average, crop-wild hybrids had 46% of seeds eaten, compared to less than 2% predation for wild plants.

Small mammals can also contribute to pre-dispersal crop seed granivory, with impacts varying by crop. Pre-dispersal seed predation of maize by coati and raccoon in fields adjacent to the Natural Protected Area of Agua Blanca in Mexico was substantially lower than that caused by birds (Can-Hernandez et al., 2019). Small rodents can be particularly problematic pre-dispersal predators in short-statured crops such as rice (John, 2014), wheat (Kaboodvandpour and Leung, 2012), and soybean (Neelanarayanan et al., 1997). As with pre-dispersal crop seed predation by birds, variation in seed traits can also influence the rate of crop seed predation by vertebrates, with valuable traits for predation reduction found in wild plant populations. For example, high glucosinolate levels protected wild *Brassica napa* seeds from wood mouse predation compared to cultivated oilseed rape (de Jong et al., 2016).

2.2 Pre-dispersal pathogen infection and the transmission of beneficial endophytes

Numerous fungal diseases from diverse fungal phyla infect flowers and cause significant economic losses to crop plants through destruction of the infructescence. These include fusarium head blight, responsible for large losses in wheat and barley production due to mycotoxin production, sorghum ergot affecting cultivations in the Americas and Australia, and smut diseases affecting maize, barley, and wheat (Goswami and Kistler, 2004, Ngugi and Scherm, 2006, Ghimire et al., 2020). In addition to these economically important pathogens, developing seeds associate with diverse communities of both endophytic microbes (present in the internal tissues of seeds) and epiphytic microbes (present on external seed or maternally derived tissues; Shade et al., 2017, Nelson, 2018). While these microbial communities include bacterial and fungal pathogens, many microbial infections show no symptoms of disease, or may protect against pathogens at the post-dispersal stage. Still others may have growth-promoting effects on developing seedlings. The seed therefore represents a critical nexus linking the multiple infection pathways by which microbes (especially bacteria) colonize plant maternal tissues to their subsequent proliferation in seedling tissues. More broadly, the transmission of pathogens by seeds is an important means by which disease organisms are transmitted to new planting areas (Baker and Smith, 1966, Farahani-Kofoet et al., 2012). Moreover, the potential for asymptomatic hosts creates the opportunity for seeds to carry infections to susceptible hosts (Darrasse et al., 2010).

Several studies have shown that inoculation of flowers and fruit with pathogenic or non-pathogenic bacteria can result in significant levels of seed infestation (reviewed in Nelson, 2018). For example, Dutta et al. (2015) showed that inoculation of watermelon stigmas with *Acidovorax citrulli*, the causal agent of bacterial fruit blotch in cucurbits, results in colonization of watermelon seed endocarps. Similarly, Mitter et al. (2016) showed that plant growth-promoting bacteria introduced to flowers colonize seed embryos and transfer to seedlings promoting growth. Likewise, in the absence of inoculation experiments Fürnkranz et al. (2012) showed that the bacteria (*Pseudomonas* and *Bacillus*) present in the seed cavity of cucurbit fruits are capable of colonizing seeds and providing protection to seedlings from pathogenic bacteria and fungi. However, seeds are not simply passive carriers of bacterial infections. In an elegant study Terrasson et al. (2015) compared the transcriptional profiles of *Medicago* seeds infected with a natural pathogen (*Xanthomonas alfalfae* subsp. *alfalfae*; *Xaa*) or with *Xanthomonas campestris* pv. *campestris* (*Xcc*) a bacterium with no known legume hosts. As expected *Xcc* but not *Xaa* induced an immune response. Furthermore, activation of the defense response resulted in a repression of transcripts associated with seed maturation pathways and a reduction in seed weight highlighting a potential trade-off between seed development and seed defense.

While transmission from maternal tissue to the seed embryo or storage tissues appears to be common across diverse bacterial lineages, similar 'vertical' transmission of fungi appears to be much more limited. An exception appears to be the non-pathogenic ascomycete fungus *Epichlöe*, which acts as a seed endophyte in grasses providing protection against pathogen infection and toxicity to herbivores (Saikkonen et al., 2016). More generally, a variety of culture-based studies indicate that the richness of fungi isolated from the interior of seeds is much lower than that of leaves and roots of the same plant species (Ganley and Newcombe, 2006, Sarmiento et al., 2017, Newcombe et al., 2018). Next-generation sequencing (NGS) provides additional support for limited microbial colonization. While a large fraction of seeds may be colonized by fungi, the interior of individual fresh seeds harbours very few species in most natural plant communities examined to date (see Ridout et al., 2019). Consistent observations of limited fungal diversity in individual intact, surface-sterilized seeds of dicots suggest that seeds represent a bottleneck in plant microbiomes, with important implications for how initial seed colonists might shape seed fate.

In contrast to studies of individual seeds, a review of metagenomic studies of pooled seed samples (typically consisting of ~1000 seeds), including both 'raw' and surface-disinfected seeds has revealed diverse seed microbiota (Simonin et al., 2021). Although host seed sampling in the study was predominantly from four crop species (radish, rapeseed, common bean, and wheat) the study showed that across different continents seeds supported a core microbial community of around 30 bacterial and fungal taxa, including *Pantoea*, *Pseudomonas*, *Cladosporium*, and *Alternaria*, present in most plant species. The presence of these core communities suggests that common microbial targets might exist for developing management approaches to protect against pathogen infection or increase seedling vigor.

2.3 Post-dispersal seed predation

Although weed and crop plants may grow together in the same field, the post-dispersal fates of their seeds are highly divergent due to the differential impact of harvest operations on crop and weed seeds. Weed seeds maturing in a crop canopy may either be dispersed by natural means prior to harvest or by the action of the harvest machinery (Cousens et al., 2008). Cereal crops are harvested with specialized machinery optimized to capture, thresh, and store crop seeds while passing the 'fines' fraction (weed seeds, cracked grain and small particles of plant residue) and chaff out the rear of the device (Walsh et al., 2013). The phenology of weed seed maturation and dispersal in relation to crop harvest varies by growing region and by weed-crop species combination (Schwartz-Lazaro et al., 2021a,b). The efficacy of an emerging approach to integrated management of weeds ('Harvest Weed Seed Control' or HWSC), based on capturing and destroying weed seeds during crop harvest (Shergill et al., 2020), rests on the phenological overlap between crop harvest and weed seed maturation. In such a system, weed seeds that are not dispersed prior to crop harvest may be destroyed by a mechanical analogue to post-dispersal seed predation: passage through a milling device located in the harvester, whereas weed seeds that are dispersed prior to harvest then become available to granivores for post-dispersal seed predation.

Post-dispersal weed seed predation is ubiquitous, with far-reaching impacts (Bohan et al., 2011). The ecology of this process is complex, involving abiotic and biotic factors affecting the availability of dispersed weed seeds to seed predators and the activity of those predators (Long et al., 2015). Daily rates of weed seed consumption by invertebrate granivores, such as crickets and carabid beetles, are positively influenced by the population density of dispersed seeds and the local abundance of granivores, and negatively influenced by the presence of the granivores' predators, such as wolf spiders (Araneae) (Davis and Raghu, 2010). Daily rates of weed seed consumption by vertebrate granivores, such as mice and birds, increase as falling temperatures indicate the approach of winter and the need to store energy and decrease on moonlit nights, when their predators are more apt to hunt them successfully (Davis and Raghu, 2010). The presence of cover crops and other living covers can help to provide habitat and shelter to weed seed predators, supporting higher predation rates (Blubaugh and Kaplan, 2015, Heggenstaller et al., 2006).

Despite their small size, invertebrate granivores can eat enormous numbers of weed seeds over a short period of time. In a single day, male and female crickets (*Gryllus pennsylvanicus*) consumed an average of 223 and 90 redroot pigweed (*Amaranthus retroflexus*), 12 and 8 velvetleaf (*Abutilon theophrasti*), 25 and 9 giant foxtail (*Setaria faberi*), and 87 and 69 crabgrass (*Digitaria sanguinalis*) seeds, respectively (Carmona et al., 1999). Invertebrate granivore seed preference and consumption rate is influenced by seed shape and seed coat strength (Lundgren and Rosentrater, 2007) and also by concentrations of defense compounds (de Jong et al., 2016). In agricultural systems, dispersed weed seeds are a preferred energy source of native mice (Berl et al., 2017). For these vertebrate seed predators, physical seed dormancy may be an additional important mediator of seed losses; impermeable 'hard seeds' do not produce the volatile chemicals that can be used to detect buried seeds by mammalian predators (Paulsen et al., 2013).

When considered over longer periods, the short term, intense bursts of seed feeding activity by both invertebrate and vertebrate granivores can add up to considerable proportions of the total weed seed rain. Depending upon location, cropping system and management practices, season-long post-dispersal predation rates range from 10% to 90% of the total seed rain (Davis et al., 2011). Potentially, this is a sufficient reduction in the survival of newly produced seed to have profound suppressive effects on weed population growth rates (Sarabi, 2019). However, obtaining the full biocontrol benefit of post-dispersal granivores requires allowing weed seeds to remain on the soil surface, where they are most exposed to predators. Primary tillage of soil following crop harvest greatly reduces post-dispersal weed seed predation rates (Blubaugh and Kaplan, 2015).

In contrast to the sudden and continuous exposure of weed seeds to predators following crop harvest, crop seeds are carefully stored under temperature and humidity conditions designed to exclude seed predators and maintain seed viability and quality, either for later consumption or for sowing a following crop. In cases where grain storage is not secure from granivores, as in some developing economies, post-harvest crop seed losses may be substantial enough to affect food security (Htwe et al., 2017, 2021, Ognakossan et al., 2018). Apart from such cases, however, most post-dispersal predation of crop seeds occurs during the short temporal window between planting and seedling establishment.

Soil invertebrates can have large negative impacts on crop stands through seed feeding activity. Wireworms (*Agriotes* spp.) are widespread and highly effective in their predatory activity on newly planted crop seeds, with the level of damage depending upon risk factors such as very high soil organic matter (> 5%), adjacent areas of unmanaged grassland vegetation and poor drainage (Furlan et al., 2017). Another invertebrate pest of newly planted crop seeds, the seed corn maggot (*Delia platura*), also uses high soil organic matter and freshly-tilled plant residues as an oviposition cue, however allowing sufficient thermal time to elapse between tillage and planting greatly minimizes the risk of seed damage due to this species (Hammond, 1995).

Vertebrate granivores can also have negative impacts on newly planted crop seeds prior to seedling establishment. The horned lark (*Eremophila alpestris*) causes major damage to pre-emerged canola seedlings in inland Pacific Northwest, USA, resulting in complete crop losses in some cases (Schillinger and Werner, 2016). Blackbirds (*Agelaius phoeniceus*) feeding in drill-planted rice fields can cause stand reductions of over 50%; anthraquinone seed treatments were able to reduce damage to below 10% even under high blackbird population densities. Reduction in tillage, gaining popularity as a management tactic to increase soil carbon sequestration, also leads to improved habitat for vole, with concomitant increases in predation of newly planted crop seeds and emerged seedlings (Roos et al., 2019). Providing attractive semi-natural habitat in edge of field margins may lure voles away from field interiors, reducing cereal crop damage (Rodriguez-Pastor et al., 2016).

2.4 Post-dispersal pathogen infection of seeds

Once seeds reach the soil, they interact with a new microbial community. These interactions may have additive effects on seed losses beyond those attributable to pathogens acquired on the maternal plant. In general, seed losses in the soil appear to be mostly attributable to fungi (Dalling et al., 2020), as numerous experiments have shown that fungicide applications to seeds in the soil can increase the fraction of viable seeds in a cohort (reviewed by Wagner and Mitschunas, 2008). Interpretation of experimental results, however, requires caution. For many species, seed cohorts consist of both dormant and non-dormant seeds. Fungi not only kill seeds, they can also stimulate seed germination by breaking dormancy. Fungicide addition experiments may not always distinguish between these fates (Wagner and Mitschunas, 2008). Dormancy break by fungi is unlikely to provide a fitness advantage for seeds if it lacks a mechanism to ensure appropriate germination timing (Baskin and Baskin, 2000). However, inoculation experiments show that fungi present on the seed surface can greatly enhance germination of stored seeds with physiological dormancy (Delgado-Sánchez et al., 2011), or in seeds for which rupture of the pericarp is necessary for germination (Sperber et al., 2017). Our understanding of how fungi impact germination for seeds with different dormancy types, and the mechanisms underlying fungal species-specific effects in dormancy loss (Delgado-Sanchez et al., 2011, Sarmiento et al., 2017, Dominguez et al., 2019) remains incomplete but represents an exciting research area for sustainable crop management.

Exposure to the soil environment affects both epiphytic and endophytic seed microbial communities. Existing evidence suggests that epiphytic microbial communities consist mostly of host-generalist taxa. Links et al. (2014) showed that the epiphytic microbial communities of *Triticum* and *Brassica* seed washes, characterized by a combination of amplicon sequencing and qPCR of specific micro-organisms, had similar bacterial loads and similar richness and evenness of fungal and bacterial communities. However, host differences in the abundance of one fungal isolate (*Alternaria*) were attributed to inhibitory interactions between fungi and seed-associated bacteria. In contrast, endophytic microbial communities must breach seed defenses to reach internal seed tissues, a process that filters the microbial community and can result in high levels of apparent host specificity. For example, in a common-garden seed burial experiment in a tropical forest in Panama, Sarmiento et al. (2017) found that host species identity was far more important than common garden location or burial duration in determining the composition of fungal communities among nine tree species, with clear differentiation of communities present in the seeds of congeneric tree species buried <1 m apart (Zalamea et al., 2021).

The Sarmiento et al. (2017) study also showed that fungi can rapidly colonize seed interior structures, in some cases within a few weeks of burial in the soil. However, infection rates were lower for species with physical dormancy where there is an impermeable barrier that restricts microbial access (Sarmiento et al., 2017). Increased susceptibility of non-dormant seeds to fungal pathogens is apparent from studies of seed 'rot' diseases caused by *Fusarium graminearum*, *F. tricinctum*, and related species (Chen et al., 2018). In addition to causing floral diseases in wheat and other crops, these pathogens cause stand failure in the invasive grass species *Bromus tectorum* by infecting seeds in the soil either before or after germination (Franke et al., 2014). Comparison of seed banks with and without stand failure showed that *Fusarium* causes mortality of the non-dormant fraction of the seed bank. Furthermore, mortality is more severe under water stress conditions. Similar results have been reported for desert annuals in southwestern North America, where inoculation experiments showed higher seed losses for a native and invasive species under a summer temperature regime when seeds tend to lose dormancy (Li et al., 2019).

The reduced susceptibility of dormant seeds to soil pathogens or to epiphytic pathogens acquired during seed development may limit the effectiveness of soil-borne pathogens in regulating populations of weed species (or invasive species like *Bromus*). Weed infestations can be maintained through the fraction of seeds that form persistent, dormant seed banks, potentially thwarting attempts to control weeds using above-ground herbicides (Chancellor, 1981, Kremer, 1993). Nonetheless, some fungi are effective in reducing the viability of important weed species (Kirkpatrick and Bazzaz, 1979, Pitty et al., 1987, Thomas et al., 1999), and some, such as *Pyrenophora semeniperda*, a pathogen of *Bromus* and other grasses, are capable of killing dormant seeds in the soil (Meyer et al., 2007) probably through the production of cell wall degrading enzymes (Pollard, 2018). Weed species with physically dormant seeds may also invest less in chemical defenses than species with physiological dormancy (as has been shown for tropical tree species (Zalamea et al., 2018)), potentially making them more susceptible to lethal infection and the emergence stage.

Additional complications arise in understanding seed-pathogen interactions. First, pathogens may have a broader host range in terms of the ability to infect seeds than their ability to cause disease. Inoculation experiments using a subset of the soil-borne seed-infecting fungal community cultured from seeds in the Panama common-garden experiment showed that fungi were able to infect the ungerminated seeds of a range of host species, while significant seed mortality was restricted to a subset of species which differed with fungal taxon. Furthermore, some fungal inocula increased seed mortality of some host species while stimulating germination in others, a phenomenon they termed 'functional specificity' (Sarmiento et al., 2017). Additional analysis of the seed-infecting fungi in this study also revealed the presence of endohyphal bacteria (EHB; bacteria present within hyphal tissue (Shaffer et al., 2016)). Elimination of EHB using antibiotics showed that EHB can have strong impacts on seed germination, adding another poorly understood layer of context dependency to seed-fungal interactions.

More generally, researchers have questioned whether existing experimental approaches over-estimate the efficacy of seed pathogens, with important implications for the development of biocontrol agents. In a recent review, Wagner and Mitschunas (2022) noted that losses to pathogens are likely to be higher in lab inoculation experiments than under field conditions because of the use of high inoculum levels in vitro, and because seeds are typically surface-sterilized removing potentially antagonistic microbes (especially bacteria) that can protect seeds against pathogen infection (Maksimov et al., 2011). Even field assays of microbial seed losses may be biased because the high densities of seeds placed together in experimental replicates may provide greater opportunities for contagious spread of pathogens from seed to seed that would occur under natural conditions with lower seed densities (van Mourik et al., 2005). Mesh bags typically used to enclose seeds in field burial experiments may also alter the soil microenvironment in ways that favour microbial growth, for example, by wicking moisture from the soil surface to the interior of the burial bag (Teo-Sherrell et al., 1996).

3 Mitigating effects of seed losses through management

Previous sections of this review have highlighted morphological attributes of seeds, phenological patterns of seed production, interactions among microbes, and the wider spatial matrix in which interactions occur as important predictors of seed losses to pathogens and predators. Manipulation of these factors therefore provides opportunities to reduce seed losses through management.

Given their differing biology and the divergent roles that weeds and crops play in agricultural systems, it is natural that the management objectives and approaches for mitigating seed losses due to biotic stress would also be different. For weeds, their seeds not only form an important management target for weed control, but for weeds with annual life histories, reducing viable seed rain and persistence of the soil seed bank are demographic transitions whose variation is capable of creating the greatest reductions in weed population growth (Davis, 2006). Increasing seed losses due to biotic stress in weed seeds boils down to a few core principles: (a) manage the agroecosystem to improve habitat for seed predators and pathogens (Blubaugh et al., 2016, Carmona and Landis, 1999, Gallandt et al., 2005, Sarabi, 2019), (b) delay the incorporation of weed seeds into the soil for as long as possible after seed rain occurs, so they remain near the soil surface where they are most vulnerable to losses due to seed predation, fatal germination and decay (Blubaugh and Kaplan, 2015, Cordeau et al., 2017, Davis and Renner, 2007); and (c) avoid sending weed seeds deep into the soil profile, where they are likely to remain dormant, persistent and protected from predators and pathogens (Long et al., 2015).

For crop species, the goal of the manager is clearly to reduce seed losses to biotic stresses so that crop stands emerge at the intended population densities and crop yields are not lost to depredation. In contrast to weed seed bank management, which takes a holistic, ecological view of the process in order to identify management practices that alter natural processes that regulate seed bank population dynamics (Long et al., 2015, Sarabi, 2019), approaches to managing crop seed loss are almost entirely focused on methods that directly deter granivores and pathogens, with a much smaller, nascent effort to alter depredation rates through habitat alteration. Humans have likely been managing crop seed losses as long as they have been practicing agriculture. In shifting cultivation systems, a primary means of reducing seed losses may have just been to move before pathogen or predation pressure built to unacceptable levels (Bargues-Ribera and Gokhale, 2020). In modern sedentary agricultural systems, deterring granivores and pathogens relies on three major classes of management tactics: seed treatments, mechanical deterrents, and pesticides. The relative importance of these tactics depends upon both the crop and the pest species.

Kremer (1993) noted that our capacity to deplete weed seed banks using pathogens is limited. With some notable exceptions (Gogue and Emino, 1979, Beckstead et al., 2007) microbial pathogens rarely appear to be able to significantly reduce populations of dormant seeds in the soil. Nonetheless, pathogens are frequently effective at killing seeds (or germinating seedlings) once dormancy is broken (Kremer et al., 1984). Mechanical or chemical seed treatments that break dormancy of persistence weeds may therefore provide an effective initial step in reducing weed infestations (Kremer, 1993).

Microbial treatments that protect the seeds of crop species at pre-emergence and emergence stages hold more promise. Seed treatments can be designed to target specific seed pathogens, and because crop seeds typically are sown shortly before germination, they need not maintain efficacy for prolonged periods in the soil, a major challenge for microbial products (van Lenteren, 2012). Advances in seed coating techniques and in enhancing seed defenses using microbial priming agents are addressed elsewhere in the volume. However successful development of microbial biocontrols to suppress plant pathogens remains challenging (Parnell et al., 2016) given the diverse microbial interactions in the soil that influence efficacy.

4 Future trends in research

Recent advances in our understanding of seed and fruit microbiomes, and of seed and fruit surface chemistry have important implications for manipulating plant defenses. In a recent article, Hemapriya et al. (2020) noted that fruit tissues (lemma and palea) of cereals often are associated with large fungal and bacterial communities. In the wild relatives of domesticated crops, these structures constitute part of the diaspore and therefore potentially provide a beneficial microbial inoculum that protects germinating seeds or infects growing seedlings. With the advent of domestication, the function of the shattering gene that abscises seed and maternal tissues from the infructescence has been lost from cereals, and seeds are instead mechanically separated from surrounding maternal tissues during threshing. As a consequence, germinating seeds of cereal crops may have lost a key component of the microbiome that protects their wild relatives

(Hemapriya et al., 2020). Further studies are needed to determine the functional role of maternally derived microbiomes before developing appropriate seed treatments to improve seed survival and seedling vigor. Examination of seed surface microbiomes of agricultural weeds may be especially fruitful as weeds and their microbiomes have already adapted to cropping systems.

In addition to harboring microbial communities, maternally derived hulls (lemma and palea) of wild oats (*Avena fatua*) also contain enzymatic defenses in the form polyphenol oxidases that are induced in the presence of seed decaying *Fusarium* fungi (Fuerst et al., 2014). As yet little is known about the generality of these enzymatic defenses across plant species or their distribution across fruit and seed tissues. However, preliminary evidence suggests that these inducible defenses may contribute to differential susceptibility to fungal pathogens: for example, polyphenol oxidase enzyme activity can be induced by *Fusarium* strains in wild oat caryopses but inhibited by *Pythium* (Fuerst et al., 2014). Enzymatic defenses may have been lost from many crop species either through the separation of maternal tissues or through selective breeding for other desirable traits. In addition to specificity, enzymatic defenses may be especially attractive targets for incorporation into crop defense systems because while they are in inactive forms they appear to be persistent, tolerant to temperature changes, and resistant to leaching (Godwin et al., 2017, Raviv et al., 2017, Pollard, 2018).

5 Where to look for further information

Several recent reviews have examined the defenses of seeds against pathogens and predators, and the potential use and mode of action of biological control microbes as alternatives to pesticide application:

- Ab Rahman, S. F. S., Singh, E., Pieterse, C. M. J. and Schenk, P. M. 2018. Emerging microbial biocontrol strategies for plant pathogens. *Plant Science* 267:102-111.
- Dalling, J. W., Davis, A. S., Arnold, A. E., Sarmiento, C. and Zalamea, P.-C. 2020. Extending plant defense theory to seeds. *Annual Review of Ecology, Evolution and Systematics*, 51:123-141.
- Köhl, J., Kolnaar, R. and Ravensberg, W. J. 2019. Mode of Action of Microbial Biological Control Agents Against Plant Diseases: Relevance Beyond Efficacy. *Frontiers in Plant Science* 10:19.

6 References

Ahmad, S., Khan, H. A. and Khalil-Ur-Rehman, J. M. (2012). Management of maize and sunflower against the depredations of rose-ringed parakeet (*Psittacula krameri*) using

mechanical repellents in an agro-ecosystem. *International Journal of Agriculture and Biology* 14:286–290.
Baker, H. G. (1974). The evolution of weeds. *Annual Review of Ecology and Systematics* 5(1):1–24.
Baker, K. F. and Smith, S. H. (1966). Dynamics of seed transmission of plant pathogens. *Annual Review of Phytopathology* 4(1):311–332.
Bargues-Ribera, M. and Gokhale, C. S. (2020). Eco-evolutionary agriculture: Host-pathogen dynamics in crop rotations. *PLoS Computational Biology* 16(1):e1007546.
Baskin, J. M. and Baskin, C. C. (2000). Evolutionary considerations of claims for physical dormancy-break by microbial action and abrasion by soil particles. *Seed Science Research* 10(4):409–413.
Beckstead, J., Meyer, S. E., Molder, C. J. and Smith, C. (2007). A Race for survival: Can *Bromus tectorum* seeds escape *Pyrenophora semeniperda*-caused mortality by germinating quickly? *Annals of Botany* 99(5):907–914.
Bello-Bedoy, R., Cruz, L. L. and Nunez-Farfan, J. (2011). Inbreeding alters a plant-predispersal seed predator interaction. *Evolutionary Ecology* 25(4):815–829.
Bensch, C. N., Horak, M. J. and Peterson, D. (2003). Interference of redroot pigweed (*Amaranthus retroflexus*), Palmer amaranth (*A. palmeri*), and common waterhemp (*A. rudis*) in soybean. *Weed Science* 51(1):37–43.
Berl, J. L., Johnstone, H. A., Wu, J. Y., Flaherty, E. A. and Swihart, R. K. (2017). Winter preference for weed seed and waste grain by native mice in row-crop agriculture. *Weed Science* 65(3):406–412.
Blubaugh, C. K., Hagler, J. R., Machtley, S. A. and Kaplan, I. (2016). Cover crops increase foraging activity of omnivorous predators in seed patches and facilitate weed biological control. *Agriculture, Ecosystems and Environment* 231:264–270.
Blubaugh, C. K. and Kaplan, I. (2015). Tillage compromises weed seed predator activity across developmental stages. *Biological Control* 81:76–82.
Bohan, D. A., Boursault, A., Brooks, D. R. and Petit, S. (2011). National-scale regulation of the weed seedbank by carabid predators. *Journal of Applied Ecology* 48(4):888–898.
Can-Hernandez, G., Villanueva-Garcia, C., Gordillo-Chavez, E. J., Pacheco-Figueroa, C. J., Perez-Netzahual, E. and Garcia-Morales, R. (2019). Wildlife damage to crops adjacent to a protected area in southeastern Mexico: Farmers' perceptions versus actual impact. *Human-Wildlife Interactions* 13:423–438.
Carmona, D. M. and Landis, D. A. (1999). Influence of refuge habitats and cover crops on seasonal activity-density of ground beetles (Coleoptera: Carabidae) in field crops. *Environmental Entomology* 28(6):1145–1153.
Carmona, D. M., Menalled, F. D. and Landis, D. A. (1999). *Gryllus pennsylvanicus* (Orthoptera : Gryllidae): Laboratory weed seed predation and within field activity-density. *Journal of Economic Entomology* 92(4):825–829.
Chancellor, R. J. (1981). The manipulation of weed behaviour for control purposes. *Philosophical Transactions of the Royal Society of London B, Biological Sciences* 295(1076):103–110.
Chen, T., Nan, Z., Zhang, X., Hou, F., Christensen, M. and Baskin, C. (2018). Does dormancy protect seeds against attack by the pathogenic fungus *Fusarium tricinctum* in a semiarid grassland of Northwest China? *Plant and Soil* 422(1–2):155–168.
Cordeau, S., Smith, R. G., Gallandt, E. R., Brown, B., Salon, P., DiTommaso, A. and Ryan, M. R. (2017). Timing of tillage as a driver of weed communities. *Weed Science* 65(4):504–514.

Cousens, R., Dytham, C. and Law, R. (2008). *Dispersal in Plants: A Population Perspective*. Oxford: Oxford University Press.

Cummings, C. L., Alexander, H. M. and Snow, A. A. (1999). Increased pre-dispersal seed predation in sunflower crop-wild hybrids. *Oecologia* 121(3):330–338.

Dalling, J. W., Davis, A. S., Arnold, A. E., Sarmiento, C. and Zalamea, P.-C. (2020). Extending plant defense theory to seeds. *Annual Review of Ecology, Evolution, and Systematics* 51(1):123–141.

Darrasse, A., Darsonval, A., Boureau, T., Brisset, M. N., Durand, K. and Jacques, M. A. (2010). Transmission of plant-pathogenic bacteria by nonhost seeds without induction of an associated defense reaction at emergence. *Applied and Environmental Microbiology* 76(20):6787–6796. https://doi.org/10.1128/AEM.01098-10.

Davis, A. S. (2006). When does it make sense to target the weed seed bank? *Weed Science* 54(3):558–565.

Davis, A. S., Daedlow, D., Schutte, B. J. and Westerman, P. R. (2011). Temporal scaling of episodic point estimates of weed seed predation to long-term predation rates. *Methods in Ecology and Evolution* 2(6):682–890.

Davis, A. S. and Raghu, S. (2010). Weighing abiotic and biotic factors in weed seed predation. *Weed Research* 50(5):402–412.

Davis, A. S. and Renner, K. A. (2007). Influence of seed depth and pathogens on fatal germination of velvetleaf (*Abutilon theophrasti*) and giant foxtail (*Setaria faberi*). *Weed Science* 55(1):30–35.

de Jong, T. J., van Goeverden, S., Jacobs, R., Rurenga, K. and Robbers, Y. (2016). Major effects of glucosinolates and minor effects of erucic acid on predation of *Brassica* seeds by mice. *Basic and Applied Ecology* 17(8):706–713.

Delgado-Sánchez, P., Ortega-Amaro, M. A., Jiménez-Bremont, J. F. and Flores, J. (2011). Are fungi important for breaking seed dormancy in desert species? Experimental evidence in *Opuntia streptacantha* (Cactaceae). *Plant Biology* 13(1):154–159.

Dominguez, C. P., Rodríguez, M. V., Batlla, D., de Salamone, I. E. G., Mantese, A. I., Andreani, A. L. and Benech-Arnold, R. L. (2019). Sensitivity to hypoxia and microbial activity are instrumental in pericarp-imposed dormancy expression in sunflower (*Helianthus annuus* L.). *Seed Science Research* 29(2):85–96.

Dutta, B., Ha, Y., Lessl, J. T., Avci, U., Sparks, A. C., Johnson, K. L. and Walcott, R. R. (2015). Pathways of bacterial invasion and watermelon seed infection by *Acidovorax citrulli*. *Plant Pathology* 64(3):537–544. https://doi.org/10.1111/ppa.12307.

Elphick, C. S. (2015). A history of ecological studies of birds in rice fields. *Journal of Ornithology* 156(S1):239–245.

Farahani-Kofoet, R. D., Römer, P. and Grosch, R. (2012). Systemic spread of downy mildew in basil plants and detection of the pathogen in seed and plant samples. *Mycological Progress* 11(4):961–966.

Farmer, J. A., Webb, E. B., Pierce, R. A. and Bradley, K. W. (2017). Evaluating the potential for weed seed dispersal based on waterfowl consumption and seed viability. *Pest Management Science* 73(12):2592–2603.

Fenner, M. and Thompson, K. (2005). *The Ecology of Seeds*. Cambridge: Cambridge University Press.

Franke, J., Geary, B. and Meyer, S. E. (2014). Identification of the infection route of a *Fusarium* seed pathogen into nondormant *Bromus tectorum* seeds. *Phytopathology* 104(12):1306–1313. https://doi.org/10.1094/PHYTO-03-14-0077-R.

Fuerst, E. P., Okubara, P. A., Anderson, J. V. and Morris, C. F. (2014). Polyphenol oxidase as a biochemical seed defense mechanism. *Frontiers in Plant Science* 5:689. https://doi.org/10.3389/fpls.2014.00689.

Furlan, L., Vasileiadis, V. P., Chiarini, F., Huiting, H., Leskovšek, R., Razinger, J., Holb, I. J., Sartori, E., Urek, G., Verschwele, A., Benvegnù, I. and Sattin, M. (2017). Risk assessment of soil-pest damage to grain maize in Europe within the framework of Integrated Pest Management. *Crop Protection* 97:52–59.

Fürnkranz, M., Lukesch, B., Müller, H., Huss, H., Grube, M. and Berg, G. (2012). Microbial diversity inside pumpkins: Microhabitat-specific communities display a high antagonistic potential against phytopathogens. *Microbial Ecology* 63(2):418–428.

Gallandt, E. R., Molloy, T., Lynch, R. P. and Drummond, F. A. (2005). Effect of cover-cropping systems on invertebrate seed predation. *Weed Science* 53(1):69–76.

Ganley, R. J. and Newcombe, G. (2006). Fungal endophytes in seeds and needles of *Pinus monticola*. *Mycological Research* 110(3):318–327.

Gebhardt, K., Anderson, A. M., Kirkpatrick, K. N. and Shwiff, S. A. (2011). A review and synthesis of bird and rodent damage estimates to select California crops. *Crop Protection* 30(9):1109–1116.

Ghimire, B., Sapkota, S., Bahri, B. A., Martinez-Espinoza, A. D., Buck, J. W. and Mergoum, M. (2020). *Fusarium* head blight and rust diseases in soft red winter wheat in the southeast United States: State of the art, challenges and future perspective for breeding. *Frontiers in Plant Science* 11:1080. https://doi.org/10.3389/fpls.2020.01080.

Godwin, J., Raviv, B. and Grafi, G. (2017). Dead pericarps of dry fruits function as long-term storage for active hydrolytic enzymes and other substances that affect germination and microbial growth. *Plants* 6(4):64. https://doi.org/10.3390/plants6040064.

Goggin, D. E., Emery, R. J. N., Kurepin, L. V. and Powles, S. B. (2015). A potential role for endogenous microflora in dormancy release, cytokinin metabolism and the response to fluridone in *Lolium rigidum* seeds. *Annals of Botany* 115(2):293–301. https://doi.org/10.1093/aob/mcu231.

Gogue, G. J. and Emino, E. R. (1979). Seed coat scarification of *Albizia julibrissin* Durazz, by natural mechanisms. *Journal of American Society of Horticultural Science* 104:421–423.

Goswami, R. S. and Kistler, H. C. (2004). Heading for disaster: *Fusarium graminearum* on cereal crops. *Molecular Plant Pathology* 5(6):515–525. https://doi.org/10.1111/j.1364-3703.2004.00252.x.

Gross, P. L. and Hanzel, J. J. (1991). Stability of morphological traits conferring bird resistance to sunflower across different environments. *Crop Science* 31(4):997–1000.

Hammond, R. B. (1995). Timing of plowing and planting - Effects on seedcorn maggot populations in soybean. *Crop Protection* 14(6):471–477.

Heggenstaller, A. H., Menalled, F. D., Liebman, M. and Westerman, P. R. (2006). Seasonal patterns in post-dispersal seed predation of *Abutilon theophrasti* and *Setaria faberi* in three cropping systems. *Journal of Applied Ecology* 43(5):999–1010.

Hemapriya, M., Nataraja, K. N., Suryanarayanan, T. S. and Shaanker, R. U. (2020). Threshing yards: Graveyard of maternally borne seed microbiome? *Trends in Ecology and Evolution* 35(11):965–968. https://doi.org/10.1016/j.tree.2020.08.010.

Holm, L., Doll, J., Holm, E., Pancho, J. and Herberger, J. (1997). *World Weeds: Natural Histories and Distribution*. Hoboken: John Wiley and Sons, Inc.

Htwe, N. M., Sarathchandra, S. R., Sluydts, V., Nugaliyadde, L., Singleton, G. R. and Jacob, J. (2021). Small mammal communities, associated damage to rice and damage prevention in smallholder rice storage facilities in Sri Lanka. *Crop Protection* 145:105638.

Htwe, N. M., Singleton, G. R. and Maw, P. P. (2017). Post-harvest impacts of rodents in Myanmar; how much rice do they eat and damage? *Pest Management Science* 73(2):318–324.

John, A. (2014). Rodent outbreaks and rice pre-harvest losses in Southeast Asia. *Food Security* 6(2):249–260.

Kaboodvandpour, S. and Leung, L. K.-P. (2012). Modelling density thresholds for managing mouse damage to maturing wheat. *Crop Protection* 42:134–140.

Khaleghizadeh, A. (2011). Effect of morphological traits of plant, head and seed of sunflower hybrids on house sparrow damage rate. *Crop Protection* 30(3):360–367.

Kirkpatrick, B. L. and Bazzaz, F. A. (1979). Influence of certain fungi on seed germination and seedling removal of four colonizing annuals. *Journal of Applied Ecology* 16:515–527.

Kivilaan, A. and Bandurski, R. S. (1981). The one hundred-year period for Dr. Beal's seed viability experiment. *American Journal of Botany* 68(9):1290–1292.

Kremer, R. J. (1993). Management of weed seed banks with microorganisms. *Ecological Applications: A Publication of the Ecological Society of America* 3(1):42–52. https://doi.org/10.2307/1941791.

Kremer, R. J., Hughes, L. B. and Aldrich, R. J. (1984). Examination of microorganisms and deterioration resistance mechanisms associated with velvetleaf seed 1. *Agronomy Journal* 76(5):745–749. https://doi.org/10.2134/agronj1984.00021962007600050009x.

Kremer, R. J. and Spencer, N. R. (1989). Impact of a seed-feeding insect and microorganisms on velvetleaf (*Abutilon theophrasti*) seed viability. *Weed Science* 37(2):211–216.

Li, Y. M., Shaffer, J. P., Hall, B., Ko, H. (2019) Soil-borne fungi influence seed germination and mortality, with implications for coexistence of desert winter annual plants. *PLoS ONE* 14(10): e0224417. https://doi.org/10.1371/journal.

Links, M. G., Demeke, T., Gräfenhan, T., Hill, J. E., Hemmingsen, S. M. and Dumonceaux, T. J. (2014). Simultaneous profiling of seed-associated bacteria and fungi reveals antagonistic interactions between microorganisms within a shared epiphytic microbiome on *Triticum* and *Brassica* seeds. *The New Phytologist* 202(2):542–553. https://doi.org/10.1111/nph.12693.

Long, R. L., Gorecki, M. J., Renton, M., Scott, J. K., Colville, L., Goggin, D. E., Commander, L. E., Westcott, D. A., Cherry, H. and Finch-Savage, W. E. (2015). The ecophysiology of seed persistence: A mechanistic view of the journey to germination or demise. *Biological Reviews of the Cambridge Philosophical Society* 90(1):31–59.

Lundgren, J. G. and Rosentrater, K. A. (2007). The strength of seeds and their destruction by granivorous insects. *Arthropod-Plant Interactions* 1(2):93–99.

Maksimov, I. V., Abizgil'dina, R. R. and Pusenkova, L. I. (2011). Plant growth promoting rhizobacteria as alternative to chemical crop protectors from pathogens (review). *Applied Biochemistry and Microbiology* 47(4):333–345. https://doi.org/10.1134/S0003683811040090.

Meyer, S. E., Quinney, D., Nelson, D. L. and Weaver, J. (2007). Impact of the pathogen Pyrenophora semeniperda on *Bromus tectorum* seedbank dynamics in North American cold deserts. *Weed Research* 47(1):54–62.

Mills, J. T. (1983). Insect-fungus associations influencing seed deterioration. *Phytopathology* 73(2):330–335.

Mitter, B., Pfaffenbichler, N. and Sessitsch, A. (2016). Plant-microbe partnerships in 2020. *Microbial Biotechnology* 9(5):635–640.

Neelanarayanan, P., Nagarajan, R. and Kanakasabai, R. (1997). Magnitude of rodent depredation in soybeans. *Journal of Environmental Biology* 18:111–113.

Nelson, E. B. (2018). The seed microbiome: origins, interactions, and impacts. *Plant and Soil* 422(1–2):7–34. https://doi.org/10.1007/s11104-017-3289-7.

Newcombe, G., Harding, A., Ridout, M. and Busby, P. E. (2018). A hypothetical bottleneck in the plant microbiome. *Frontiers in Microbiology* 9:1645. https://doi.org/10.3389/fmicb.2018.01645.

Ngugi, H. K. and Scherm, H. (2006). Biology of flower-infecting fungi. *Annual Review of Phytopathology* 44:261–282. https://doi.org/10.1146/annurev.phyto.44.070505.143405.

Oerke, E.-C. (2006). Crop losses to pests. *The Journal of Agricultural Science* 144(1):31–43. https://doi.org/10.1017/S0021859605005708.

Ognakossan, K. E., Mutungi, C. M., Otieno, T. O., Affognon, H. D., Sila, D. N. and Owino, W. O. (2018). Quantitative and quality losses caused by rodents in on-farm stored maize: A case study in the low land tropical zone of Kenya. *Food Security* 10(6):1525–1537.

Parnell, J. J., Berka, R., Young, H. A., Sturino, J. M., Kang, Y. W., Barnhart, D. M. and DiLeo, M. V. (2016). From the lab to the farm: An industrial perspective of plant beneficial microorganisms. *Frontiers in Plant Science* 7:1110.

Paulsen, T. R., Colville, L., Kranner, I., Daws, M. I., Högstedt, G., Vandvik, V. and Thompson, K. (2013). Physical dormancy in seeds: A game of hide and seek? *The New Phytologist* 198(2):496–503. https://doi.org/10.1111/nph.12191.

Pejchar, L., Clough, Y., Ekroos, J., Nicholas, K. A., Olsson, O., Ram, D., Tschumi, M. and Smith, H. G. (2018). Net effects of birds in agroecosystems. *BioScience* 68:896–904.

Pitty, A., Staniforth, D. W. and Tiffany, L. H. (1987). Fungi associated with caryopses of Setaria species from field-harvested seeds and from soil under two tillage systems. *Weed Science* 35(3):319–323.

Pollard, A. T. (2018). Seeds vs fungi: An enzymatic battle in the soil seedbank. *Seed Science Research* 28(3):197–214. https://doi.org/10.1017/S0960258518000181.

Raviv, B., Aghajanyan, L., Granot, G., Makover, V., Frenkel, O., Gutterman, Y. and Grafi, G. (2017). The dead seed coat functions as a long-term storage for active hydrolytic enzymes. *PLoS ONE* 12(7):e0181102. https://doi.org/10.1371/journal.pone.0181102.

Ridout, M. E., Schroeder, K. L., Hunter, S. S., Styer, J. and Newcombe, G. (2019). Priority effects of wheat seed endophytes on a rhizosphere symbiosis. *Symbiosis* 78(1):19–31. https://doi.org/10.1007/s13199-019-00606-6.

Roberts, H. A. and Feast, P. M. (1973). Emergence and longevity of seeds of annual weeds in cultivated and undisturbed soil. *The Journal of Applied Ecology* 10(1):133–143.

Rodriguez-Pastor, R., Luque-Larena, J. J., Lambin, X. and Mougeot, F. (2016). "Living on the edge": The role of field margins for common vole (*Microtus arvalis*) populations in recently colonised Mediterranean farmland. *Agriculture, Ecosystems and Environment* 231:206–217.

Roos, D., Saldana, C. C., Arroyo, B., Mougeot, F., Luque-Larena, J. J. and Lambin, X. (2019). Unintentional effects of environmentally-friendly farming practices: arising conflicts between zero-tillage and a crop pest, the common vole (*Microtus arvalis*). *Agriculture, Ecosystems and Environment* 272:105–113.

Saikkonen, K., Young, C. A., Helander, M. and Schardl, C. L. (2016). Endophytic *Epichloë* species and their grass hosts: From evolution to applications. *Plant Molecular Biology* 90(6):665–675. https://doi.org/10.1007/s11103-015-0399-6.

Sarabi, V. (2019). Factors that influence the level of weed seed predation: A review. *Weed Biology and Management* 19(3):61–74.

Sarmiento, C., Zalamea, P. C., Dalling, J. W., Davis, A. S., Stump, S. M., U'Ren, J. M. and Arnold, A. E. (2017). Soilborne fungi have host affinity and host-specific effects on seed germination and survival in a lowland tropical forest. *Proceedings of the National Academy of Sciences of the United States of America* 114(43):11458–11463. https://doi.org/10.1073/pnas.1706324114.

Savary, S., Willocquet, L., Pethybridge, S. J., Esker, P., McRoberts, N. and Nelson, A. (2019). The global burden of pathogens and pests on major food crops. *Nature Ecology and Evolution* 3(3):430–439. https://doi.org/10.1038/s41559-018-0793-y.

Schillinger, W. F. and Werner, S. J. (2016). Horned lark damage to pre-emerged canola seedlings. *Industrial Crops and Products* 89:465–467.

Schwartz-Lazaro, L. M., Shergill, L. S., Evans, J. A., Bagavathiannan, M. V., Beam, S. C., Bish, M. D., Bond, J. A., Bradley, K. W., Curran, W. S., Davis, A. S., Everman, W. J., Flessner, M. L., Haring, S. C., Jordan, N. R., Korres, N. E., Lindquist, J. L., Norsworthy, J. K., Sanders, T. L., Steckel, L. E., VanGessel, M. J., Young, B. and Mirsky, S. B. (2021a). Seed-shattering phenology at soybean harvest of economically important weeds in multiple regions of the United States. Part 1: Broadleaf species. *Weed Science* 69(1):95–103.

Schwartz-Lazaro, L. M., Shergill, L. S., Evans, J. A., Bagavathiannan, M. V., Beam, S. C., Bish, M. D., Bond, J. A., Bradley, K. W., Curran, W. S., Davis, A. S., Everman, W. J., Flessner, M. L., Haring, S. C., Jordan, N. R., Korres, N. E., Lindquist, J. L., Norsworthy, J. K., Sanders, T. L., Steckel, L. E., VanGessel, M. J., Young, B. and Mirsky, S. B. (2021b). Seed-shattering phenology at soybean harvest of economically important weeds in multiple regions of the United States. Part 2: Grass species. *Weed Science* 69(1):104–110.

Shade, A., Jacques, M. A. and Barret, M. (2017). Ecological patterns of seed microbiome diversity, transmission, and assembly. *Current Opinion in Microbiology* 37:15–22. https://doi.org/10.1016/j.mib.2017.03.010.

Shaffer, J. P., Sarmiento, C., Zalamea, P.-C., Gallery, R. E., Davis, A. S., Baltrus, D. A. and Arnold, A. E. (2016). Diversity, specificity, and phylogenetic relationships of endohyphal bacteria in fungi that inhabit tropical seeds and leaves. *Frontiers in Ecology and Evolution* 4:116. https://doi.org/10.3389/fevo.2016.00116.

Shergill, L. S., Schwartz-Lazaro, L. M., Leon, R., Ackroyd, V. J., Flessner, M. L., Bagavathiannan, M., Everman, W., Norsworthy, J. K., VanGessel, M. J. and Mirsky, S. B. (2020). Current outlook and future research needs for harvest weed seed control in North American cropping systems. *Pest Management Science* 76(12):3887–3895.

Simonin, M., Briand, M., Chesneau, G., Rochefort, A., Marais, C., Sarniguet, A. and Barret, M. (2021). Seed microbiota revealed by a large-scale meta-analysis including 50 plant species. *bioRxiv*.

Sperber, K., Steinbrecher, T., Graeber, K., Scherer, G., Clausing, S., Wiegand, N., Hourston, E., Kurre, R., Luebner-Metzger, G. and Mummenhoff, K. (2017). Fruit fracture biomechanics and the release of Lepidium didymum pericarp-imposed mechanical dormancy by fungi. *Nature Communications* 8:1.

Teo-Sherrell, C. P. A., Mortensen, D. A. and Keaton, M. E. (1996). Fates of weed seeds in soil: A seeded core method of study. *Journal of Applied Ecology* 33(5):1107–1113.

Terrasson, E., Darrasse, A., Righetti, K., Buitink, J., Lalanne, D., Ly Vu, B., Pelletier, S., Bolingue, W., Jacques, M. A., Leprince, O. (2015). Identification of a molecular dialogue between developing seeds of Medicago truncatula and seedborne xanthomonads. *Journal of Experimental Botany* 66(13):3737–3752. doi: 10.1093/jxb/erv167. PMID: 25922487.

Thomas, H., Heller, A., Sauerborn, J. and Müller-Stöver, D. (1999). *Fusarium oxysporum* f. sp. *orthoceras*, a potential mycoherbicide, parasitizes seeds of *Orobanche cumana* (sunflower broomrape): A cytological study. *Annals of Botany* 83(4):453–458.

van Lenteren, J. C. (2012). The state of commercial augmentative biological control: Plenty of natural enemies, but a frustrating lack of uptake. *BioControl* 57(1):1–20. https://doi.org/10.1007/s10526-011-9395-1.

van Mourik, T. A., Stomph, T. J. and Murdoch, A. J. (2005). Why high seed densities within buried mesh bags may overestimate depletion rates of soil seed banks. *Journal of Applied Ecology* 42(2):299–305. https://doi.org/10.1111/j.1365-2664.2005.01016.x.

Wagner, M. and Mitschunas, N. (2008). Fungal effects on seed bank persistence and potential applications in weed biocontrol: A review. *Basic and Applied Ecology* 9(3):191–203. https://doi.org/10.1016/j.baae.2007.02.003.

Wagner, M. and Mitschunas, N. (2022). Soil microbial effects on weed seed bank persistence: Current knowledge and applications for weed management. In: Upadhyaya, M. K., Clements, D. R. and Shrestha, A. (Eds). *Persistence Strategies of Weeds in Agriculture*. Wiley Blackwell in Press. https://doi.org/10.1002/9781119525622.ch13

Walsh, M., Newman, P. and Powles, S. (2013). Targeting weed seeds in-crop: A new weed control paradigm for global agriculture. *Weed Technology* 27(3):431–436.

Wilson, B. J. and Lawson, H. M. (1992). Seedbank persistence and seedling emergence of seven weed species in autumn-sown crops following a single year's seeding. *Annals of Applied Biology* 120(1):105–116.

Zalamea, P. C., Dalling, J. W., Sarmiento, C., Arnold, A. E., Delevich, C., Berhow, M. A., Ndobegang, A., Gripenberg, S. and Davis, A. S. (2018). Dormancy-defense syndromes and tradeoffs between physical and chemical defenses in seeds of pioneer species. Ecology 99(9): 1988–1998.

Zalamea, P. C., Sarmiento, C., Arnold, A. E., Davis, A. S., Ferrer, A. and Dalling, J. W. (2021). Closely related tree species support distinct communities of seed-associated fungi in a lowland tropical forest. *Journal of Ecology* 109(4):1858–1872.

Chapter 5

Advances in understanding the genetic and environmental factors determining seed longevity

Julia Buitink and Olivier Leprince, Institut Agro, Univ Angers, INRAE, IRHS, France

1 Introduction

Longevity is defined as the extent to which seeds can remain alive during storage. This property is progressively acquired during the second half of seed development and serves to bridge the final step of maturation drying and seed abscission from the mother plant until conditions are conducive to germination. Together with seed dormancy, it is an essential strategy to allow seed dispersion both in time and in space and therefore impacts seed persistence in the soil (Long et al., 2015). Seed longevity also underpins global agriculture and plant species conservation (Li and Pritchard, 2009; Colville and Pritchard, 2019). This was recognized as early as antiquity as a method to preserve seeds for the next sowing season as Theophrastus (-370 BC) mentioned in his book *Enquiry into Plants*, translated by Sir Arthur Hort (Fig. 1).

Prerequisites for seed longevity are the acquisition of desiccation tolerance and the capacity to resist freezing in a dry state. Desiccation tolerance is the capacity to remain alive after drying in ambient air or at low relative humidity (RH), which practically brings the seed moisture to around 5–10% of the dry weight.

http://dx.doi.org/10.19103/AS.2022.0105.07

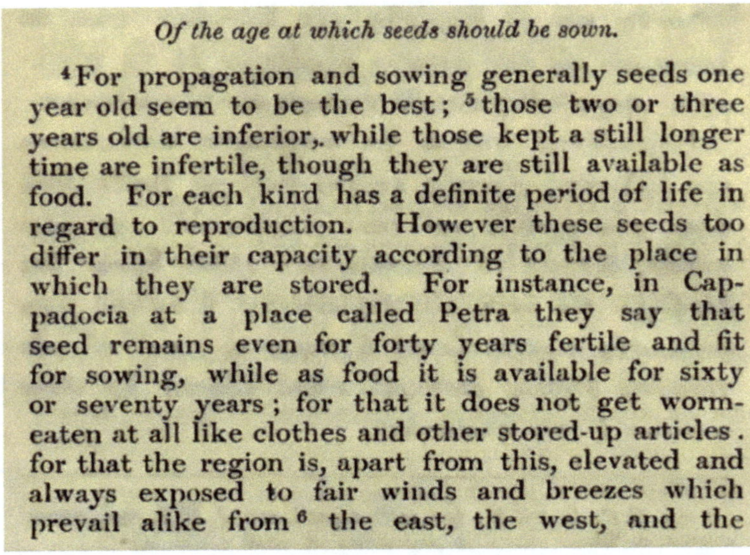

Of the age at which seeds should be sown.

[4]For propagation and sowing generally seeds one year old seem to be the best; [5]those two or three years old are inferior,. while those kept a still longer time are infertile, though they are still available as food. For each kind has a definite period of life in regard to reproduction. However these seeds too differ in their capacity according to the place in which they are stored. For instance, in Cappadocia at a place called Petra they say that seed remains even for forty years fertile and fit for sowing, while as food it is available for sixty or seventy years; for that it does not get worm-eaten at all like clothes and other stored-up articles. for that the region is, apart from this, elevated and always exposed to fair winds and breezes which prevail alike from [6] the east, the west, and the

Figure 1 Insert from the section *of the age at which seeds should be sown* by Theophrastus, *Enquiry into plants and minor works on odours and whether signs*, with an English translation by Sir Arthur Hort, volume II. William Heinemann Editor, 1916. Downloaded from: https://archive.org/details/enquiryintoplant02theouoft.

Seeds with these characteristics are called 'orthodox' and will be the focus of this chapter. In contrast, seeds of up to 2–10% of herbaceous and woody species lost their desiccation tolerance during evolution. They can therefore not be stored in a dry atmosphere storage or below 0°C as the water remaining in these seeds forms lethal ice crystals when stored in a domestic freezer. These seeds are called 'recalcitrant' and include crops such as cocoa, rubber tree, or mango. Models based on data from 17 539 species predict the likely seed response to desiccation and suggest that taxonomy affinity, annual precipitation, and seed mass are the main variables explaining recalcitrance in Angiosperms (Wyse and Dickie, 2018). There exists a third category, named 'intermediate', representing 10–15% of Angiosperms worldwide. Intermediate seeds, such as papaya and *Coffea* ssp., can survive drying when exposed to 45–60% RH but deteriorate faster when stored in a domestic freezer (−18°C) compared to 5°C. In comparison with orthodox seeds, intermediate seeds exhibit much shorter longevity and usually die within 5 years. However, it is argued that there is a continuum of seed response to desiccation from highly desiccation-sensitive to intermediate rather than distinct categories (Walters, 2015). Further information on this continuum and the challenges associated with predicting recalcitrant versus intermediate seeds and how to store these seeds have recently been reviewed elsewhere (Berjak and Pammenter, 2013; Walters, 2015; Pence et al., 2020).

The actual time that seeds remain alive during storage, or lifespan, depends on three interacting storage conditions: temperature, water content (or RH of the storage atmosphere since the seed water content always equilibrates with the environment), and oxygen. Temperature and water interact with each other to influence the physical properties of the cytoplasmic matrix. One of the fundamental outcomes of this influence is that the cytoplasm of dry seeds is in a solid-like state that severely reduces the molecular motions and thereby decreases the rate of the deteriorative reactions. Both parameters also affect the nature of these reactions (Mira et al., 2016; Gerna et al., 2020; Michalak et al., 2021). Reviews regarding the different concepts of the glassy state applied to seeds (its detection and characteristics such as molecular mobility and structural aspects) and the implication of these concepts to seed storage can be found in Buitink and Leprince (2008); Walters et al. (2010); Ballesteros et al. (2019); Ballesteros et al. (2020). Temperature also affects the physical properties of lipids. Above 0°C, storage lipids are in a liquid state, which allows the diffusion of molecules including oxygen and activities of lipophilic enzymes (Oenel et al., 2017). Depending on the species and nature of triacylglycerols, lipids start to solidify between 0°C and −50°C in a complex manner that could influence survival during storage in the freezer (Mira et al., 2019). Oxygen is deleterious for long-term storage of seeds because oxidation is a major degradative reaction still occurring in the dry state that leads to cell death (Groot et al., 2012).

Mechanisms inducing seed longevity have been reviewed recently and will be only briefly mentioned here (Sano et al., 2016; Ballesteros et al., 2020; Zhou et al., 2020; Zinsmeister et al., 2020b). They can be categorized by two main functions: protection during drying and storage and repair during imbibition after storage. Protection is needed to avoid oxidation of proteins, lipids, and nucleic acids and to prevent structural collapse and dehydration-induced structural changes in proteins and membrane phospholipids. The seed coat protects against oxidation via a thin layer of lipid polyesters such as cutin and suberin that seals off the embryo from the outside oxygen (De Giorgi et al., 2015; Renard et al., 2021). The seed coat also contains antioxidant molecules such as anthocyanins. Within the embryo, protection against oxidation is provided by lipophilic antioxidants such as lipocalin and molecules from the vitamin E family and by the pool of hydrophilic antioxidants such as reduced glutathione (Walters et al., 2005; Kranner et al., 2006; Mène-Saffrané et al., 2010; Boca et al., 2013; Chen et al., 2016; Roach et al., 2018). These molecules scavenge reactive oxygen species (ROS) that are formed during storage, thereby preventing further oxidative degradation of the cellular compounds. They need an enzymatic machinery to regenerate their antioxidant efficiency such as the ascorbate/glutathione cycle or superoxide dismutase/catalase and are therefore only functional at higher water contents during storage

(Fernández-Marín et al., 2013). Structural protection is brought by the formation of the glassy state during drying (reviewed in (Buitink and Leprince, 2004; Walters et al., 2010). Its composition remains to be established but likely involves non-reducing sugars and late embryogenesis abundant (LEA) proteins. Next to forming a glass, *in vitro* studies showed that these molecules can stabilize membranes and proteins during drying and in the dry state, by direct binding and/or providing hydrogen bonds to replace the water molecules (Hincha et al., 2021) and references therein).

Repair processes are necessary to restore genome integrity that can be compromised during ageing and to eliminate aggregated proteins and oxidized lipids and proteins that are formed during storage (reviewed in Dirk and Downie, 2018; Waterworth et al., 2019). Molecular chaperones eliminating aggregated proteins include LEA proteins (Boucher et al., 2010; Furuki et al., 2020) and small heat shock proteins (sHSP) Haslbeck and Vierling, 2015). Their importance in longevity has been demonstrated in seeds stored at high temperature and high moisture content (Hundertmark et al., 2011; Personat et al., 2014; Kaur et al., 2015; Zinsmeister et al., 2016, 2020a), but their precise function as repair proteins remains to be established *in vivo*. Repair enzymes such as PROTEIN ISOASPARTYL METHYLTRANSFERASE (PMIT) and methionine sulfoxide reductase target-specific amino acids in proteins that are prone to damage through oxidation, isomerization, and spontaneous conversion during storage (Petla et al., 2016; Dirk and Downie, 2018 and references therein). The list of client proteins of these above-mentioned chaperones as well as PIMT is skewed towards proteins involved in translation, suggesting that maintaining a high translational efficiency might be pivotal to seed longevity (Dirk and Downie, 2018). Another type of repair is related to detoxification. Lipid hydroperoxides and peroxy radicals that accumulate during storage must be rapidly eliminated because they can fragment and produce reactive intermediates, such as unsaturated carbonyls and alkanes. These molecules have been detected during storage (Colville et al., 2012; Shenzao et al., 2018; Han et al., 2021). Aldehyde dehydrogenases are a large family of enzymes that catalyze the oxidation of reactive aldehydes to their corresponding carboxylic acids. Their role in seed longevity has been demonstrated during storage at high moisture contents that allow for the enzymatic reactions to take place. For example, in rice and tobacco, seeds over-expressing *ALDO-KETOREDUCTASE1* showed lower levels of cytotoxic compounds and glycation products that resulted in improved seed viability and vigour (Narayana et al., 2017).

2 Distribution of seed lifespans among species

Since the creation of dedicated seed banks in the 1950s, longevity data for seeds stored >20 years become increasingly available for predictions and meta-analyses. Longevity varies greatly between species, ranging from months to centuries under seed bank storage conditions, and is summarized in Table 1 and Fig. 2 (Walters et al., 2004, 2005; Probert et al., 2009; Colville and Pritchard, 2019; Solberg et al., 2020). Altogether, the published meta-analyses of seed longevity are based on only 0.2% of the estimated 370 000 species, which corresponds to 732 orthodox species from 94 families. They demonstrate that over 50% of species are likely to have relatively short lifespans (Colville and Pritchard, 2019). Lifespan can be roughly categorized as short, medium, or long, based on the comparison of species P50 (the time to 50% viability during storage) with the median P50 for all species for a particular storage condition, although there is a large variation within species (Table 1). Species with shorter-lived seeds are more common in the Asteraceae and Brassicaceae families and in the grasses such as Poaceae (Table 1). Relatively longer lifespans are mainly found in the rosid clade of the core eudicots, while relatively short to medium lifespans predominate in all other major groups including gymnosperms, monocots, basal eudicots, and asterids (Colville and Pritchard, 2019). Of the 172 species that were classified as having relatively long-lived seeds, the legume (Fabaceae) and myrtle (Myrtaceae) families are well represented (Colville and Pritchard, 2019). Of the 67 species that were stored under different storage conditions, 25 were attributed to the same relative lifespan category in all storage conditions and 36 were assigned to 2 adjacent lifespan categories, i.e. long-/medium-lived or short-/medium-lived, suggesting that some cellular components are more sensitive to deterioration depending on the storage conditions. The comparison between wild species and crops does not show a significant difference in lifespan during ambient and cold storage (Fig. 2 Solberg et al., 2020). Another comparative study involving 195 species indicated that endospermic seeds of early Angiosperms are relatively shorter-lived than non-endospermic seeds (Probert et al., 2009). These authors predicted that seeds from hot and dry environments were longer lived than those from cool, wet conditions, consistent with the idea that longevity is an adaptive strategy to disseminate the species (Leprince et al., 2017). This will be reviewed in Section 4.

The mechanistic basis for the large differences in seed lifespan between certain species is still unclear. Based on the biochemical composition and physical-chemical properties of dried materials, Ballesteros et al. (2020) proposed three categories of seeds to explain the variation in lifespans: (1) seeds that lack chlorophyll and contain few storage lipids may exhibit long shelf life, with deterioration during storage probably occurring through slow autoxidative processes within the glassy, solid-like cytoplasm as it relaxes; (2) seeds that lack chloroplasts but contain high amounts storage lipids may die quickly during storage at −20°C, possibly because lipids crystallize and destabilize the glassy matrix; and (3) seeds that contain chlorophyll

Table 1 Seed lifespans (P50) of selected plant families stored under the indicated storage conditions. Numbers in the middle indicate the median P50 value and those in parenthesis indicate the 25th (left) and 75th (right) percentile value. Italics and bold characters correspond respectively to short and long lifespans for the indicated storage condition.

	Storage conditions					
	60% RH, 45°C		20% RH, 5°C		4-5% moisture, 5°C then transferred to −18°C	
	0.1 year		7.3 years		54 years	
Average lifespan	Number of species	P50 (days)	Number of species	P50 (year)	Number of species	P50 (year)
Asteraceae	42	(*21*) 32 (54)	0		18	(*24*) 52 (66)
Brassicaceae	0		0		14	(*27*) *33* (47)
Fabaceae	42	(**91**) **167** (**202**)	30	(6.8) 19.5 (28.5)	47	(52) 80 (**208**)
Malvaceae	0		0		37	(**89**) **158** (**260**)
Poaceae	14	(*16*) *30* (49)	24	(4.2) 6.7 (9.7)	84	(33) 62 (**86**)
Solanaceae	11	(43) **70** (**98**)	1[a]	10	32	(*28*) 61 (80)

[a]*Nicotiana tabacum*. Data are from Colville and Pritchard (2019)

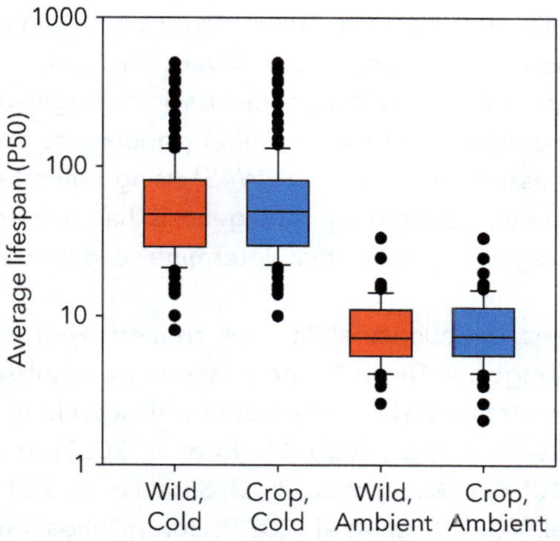

Figure 2 Average lifespan (P50) of wild and crop species that were stored in ambient conditions and cold storage conditions in gene banks. Data are from Solberg et al. (2020).

molecules may die quickly, possibly because they are prone to oxidative stress promoted by the photosynthetic pigments in the absence of water available for photosynthesis. This attractive hypothesis needs further investigation since seeds are stored in darkness, thereby eliminating the possibil ty to excite chlorophyll molecules that are the source of ROS.

The conclusions of these meta-analyses have several important implications for *ex situ* conservation of plant biodiversity. How species evolution, within-species adaptation to local climate, and ratios between seed organs have shaped seed longevity deserve further attention. In addition, current storage conditions and/or procedures prior to storage such as post-harvest treatments may not be sufficient for long-term preservation without resorting to regular regenerations of accessions, which poses a risk of genetic erosion (Richards et al., 2010). Equally alarmingly is the analysis of historical viability data obtained between 2013 and 2017 from the 7 CGIAR genebanks, representing 874 427 observations of 822 different species spanning an average of 37 years of storage. This study showed that these data are unfortunately not as valuable as expected to derive reliable estimates of survival in order to better understand and quantify seed longevity in long-term gene bank storage conditions (Hay et al., 2021). These authors made several suggestions to improve seed bank management procedures.

3 Advances in the understanding of longevity using genetic diversity

The previous section indicated that longevity varies considerably between species under the same storage conditions (Fig. 2) (Walters et al., 2005; Probert et al., 2009; Colville and Pritchard, 2019). However, considerable variation in lifespan is also found within species or cultivars. This variation has been used over the past years to decipher the genetic basis of longevity, via the analysis of bi-parental recombinant inbred line (RIL) populations and more recently by genome-wide association studies (GWAS) using natural genetic diversity. Also, induced genetic variation by mutagenesis has helped identifying the mechanisms and regulatory genes that determine seed longevity, as discussed below.

In most studies, genetic variability was created by crossing two parents with contrasting longevity. These RIL populations were subsequently used to identify quantitative trait loci (QTL) correlated with seed longevity, for instance on Arabidopsis (Bentsink et al., 2000; Clerkx et al., 2003), lettuce (Schwember and Bradford, 2010), rice (Miura et al., 2002; Sasaki et al., 2015; Thi Hang et al., 2015; Raquid et al., 2021; Yuan et al., 2021), durum wheat (Rehman Arif et al., 2012), *Medicago truncatula* (Zinsmeister et al., 2016), and soybean (Zhang et al., 2019). All these studies identified many longevity QTL under different storage conditions with each one explaining relatively small variations in longevity, showing the true polygenic nature of this trait. For most of these studies, the underlying genes or loci explaining the variation in longevity remain to be discovered, although in some cases, putative candidate genes were suggested. For example, fine-mapping of the qLG-9 of seed longevity that was identified using a backcross inbred line from a cross between a temperate japonica × indica rice population resulted in the identification of a 30-kb region explaining 59.5% of the variation (Miura et al., 2002; Sasaki et al., 2015). This region includes two candidate genes, one encoding trehalose-6-phosphate phosphatase (TPP, Os09g0369400) and the other encoding an unknown protein (Os09g0369500). The same TPP was also identified in another QTL study using a population generated from a cross of temperate japonica and indica genotypes that was evaluated under both natural storage and accelerating ageing (Yuan et al., 2021). In contrast, the qLG-9 region was not found in a QTL study on tropical rice population (Raquid et al., 2021), showing the limits of using biparental populations. Combining the list of 418 genes in the longevity QTL region with public gene network data sets, these authors identified eight candidate genes involved in oxidative stress response and DNA repair.

In *M. truncatula*, a model legume, ten longevity QTLs were detected during accelerated ageing and five during moderate ageing (Zinsmeister et al., 2016). Within the confidence interval of a QTL for longevity on LG7 that was found for both storage conditions, these authors identified *MtABI5*, a b-Zip transcription factor involved in abscisic acid (ABA) signalling. This chromosomal region also harboured a QTL for the composition of sucrose and stachyose. The functional analysis of *Mtabi5* mutant seeds demonstrated that longevity and RFO accumulation were affected together with chlorophyll degradation (Zinsmeister et al., 2016). The analysis of six RIL populations in Arabidopsis revealed five loci involved in longevity, named GAAS1 to GAAS5 (germination ability after storage) (Nguyen et al., 2012). Using near-isogenic lines carrying single genomic fragments of different accessions into the Ler genetic background, it was shown that *GAAS5* was the underlying *DELAY OF GERMINATION 1* (*DOG1*), one of the major regulators of seed dormancy. The role of DOG1 in longevity was later confirmed by Dekkers et al. (2016) who showed that *dog1-1* seeds exhibited reduced seed longevity by affecting directly or indirectly the expression of *ABI5* and protective genes such as LEA and HSPs. Characterization of these six RIL populations also led to the hypothesis that dormancy and longevity were inversely correlated in Arabidopsis. Such correlation, albeit weak ($r = -0.28$ to -0.48), was also experimentally verified in rice using a genetic approach consisting of pyramiding dormancy alleles from weedy rice into cultivated rice and creating 16 homozygous lines selected as a tetragenic system (Pipatpongpinyo et al., 2020). Interestingly, the negative correlation between longevity and dormancy in Arabidopsis was based on *ex situ* storage, in the laboratory (Nguyen et al., 2012), whereas that in rice was based on survival after burial in the soil (Pipatpongpinyo et al., 2020). No significant correlation between both traits was found when 270 natural accessions of Arabidopsis were examined (Renard et al., 2020) or in genetic variants of Arabidopsis originating from the cytoplasmic genomes (Boussardon et al., 2019). Such discrepancy might highlight the complexity of the genetic architecture of longevity. They could also be explained by differences in post-harvest treatments and methods to assess longevity and dormancy. Further research is therefore needed to better understand whether there is a trade-off between dormancy and longevity. This is crucial for seed bank curators in order not to confound the lack of germination after storage as a result of seed death whereas seeds could be dormant instead (Hay et al., 2021). It is also crucial for breeding programmes aiming at creating varieties adapted for no-tilling practices.

The above-mentioned biparental crosses to map QTL are restricted in allelic diversity and have a limited genomic resolution. In recent years, the focus on using genetic diversity has shifted to GWAS. This approach, which provides higher resolution than traditional gene mapping, makes use of the natural genetic diversity that exists in species from diverse geographic origins to identify loci underlying the natural phenotypic variation in traits. The occurrence of high-density single-nucleotide polymorphism (SNP) typing allows whole-genome scans to identify often small haplotype blocks that are significantly correlated with quantitative trait variation. Although it remains challenging to identify true causative SNPs because they are in linkage disequilibrium (in non-random association) with many loci in the genome, the size of the regions with gene candidates is considerably reduced, often restricted to one to ten genes, making it a more amenable list for functional validation.

GWAS has been used to characterize genetic and environmental factors that affect ageing of barley seeds from a historic germplasm collection stored since 1974 and from 175 genotypes grown in field plots with 2 different nutrient supplies stored under more deleterious conditions (Nagel et al., 2015). A total of 107 marker–trait associations with longevity were found. Nine genetic markers were identified that were linked to major QTL of other agronomic traits such as germination speed, seed size, dormancy, and days to heading and pointed to candidate genes involved in biotic and biotic stress (Nagel et al., 2015).

Several GWAS studies performed on different common wheat (*Triticum aestivum* L.) populations also identified potential novel loci linked to ageing (Rehman Arif et al., 2012; Zuo et al., 2018, 2020; Rehman Arif and Börner, 2020). A GWAS analysis based on seeds that were aged at 82% RH and 44°C from 166 wheat accessions using the 90K SNP array identified 23 loci associated with seed longevity (Zuo et al., 2020). One of the candidate genes was a homologue of *DELTA1-PYRROLINE-5-CARBOXYLATE SYNTHASE 1*, a rate-limiting enzyme involved in proline biosynthesis and in drought stress. A comparison of the candidate genes with another QTL study performed on seeds from a 248 RIL population of Chinese wheat that were stored at 82% RH, 44°C (Zuo et al., 2018) revealed 2 common candidate genes involved in biotic stress response, including a *STEM RUST RESISTANCE PROTEIN RPG1* and *NBS-LRR RESISTANCE-LIKE PROTEIN*. Resistance genes-associated longevity was also found in the GWAS analysis (Nagel et al., 2015). These findings reinforce the idea that seed longevity evolved by co-opting the existing genetic pathways regulating the activation of defence against pathogens (Righetti et al., 2015). Recently, a putative link between defence and longevity was also found in rice (Yuan et al., 2021).

GWAS performed on rice has been very successful in identifying and validating candidate genes involved in longevity (Lee et al., 2019, 2020; Yuan et al., 2021). Using P50 values obtained during storage either at 25°C/65% RH or 43°C/88% RH, a GWAS on a panel including indica and temperate and tropical japonica accessions led to the identification of several candidate genes, including *T6PP* as mentioned above and an *HSP20* (Os02g03570) (Yuan et al., 2021). These authors also identified *GRETCHEN HAGEN3-2* (*GH3-2*), a gene encoding indole-3-acetic acid-amido synthetase, a conjugating enzyme that removes free auxins. Functional validation showed that GH3-2 is a negative regulator of seed longevity, acting on ABA synthesis and signalling pathways (Yuan et al., 2021). This is consistent with the demonstration that in Arabidopsis, auxin activity enhances longevity via the regulation of *ABI3* (Pellizzaro et al., 2020). Using seeds stored at 45°C, 60% RH (10% moisture content), another GWAS on a panel of 299 indica accessions identified 8 major loci using the harvest moisture content as a covariate (Lee et al., 2019). Interestingly, not only P50 but also the Ki (initial seed quality before storage) and speed of ageing (σ, the standard deviation of the normal distribution of seed deaths over time) were phenotyped. Association mapping identified loci of these three longevity parameters in different genome locations, but functions of the candidate genes surrounding the SNPs suggested common mechanisms of longevity (Lee et al., 2019). They were related to DNA-dependent transcription and repair, sugar metabolism, and ROS scavenging, as well as auxin. Haplotype mapping identified a set of loci that is linked to improvement in P50 (between 53% and 186%) and is potentially interesting to use in breeding programmes (Lee et al., 2019). In a follow-up study exploring the diversity of longevity in 185 Aus (semi-wild rice) varieties (Lee et al., 2020) tocol homologues were quantified including tocopherols, which are member of the vitamin E family of lipophilic antioxidants involved in seed longevity (Mène-Saffrané et al., 2010). They found that Ki was positively correlated with γ- and δ-tocopherols and negatively correlated with α-tocotrienol, while P50 was positively correlated with δ-tocopherol (Lee et al., 2020). From this study, it was suggested that a specific ratio of tocol homologues rather than total tocol content plays an important role in the seed longevity mechanisms. The role of these molecules is likely to be complex as in rice seeds, high levels of tocopherols are only found in the embryo, whereas tocotrienols are exclusively present in the pericarp and endosperm (Chen et al., 2016). The latter do not reduce rice seed deterioration during storage at 100% RH (Chen et al., 2016).

In Arabidopsis, a GWAS of longevity was performed on 270 ecotypes using natural ageing and different deleterious storage conditions to accelerate deterioration (Renard et al., 2020). This study identified multiple genomic regions associated with variation in seed longevity, demonstrating the polygenic nature of seed longevity and its dependence on storage conditions. Functional analysis of 20 candidate genes in the Columbia ecotype identified seven genes with positive effects on seed longevity (*PSAD1, SSLEA, SSTPR, DHAR1, CYP86A8, MYB47, and SPCH*) and five genes with negative effects (*RBOHD*, *RBOHE*, *RBOHF*, *KNAT7*, and *SEP3*), indicating the existence of activators and repressors of longevity. Out of these 12 genes, 7 (*CYP86A8*, *MYB47*, *RBOHD*, *RBOHE*, *RBOHF*, *KNAT7*, and *SEP3*) were associated with the seed coat and/or endosperm, highlighting the importance of the testa and maternal tissues on seed longevity. Among them, *RBOHE* and *RBOHF*, which are NAPDH oxidases generating extracellular superoxide upon imbibition, are highly expressed in the seed coat, whereas *RBOHD* is expressed in the endosperm. Seeds of *rbohd*, *rbohe*, and *rbohf* mutants exhibited increased viability after 24 months of ambient storage, whereas the digenic mutant *rbohd,f* produced seeds with an even higher lifespan (Renard et al., 2020). The remaining genes, *CYP86A8*, *MYB47*, *KNAT7*, and *SEP3*, were associated with the regulation of seed coat composition and permeability during seed development. The presence of dehydroascorbate reductase (*DHAR1*) reinforces the importance of oxidative stress on seed ageing during storage. While *dhar1* seeds exhibited reduced germination after 1 year of ambient storage, *dhar2* and *dhar3* mutants were not affected in seed longevity (Renard et al., 2020). This suggests that ascorbate may play a very specific function against seed deterioration.

Another way to decipher the polygenic nature of seed longevity is the creation of mutant populations that can be screened for altered lifespan after storage. Using an activation-tagging mutant collection of Arabidopsis, a role for gibberellins in longevity was established. First, the transcription factor ARABIDOPSIS THALIANA HOMEOBOX25 (ATHB25) was identified, which was shown to increase gibberellin levels and mucilage formation at the seed surface, and it was proposed that GA may act by reinforcing the seed coat (Bueso et al., 2014). Next, using the same activation-tagging mutant collection, two homologous DOF transcription factors, COG1/DOF1.5 and CDF4/DOF2.3, were identified (Bueso et al., 2016). When overexpressed, these genes increased the seed tolerance against deterioration in hot and humid conditions or ambient storage conditions. Reciprocal crosses showed that the role of COG1-2D in seed ageing was from the maternal origin and attributed to a thicker suberin layer and reduced permeability of the seed coat. As for *ATHB25*, the overexpression of *COG1* increased the levels of both gibberellin GA1 and ABA in siliques (Bueso et al., 2016).

Another recently discovered source of genetic variability of seed longevity is the variation in the mitochondrial and chloroplast genomes (Boussardon et al., 2019). Such variation is termed 'natural cytoplasmic variation' and has been studied in Arabidopsis by creating cytolines that combine the nuclear genome of a natural variant with the cytoplasmic genomes of a different variant (Roux et al., 2016). Seeds of the genotype $[\text{Sha}]^{\text{Cyt}}\text{Ct-1}^{\text{Nuc}}$ and $[\text{Blh-1}]^{\text{Cyt}}\text{Sha}^{\text{Nuc}}$ exhibited higher resistance to deterioration at 75% RH, 35°C compared to their respective parents (Boussardon et al., 2019). Additional vigour traits were also affected such as germination and stress tolerance. Surprisingly, the cytonuclear combination was associated with more efficient endogenous ABA degradation and lower exogenous ABA sensitivity. This is counter-intuitive because ABA signalling is necessary to induce seed longevity (see, for example, Zinsmeister et al., 2016, 2020b).

The use of natural and induced genetic diversity will remain a useful tool in the next years to increase our understanding of the molecular actors in seed longevity. Recent results obtained by GWAS indicate that this method is very promising, especially with the advance in genome-wide sequencing and the availability of mutant databases to phenotype potential candidate genes. In addition, with the generation of the big datasets, comparative analyses between species to explore the existing variability under orthologous regions have now become possible.

4 Environmental effects on longevity: plasticity and adaptation

Besides the genetic differences that make lifespan shorter or longer, the growth conditions during seed development also strongly contribute to the extent of lifespan. This is referred to as phenotypic plasticity, which describes the phenotypic variation of a trait when a genotype is exposed to different environments. Environmental cues include temperature, drought, nutrition, and light. They can be both perceived by the maternal and zygotic tissues, as also demonstrated for dormancy (Chen et al., 2014). In addition to environmental effects when seeds are attached to the mother plant, longevity can also be modulated by processing treatments after harvesting before the seeds are totally dry, such as the temperature or rate of drying (Probert et al., 2007; Whitehouse et al., 2015, 2017, 2018; Hay et al., 2018). Harvest protocols can influence seed longevity in species for which flowers appear sequentially during plant development. As a result, there is a maturation gradient among seeds from the bottom (mature) to the top (immature) plants or fruits, according to the architecture of the plant or infructescence. In this case, the way seeds are harvested can introduce bias in the interpretation of the origin of variation in longevity as illustrated for Arabidopsis in Fig. 3. In this experiment, two harvesting protocols were tested that are common in the Arabidopsis

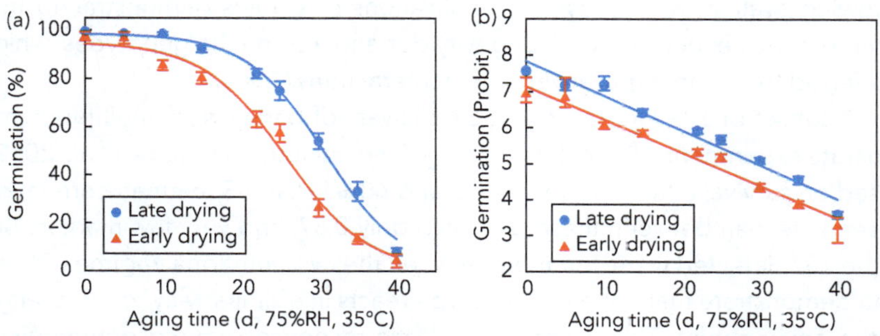

Figure 3 Loss of germination during storage of Arabidopsis seeds harvested following two protocols: early and late drying. Early drying corresponds to seeds harvested from plants for which watering was stopped when the bottom half of the siliques on the plant were brown and contained dried seeds and the top half was yellowish. Plants were left drying 2 weeks prior to harvest. Late drying corresponds to harvest when all the siliques were brown. After harvest, seeds were equilibrated to 43% RH for 3 days and then incubated at 75% RH, 35°C over a saturated NaCl solution. To test germination, seeds were imbibed on filter paper at 20°C, 16 h light for 7 days. Data represent triplicates of 100–150 seeds harvested from 4 plants each.

community, one for which watering was stopped when the bottom half of the siliques on the plant were brown and dried and the rest was yellowish, and the seeds were harvested 2 weeks later (early drying), and a second protocol where plants were irrigated until all the siliques were brown and the seed lot was fully mature (late drying). Late drying increased P50 values by 25% (from 23 to 30 days, Fig. 3A) compared to early drying. When germination data are plotted on a probit scale (Fig. 3B), the parallel lines indicate that the early drying treatment affected Ki (i.e. the longevity potential which corresponds to the intercept with the *Y*-axis) but not the speed of ageing (σ, given by the slope of the regression line). In rice, high temperature drying after harvest was found to increase longevity, but this was also dependent on the maturity level (Whitehouse et al., 2015). The benefits of high temperature drying gradually decreased as the seed water content at harvest decreased from 28% to 17%, below which no significant effect was observed (Whitehouse et al., 2017). While the above studies show that the late stages of seed maturation improve longevity, there are also species such as soybean and *M. truncatula* where longevity decreases when seeds are kept too long in their infructescences after abscission (reviewed in (Probert et al., 2007). Another post-harvest factor influencing longevity is the successive transfer of stored seeds to lower temperatures during long-term storage (Walters et al., 2004).

There is accumulating evidence that longevity is a plastic trait that is influenced by a range of environmental conditions of the mother plant. Comprehensive reviews have been recently published that document the effect of the environment on longevity, and we refer the reader to these articles for detailed information (Zhou et al., 2020; Zinsmeister et al., 2020b). In the following section, we discuss the implications of studies demonstrating that environmental effects on seed longevity depend also on the genotypes, which is referred to as genotype–environment interactions (G × E).

A subset of 122 landraces and 53 cultivars of barley were multiplied in 2 separate experimental field plots, 1 being richer in nutrients (Nagel et al., 2015). Based on survival after storage at 45°C and 60% RH, the Spearman correlation of seed lifespan between these plots was only 0.57, and very few markers out of the 107 detected were common between the two nutritional regimes. These data demonstrate that not each genotype reacts in a similar way to the change in the environment. This variation in plastic responses among genotypes is described as G × E. Likewise, in lettuce, lifespans of seeds from a RIL population produced in different years and stored either at 30% RH, 37°C or 75% RH, 50°C did not exhibit significant correlation between years of production ($r = 0.096$), and no similar QTLs were identified (Schwember and Bradford, 2010). In *M. truncatula*, a study on 159 seed lots from a RIL population (LR4) that were produced in 2 different locations in France (Angers and Dijon), where soil substrate, nutrition, as well as climate conditions, were different, showed that the correlation of seed viability between both locations after ageing at 75% RH and 35°C was very weak (*r*=0.213, Fig. 4). When longevity values were compared for accessions that were grown in the same greenhouse in Angers but in 2 different years, the Pearson's correlation between both years was 0.523 (Fig. 4, Buitink and Vandecasteele, unpublished data, Zinsmeister et al., 2016)). Since sowing, soil substrate, nutrition, watering, and harvest protocols were identical for both years, the only environmental conditions explaining these variations are light intensity, temperature, and RH during seed production.

That light and temperature during seed development affect longevity differently according to the genotype was also observed in Arabidopsis. Increasing temperature from 15°C to 25°C had a positive effect on the survival of both Col-0 and Ler ecotypes during ageing at 75% RH and 40°C, with Col-0

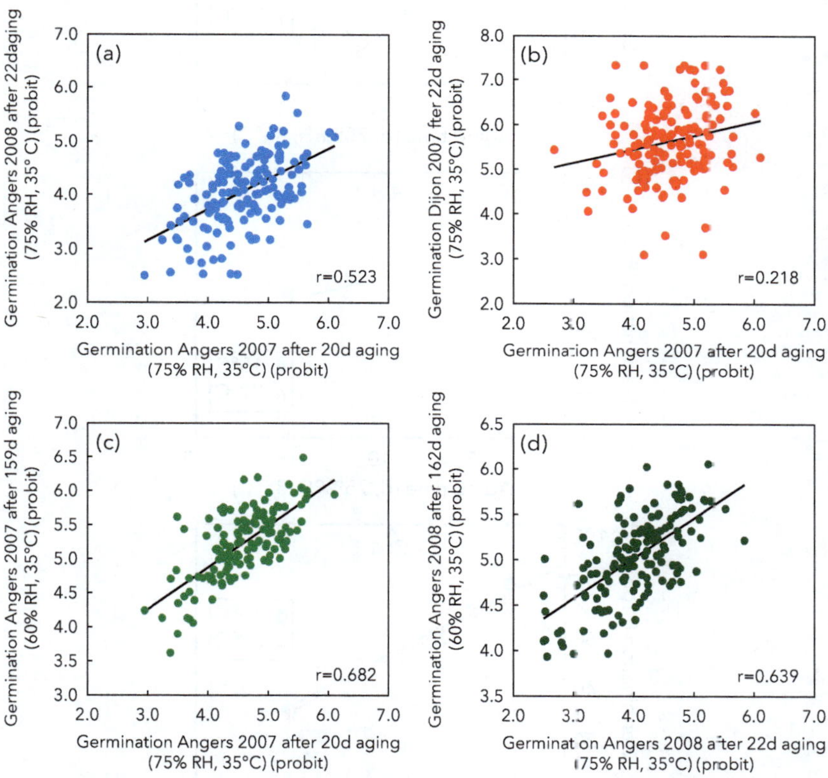

Figure 4 The relationship between the longevity of *Medicago truncatula* seeds obtained from greenhouse experiments in 2007 and 2008 in Angers and Dijon, France. Panels (a) and (b) compare the location of production, whereas panels (c) and (d) compare the storage conditions for seed lots obtained in Angers. Longevity was assessed using germination expressed as probit after the indicated time point during storage. Each data point represents 4 repetitions of 50 seeds (Zinsmeister et al., 2016). The indicated *r* value represents the Pearson's coefficient correlation.

exhibiting a higher lifespan than Ler (He et al., 2014). However, the light regime only affected Col-0 seeds with longer light duration having a positive effect on longevity (He et al., 2014). Whereas seed lifespan increased in a similar way for both Ler and Col-0 when the temperature was increased, other ecotypes showed a different response (Buitink, Penfield, unpublished data, Fig. 5). Plants of Gr-1, Rld-2, and Ct-1 genotypes were grown at 15°C and 22°C, and longevity

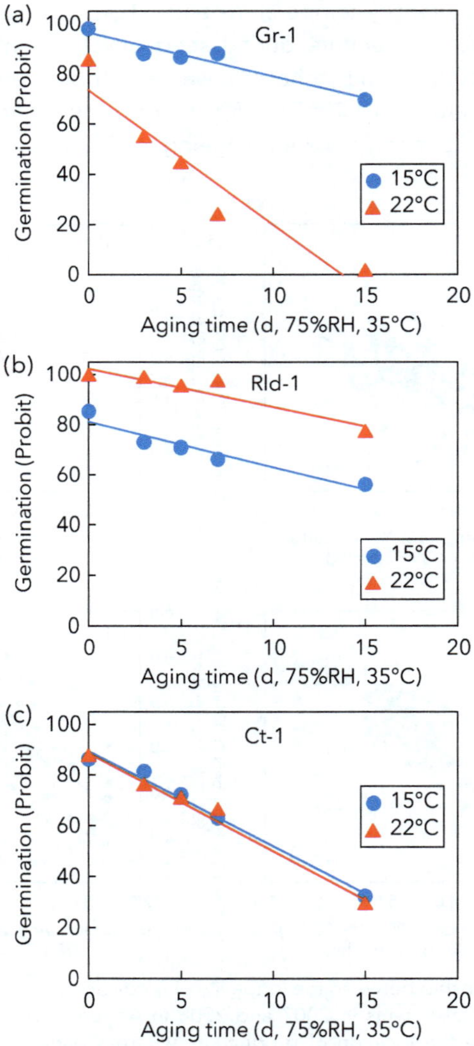

Figure 5 Loss of viability during storage of three Arabidopsis ecotypes grown at 15°C (dark red triangles) or 22°C (green circles). (a) Gr-1; (b) Rld-1; (c) Ct-1. Seeds were stratified for 7 days at 4°C. Each data point represents 100–150 seeds. Growth conditions were as described in (Springthorpe and Penfield, 2015).

was determined on the different seed lots by accelerated ageing at 75% RH, 35°C. In Gr-1, longevity was much higher when plants were grown at 15°C compared to 22°C, whereas the opposite was found for Rld-1 (Fig. 5a,b). The longevity of seeds from Ct-1 was not affected by the growth temperature. The representation as on a probit scale shows that the temperature affects the Ki of Rld-1, whereas it affects both the Ki and σ for Gr-1. Altogether, this shows that there is a large genetic variability in the environmental responses to the same ageing conditions.

These data demonstrate that the plasticity of seed longevity is under genetic control. Yet, little is known about the nature of the underlying genes. According to current models for the control of plasticity, the variation in longevity could be due to (1) heterozygosity at genes; (2) differentially environmentally sensitive alleles of genes, or (3) regulatory genes. A powerful way of identifying genes underlying the plastic response is the use of genetic diversity. As shown above, not all genotypes respond in a comparable way, and the determination of plasticity indexes that quantify the effect of the environment on seed longevity can be used for QTL and GWAS analyses. Information on the genetic basis of plasticity and methods to assess it has been recently reviewed in Laitinen and Nikoloski (2019). Nonetheless, applying GWAS to understand the genetic architecture of plasticity of longevity will be confronted to a logistic challenge. These studies require a large number of accessions, biological replicates, and a number of germination tests during storage to establish P50 values. In addition, these plants will need to be grown in different environmental conditions (e.g. a range of nutrient levels or light intensities) to adequately characterize the relationship between longevity and the said environment. It will also be necessary to be careful about the confounding effects of harvest protocols and differences in maturity levels at harvest.

The large environmental effects on seed longevity can be construed as an adaptive mechanism to pedoclimatic conditions. This comes mainly from studies on wild species showing that environmental conditions can influence seed longevity even before meiosis and seed set and that the responses to the environment of both maternal and paternal plants can be passed on to the next generation to influence seed longevity (Kochanek et al., 2011; Mondoni et al., 2014). For example, in populations of *Silene vulgaris* from alpine and lowland locations, parents that were grown in warm and wet conditions before flowering and then moved to cold and dry conditions produced seeds with a two-fold decrease in longevity compared to seeds obtained from parents grown in cold and dry conditions prior to flowering and warm and wet conditions afterwards (Mondoni et al., 2014). Likewise, in *Wahlenbergia tumidifructa* seeds, a semi-arid species from Australia, lifespan was similar for seeds produced by plants that grew in warm-wet, warm-dry, and cool-dry conditions before returning to optimal conditions after fecundation but doubled for plants grown in cool-wet conditions (Kochanek et al., 2011). That seed longevity can be adjusted according to the parental growing conditions was interpreted as an adaptive mechanism for the survival and persistence of the species with the consequence that it would exert a cascading selective pressure on future generations (Mondoni et al., 2014). Consistent with this, a longevity survey of 195 wild species stored at 60% RH/45°C showed that seeds originating from species collected in hot, dry environments tended to have a longer lifespan than those found in cool, wet conditions, regardless of the phylogenetic relatedness, suggesting that species have adapted to dry climates by increasing seed longevity (Probert et al., 2009; Mondoni et al., 2011; Zani and Müller, 2017). Thus, it appears that longevity might be considered as a long-term adaptation to the dry, arid environment. Interestingly, in *Brassica rapa*, artificial ageing under very deleterious storage conditions (40°C/90% RH) significantly increased the flowering time of the offspring (Franks et al., 2019). This was not attributed to a cascade of events that would delay the development of plants originating from aged seeds, because the increase in flowering time was also found in the second generation that was produced from seeds that did not undergo ageing like the first generation (Franks et al., 2019). This suggests that the selection of genotypes through different ageing rates among a population could impact the evolution of adult plants. The negative correlation between dormancy and longevity is also taken as evidence that selective pressure due to climate conditions allowed species to evolve different mechanisms of seed survival after abscission to ensure the dissemination of the offspring. Whether this is also valid for crops remains to be assessed. For barley, genotypic variation in seed longevity was independent of the geographical origin of the accession or its breeding status (Nagel et al., 2015). How the mother plants pass on the information of their growing environment to their offspring remains to be assessed. Identifying and characterizing the causal genes implicated in long-term adaptation and/or plasticity of longevity and whether transgenerational effects are at play represent an exciting new area of research. Likewise, there are exciting prospects in investigating whether the plasticity of longevity should be considered as adaptive.

5 Ageing protocols and the limits of extrapolating accelerated ageing conditions

Seeds can survive from several years to decades before showing symptoms of deterioration when they are stored *ex situ* in ambient conditions such as on the shelf in the laboratory (50–70% RH, 22°C) or in gene banks (<50% RH, at 5°C or at −18°C). To reduce the lifespan to experimentally tractable time scales (days to weeks) for research purposes, seeds are stored at higher temperature and/or water content (practically >40°C and RH > 75%). Claims have been made that these methods of ageing at high moisture/high temperature mimic deterioration in laboratory conditions or below 0°C. However, a survey of the data in the literature shows that longevity values are only correlated between different storage conditions that are close to each other. For example, a comparison of seed viability of 270 Arabidopsis accessions stored at 75%RH/38°C with storage at 40-60%RH/20–25°C (ambient conditions) shows a highly significant correlation ($r = 0.678$, (Renard et al., 2020). Likewise, for *M. truncatula*, 159 seed lots from the LR4 RIL population that were stored either at 60% RH/35°C or 75% RH/35°C showed a strong correlation, although there was a slight dependency on the year of harvest ($r = 0.682$ and $r = 0.639$ (Fig. 4, Buitink and Vandecasteele, unpublished data; (Zinsmeister et al., 2016). Three out of the five QTLs that were detected for seed viability at 60% RH storage were also found when seed viability was determined at 75% RH (Zinsmeister et al., 2016). Using a storage experiment involving 2 laboratories and 56 seed lots of rye, wheat, and triticale that were incubated at 35°C and a range of water contents, the correlation between P50 values at 75% RH (water content, 14% DWB) with those obtained after storage at lower water contents (RH between 14% and 65%) remained high until 0.08 g H_2O/g DW (Fig. 6) (Niedzielski et al , 2009). A comparison of the P50 values obtained during storage at 75% RH with P50 values obtained from seeds stored at lower RH, leading to seed water contents below 8%, showed a much lower correlation (Fig. 6). In lettuce, lifespans of seeds from a RIL population stored either at 30% RH/9°C (MC = 7.1%) or 75% RH/50°C (9.1%) did not exhibit any significant correlation, and no similar QTLs were identified (Schwember and Bradford, 2010). Likewise, when the viability of seeds of the 270 Arabidopsis accessions stored at 75% RH/38°C was compared with that of seeds stored at 100% RH/39°C (imbibition in water for 2 days), no significant correlation existed between both datasets (Renard et al., 2020). This indicates that longevity values do not correlate when they are obtained from storage conditions far apart from each other.

That high temperature and humidity cannot mimic ambient storage has also been verified experimentally (Roach et al., 2018; Zinsmeister et al., 2020a). For example, the transcription factor *HaHSFA9* was suggested to play a role in seed longevity, based on the observation that its repression leads to a reduced lifespan of seeds stored at 40°C, 100% RH (Tejedor-Cano et al., 2010). We confirmed these observations using seeds of Arabidopsis and *M. truncatula hsfa9* mutants ((Zinsmeister et al., 2020a). However, when seeds of these mutants were stored at less deleterious conditions, such as 75% RH and 35°C or lower RH, longevity was identical to that of wild type seeds. In barley, the levels of tocochromanols, a lipid-soluble antioxidant from the vitamin E family, decreased during storage at 20°C/ 51% RH over a 15-year period that led to a reduction in germination by 50% (Roach et al., 2018). However, after storage at 45°C/75% RH resulting in the complete loss of viability, no difference in tocochromanols levels was detected,

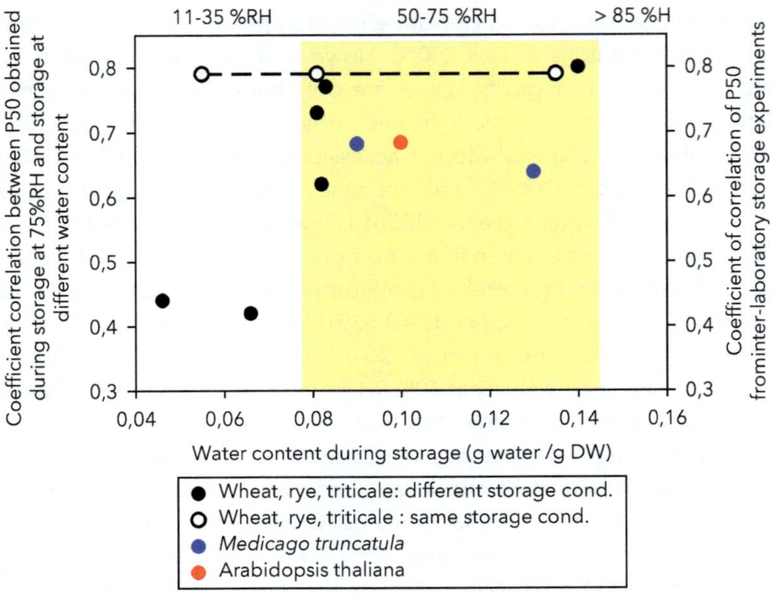

Figure 6 The effect of water content during storage at 35°C on the correlation between P50 values obtained at 75%RH/35°C and P50 obtained at these water contents (closed symbols). For wheat, rye, and triticale, data were obtained from two different laboratories using the same seed lots. The open symbols represent the correlations between the P50 obtained from each laboratory using the same storage conditions. The section in yellow represents the range of RH where extrapolations to lower water contents are valid. Data are from (Niedzielski et al., 2009). Data for *M. truncatula* and Arabidopsis are from Zinsmeister et al. (2016) and Renard et al. (2020), respectively.

suggesting that different mechanisms of deterioration occur. Altogether, these data indicate that great care should be taken in extrapolating storage conditions at higher RH to those occurring during storage in gene banks. We recommend that comparisons should be made within three ranges of RH: 11–35, 50–75, and>85%. The latter should be avoided as a means to assess longevity but can be related to seed persistence in the soil bank.

The fundamental reason that seed ageing at high humidity and temperature cannot be extrapolated to low humidity lies in the dependence of these conditions on glass-forming properties within the seeds. This has been reviewed in Buitink and Leprince (2008) and Ballesteros et al. (2020). In short, at high humidity, there are ample molecular motions within the long and short range to allow some metabolic pathway to operate, such as fermentation and glycolysis. Yet, there is not enough water and probably too low oxygen diffusion to sustain the production of ATP and respiration (Leprince and Hoekstra, 1998). This results in an imbalanced metabolism, leading to the production of molecules such as methanol, and acetaldehydes, along with a complex array of down products of metabolism whose origin is unknown (Colville et al.,

2012; Mira et al., 2016; Han et al., 2021; Michalak et al., 2021). During storage, concentrations might raise to toxic levels over time. When small amounts of methanol or acetone are added to the atmosphere of pea seeds during storage, viability is significantly decreased after 10 days at 60% RH, 50°C (Han et al., 2021). At low water content and low temperature, the cytoplasm forms a glassy, solid-like matrix with high viscosity and severely restricted molecular motion and diffusion of the cytoplasm. Under these conditions, the profile of volatiles that is emitted from seeds during ageing is different and characterized by down products of lipid oxidation (Mira et al., 2016; Michalak et al., 2021).

Oxidation is considered the main cause of seed deterioration. In Arabidopsis, barley, lettuce, and soybean, a high O_2 pressure applied during the storage of dry seeds leads to a faster deterioration compared to ambient atmospheric pressure (Groot et al., 2012; Nagel et al., 2016; Buijs et al., 2020b; Renard et al., 2020). In an effort to accelerate seed ageing without having the disadvantages of storing seeds at high MC and temperature, a new ageing condition was designed, consisting of storing dry seeds in scuba tanks at 8–20 MPa O_2 while its concentration remains identical to air (the Elevated Partial Pressure of Oxygen (EPPO) method, (Groot et al., 2012; Buijs et al., 2020b)). In Arabidopsis, the comparison of this method with storage under laboratory conditions, both being in the same range of RH (40–60%), exhibited a Spearman correlation coefficient of 0.503 (Buijs et al., 2020b; Renard et al., 2020). Intriguingly, this coefficient was similar (0.487) for seeds stored at 85% RH at normal atmospheric pressure. The correlation between survival in atmospheric storage with EPPO at 33% RH decreases to only 0.35 (Buijs et al., 2020a). Surprisingly, lifespan under EPPO conditions was lower at 33% RH than at 55 % RH, which is opposite to what has been observed under atmospheric storage conditions. Therefore, how the EPPO treatment can be extrapolated to ambient or gene bank conditions remains to be further assessed.

Another issue to address with the EPPO method is to which extent this treatment will be applicable to all Angiosperms. Indeed, the lifespan of onion seeds does not seem to be affected by oxygen in the same way and to the same extent as lettuce seeds (Schwember and Bradford, 2011). Furthermore, the EPPO also affects the incidence of abnormal seedlings originating from the stored viable seeds (Groot et al., 2012). However, in onions, this was not observed using a high pO_2 treatment with ten-fold elevated pC_2 (0.21 MPa O_2) (Hourston et al., 2020). It is unlikely that onion seed would constitute an exception. A comparison of the effects of hypoxia on the longevity and deterioration of seeds from three species that differed in their size, phylogeny, and ratio embryo:seed also revealed differences in oxygen sensitivity when stored under atmospheric conditions (Han et al., 2021). The presence of oxygen at 60%RH/50°C accelerated the loss of viability of *Agrostemma githago* but not that of *Lolium perenne* (both being small and endospermic) and pea.

6 Conclusions and future research trends

6.1 Phenotyping seed ageing and longevity

There is no silver bullet for designing storage experiments aiming at understanding seed longevity from a holistic point of view. None of the storage methods that accelerate seed deterioration will be able to correctly extrapolate ageing conditions encountered in gene banks. Therefore, it is proposed to use ageing conditions that are appropriate to the potential downstream use of the findings (Hay et al., 2018). Ideally, such conditions need to be standardized to enable comparisons across studies and species. Historical collections available in gene banks will be an even more valuable resource to understand the long-term effect of seed storage in the cold, dry state and to link the loss of viability with structural and biochemical changes within the seeds (Fleming et al., 2017; Nagel et al., 2018).

For ambient storage conditions (50–60% RH and 15–20°C), ageing can be accelerated using slightly higher temperatures and/or RH. However, several studies show that practically, storage conditions above 75% RH and/or 40°C should be avoided. We suggest that to allow future meta-studies, an RH of 60% is used and the temperature should be adjusted based on the lifespan of the species. Finally, the most difficult test to set up is to mimic the deterioration of soil seed banks. In these conditions, seeds will imbibe and dry out slowly according to the rain regimes. They may remain hydrated for long periods as their dormant state will prevent them from germinating, allowing repair mechanisms to proceed. In the laboratory, ageing conditions will need to include cycles of high and low water contents during storage.

The ever-decreasing costs of sequencing and metabolomic techniques will enable to screen large populations representing the genetic diversity of longevity, with the aim of elucidating the underlying factors leading to seed survival. Such large-scale studies will generate complex datasets for which the advances in computational statistics, deep learning, and artificial intelligence will enable us to untangle cause–effect relationships between seed composition and lifespan, decipher the complex molecular networks, and predict seed lifespan during storage. For example, in an attempt to correlate seed composition and lifespan of oil seed rape, introducing a matrix of several input vectors such as fatty acid and thousand seed weight in the regression analysis showed better predictability of seed storage (Nagel et al., 2018). This approach also identified glucosinolates as a new factor influencing seed longevity that could not have been deduced with Pearson's correlations.

For correct and successful interpretation of the data, it will be necessary to consider the following issues: (1) seed viability needs to be characterized prior to storage. Indeed, both the level of maturity, post-harvest treatments and genotype influence the synthesis of protective compounds and thus the potential longevity (Ki); (2) to characterize the asymptomatic phase of seed ageing that does not yet lead to a decrease in germination, more sensitive parameters are needed to monitor the decrease in seed vigour, such as germination speed. This will also allow to apply population-based threshold models to quantify seed ageing (Bradford et al., 1993); (3) seeds should preferably not have been manipulated during storage, such as moving from the fridge to −18°C at some point or alternatively, this information should be adequately documented (Hay et al., 2021); (4) instead of assessing the percentages of germination at a single time point during storage, it is advised to monitor the progressive decrease in viability over time, which allows for the determination of the P50 (and σ and Ki); (5) the time of death during storage should be known. Indeed, this is not because seeds are dead that deterioration stops over time (see Zinsmeister et al., 2020b) for further explanations).

6.2 Enhancing lifespan

Enhancing the seed's lifespan should be a focus of future research. Considering the impact of oxidation, research using methods to reduce oxygen content during storage should be promoted, along with indicators of deterioration. While hermetic sealing with oxygen absorbers and detectors appears to be a promising way of extending the seed's lifespan (Han et al., 2021), research is necessary to understand the relationship between glass properties, oxygen diffusion, and oxidation. Another issue that received little attention is whether the localization or local concentrations of certain protective molecules and/or toxic molecules within the different tissues and organs are important in determining the rate of deteriorative processes during storage. Imaging technique analysis such as matrix-assisted laser desorption/ionization mass spectrometry has revealed differential localization of phospholipids and triacylglycerols within the embryo. In oilseed rape, metabolite imaging using Fourier transform infrared microscopy imaging revealed that raffinose, an oligosaccharide acting as a protective molecule against deterioration during storage at high humidity, was located in the hypocotyl and radicle only (Munz et al., 2017). Other factors that need more attention to enhance longevity are the timing of harvest and drying methods of the harvested seeds (Whitehouse et al., 2015, 2017).

6.3 Identification of the regulatory network of longevity acquisition

GWAS analysis and co-expressed gene network analysis are powerful tools to identify transcription factors that regulate the acquisition of seed longevity (Righetti et al., 2015; Renard et al., 2020; Bizouerne et al., 2021). Further research is needed to decipher the complexity of the regulatory networks leading to the acquisition of longevity, including the interaction between the regulatory pathways that have been identified so far. In seeds, post-transcriptional regulation controls the expression of thousands of genes, as well as mRNA splicing, translation, and stability. In *M. truncatula*, three splicing isoforms of MtABI3 were identified that are temporally regulated during seed development and each of them regulated a subset of the maturation programme (Lalanne et al., 2021). Epigenetic regulation includes the modification of histones and methylation of DNA, RNA splicing, chromatin remodelling, and the activity of microRNAs. The role of these processes in the regulation of dormancy is progressively unravelled (reviewed in Tognacca and Botto, 2021) and deserved to be investigated for longevity. Evidence that epigenetic regulation might be important for seed longevity has recently been suggested for *Mentha aquatica* and rye seeds (Mira et al., 2020). In both species, methylation-sensitive amplification polymorphism revealed changes in the cytosine methylation profile in seeds after storage at 35°C, 75% RH and in a higher proportion in seedlings produced from aged seeds.

7 Where to look for further information

- Recommendations about designing seed storage experiment and modelling loss of seed viability: (Hay et al., 2018, 2021).
- Role of the maternal environment in seed longevity: (Probert et al., 2007; Righetti et al., 2015; Zinsmeister et al., 2020b).
- Comparative analysis of seed lifespans : (Walters et al., 2005; Probert et al., 2009; Merritt et al., 2014; Colville and Pritchard, 2019; Solberg et al., 2020).
- Mechanisms of seed ageing and protective compounds: (Sano et al., 2016; Ballesteros et al., 2020; Zinsmeister et al., 2020b).
- Signalling pathway governing the acquisition of seed longevity: (Sano et al., 2016; Leprince et al., 2017).

8 Acknowledgements

The authors thank Ms Martine Neveu and Joseph Ly Vu for technical support on Arabidopsis seed ageing experiments. The authors thank Dr Steven Penfield, John Innes Centre, for the generous gift of the Arabidopsis accessions produced at different temperatures. This research was conducted in the framework of the regional programme 'Objectif Végétal, Research, Education and Innovation in Pays de la Loire', supported by the French Region Pays de la Loire, Angers Loire Métropole, and the European Regional Development Fund. This study was also supported by the ANR grant REGULEG (no. ANR-15-CE20-0001).

9 References

Ballesteros, D., Pritchard, H. W. and Walters, C. (2020). Dry architecture: towards the understanding of the variation of longevity in desiccation-tolerant germplasm. *Seed Sci. Res.* 30(2): 142–155.

Ballesteros, D., Walters, C. and Pritchard, H. W. (2019). Probing life and death in the glassy state of cryopreserved seeds, fern spores and pollen: fundamental research to optimise ex situ preservation. *Cryobiology* 91: 161.

Bentsink, L., Alonso-Blanco, C., Vreugdenhil, D., Tesnier, K., Groot, S. P. C. and Koornneef, M. (2000). Genetic analysis of seed-soluble oligosaccharides in relation to seed storability of Arabidopsis. *Plant Physiol.* 124(4): 1595–1604.

Berjak, P. and Pammenter, N. W. (2013). Implications of the lack of desiccation tolerance in recalcitrant seeds. *Front. Plant Sci.* 4: 478.

Bizouerne, E., Buitink, J., Vu, B. L., Vu, J. L., Esteban, E., Pasha, A., Provart, N., Verdier, J. and Leprince, O. (2021). Gene co-expression analysis of tomato seed maturation reveals tissue-specific regulatory networks and hubs associated with the acquisition of desiccation tolerance and seed vigour. *BMC Plant Biol.* 21(1): 124.

Boca, S., Koestler, F., Ksas, B., Chevalier, A., Leymarie, J., Fekete, A., Mueller, M. J. and Havaux, M. (2013). Arabidopsis lipocalins AtCHL and AtTIL have distinct but overlapping functions essential for lipid protection and seed longevity. *Plant Cell Environ.* 37(2): 368–381. doi: 10.1111/pce.12159.

Boucher, V., Buitink, J., Lin, X., Boudet, J., Hoekstra, F. A., Hundertmark, M., Renard, D. and Leprince, O. (2010). MtPM25 is an atypical hydrophobic late embryogenesis-abundant protein that dissociates cold and desiccation-aggregated proteins. *Plant Cell Environ.* 33(3): 418–430.

Boussardon, C., Martin-Magniette, M. L., Godin, B., Benamar, A., Vittrant, B., Citerne, S., Mary-Huard, T., Macherel, D., Rajjou, L. and Budar, F. (2019). Novel cytonuclear combinations modify Arabidopsis thaliana seed physiology and vigor. *Front. Plant Sci.* 10: 32.

Bradford, K. J., Tarquis, A. M. and Duran, J. M. (1993). A population-based threshold model describing the relationship between germination rates and seed deterioration. *J. Exp. Bot.* 44(7): 1225–1234.

Bueso, E., Muñoz-Bertomeu, J., Campos, F., Brunaud, V., Martínez, L., Sayas, E., Ballester, P., Yenush, L. and Serrano, R. (2014). Arabidopsis thaliana HOMEOBOX25 uncovers a role for gibberellins in seed longevity. *Plant Physiol.* 164(2): 999–1010.

Bueso, E., Muñoz-Bertomeu, J., Campos, F., Martínez, C., Tello, C., Martínez-Almonacid, I., Ballester, P., Simón-Moya, M., Brunaud, V., Yenush, L., Ferrándiz, C. and Serrano, R. (2016). Arabidopsis COGWHEEL1 links light perception and gibberellins with seed tolerance to deterioration. *Plant J.* 87(6): 583–596.

Buijs, G., Vogelzang, A., Nijveen, H. and Bentsink, L. (2020a). Dormancy cycling: translation-related transcripts are the main difference between dormant and non-dormant seeds in the field. *Plant J.* 102(2): 327–339.

Buijs, G., Willems, L. A. J., Kodde, J., Groot, S. P. C. and Bentsink, L. (2020b). Evaluating the EPPO method for seed longevity analyses in Arabidopsis. *Plant Sci.* 301: 110644.

Buitink, J. and Leprince, O. (2004). Glass formation in plant anhydrobiotes: survival in the dry state. *Cryobiology* 48(3): 215–228.

Buitink, J. and Leprince, O. (2008). Intracellular glasses and seed survival in the dry state. *C. R. Biol.* 331(10): 788–795.

Chen, D., Li, Y., Fang, T., Shi, X. and Chen, X. (2016). Specific roles of tocopherols and tocotrienols in seed longevity and germination tolerance to abiotic stress in transgenic rice. *Plant Sci.* 244: 31–39.

Chen, M., MacGregor, D. R., Dave, A., Florance, H., Moore, K., Paszkiewicz, K., Smirnoff, N., Graham, I. A. and Penfield, S. (2014). Maternal temperature history activates Flowering Locus T in fruits to control progeny dormancy according to time of year. *Proc. Natl Acad. Sci. U. S. A.* 111(52): 18787–18792.

Clerkx, E. J. M., Vries, H. B., Ruys, G. J., Groot, S. P. and Koornneef, M. (2003). Characterization of green seed, an enhancer of abi3-1 in Arabidopsis that affects seed longevity. *Plant Physiol.* 132(2): 1077–1084.

Colville, L., Bradley, E. L., Lloyd, A. S., Pritchard, H. W., Castle, L. and Kranner, I. (2012). Volatile fingerprints of seeds of four species indicate the involvement of alcoholic fermentation, lipid peroxidation, and Maillard reactions in seed deterioration during ageing and desiccation stress. *J. Exp. Bot.* 63(18): 6519–6530.

Colville, L. and Pritchard, H. W. (2019). Seed life span and food security. *New Phytol.* 224(2): 557–562.

De Giorgi, J., Piskurewicz, U., Loubery, S., Utz-Pugin, A., Bailly, C., Mène-Saffrané, L. and Lopez-Molina, L. (2015). An endosperm-associated cuticle is required for Arabidopsis seed viability, dormancy and early control of germination. *PLoS Genet.* 11(12): e1005708.

Dekkers, B. J. W., He, H., Hanson, J., Willems, L. A. J., Jamar, D. C. L., Cueff, G., Rajjou, L., Hilhorst, H. W. M. and Bentsink, L. (2016). The arabidopsis delay of germination 1 gene affects abscisic acid insensitive 5 (ABI5) expression and genetically interacts with ABI3 during Arabidopsis seed development. *Plant J.* 85(4): 451–465.

Dirk, L. M. A. and Downie, A. B. (2018). An examination of Job's rule: protection and repair of the proteins of the translational apparatus in seeds. *Seed Sci. Res.* 28(3): 168–181. doi: 10.1017/S0960258518000284.

Fernández-Marín, B., Kranner, I., Sebastián, M. S., Artetxe, U., Laza, J. M., Vilas, J. L., Pritchard, H. W., Nadajaran, J., Míguez, F., Becerril, J. M. and García-Plazaola, J. I. (2013). Evidence for the absence of enzymatic reactions in the glassy state: a case study of xanthophyll cycle pigments in the desiccation-tolerant moss Syntrichia ruralis. *J. Exp. Bot.* 64(10): 3033–3043.

Fleming, M. B., Richards, C. M. and Walters, C. (2017). Decline in RNA integrity of dry-stored soybean seeds correlates with loss of germination potential. *J. Exp. Bot.* 68(9): 2219–2230.

Franks, S. J., Sekor, M. R., Davey, S. and Weis, A. E. (2019). Artificial seed aging reveals the invisible fraction: implications for evolution experiments using the resurrection approach. *Evol. Ecol.* 336(33): 811–824.

Furuki, T., Takahashi, Y., Hatanaka, R., Kikawada, T., Furuta, T. and Sakurai, M. (2020). Group 3 LEA protein model peptides suppress heat-induced lysozyme aggregation: elucidation of the underlying mechanism using coarse-grained molecular simulations. *J. Phys. Chem. B* 124(14): 2747–2759.

Gerna, D., Ballesteros, D., Stöggl, W., Arc, E., Seal, C. E., Na, C. S., Kranner, I. and Roach, T. (2020). Cytoplasmic physical state governs the influence of oxygen on Pinus densiflora seed ageing. *bioRxiv*: 2020.12.11.421446.

Groot, S. P. C., Surki, A. A., de Vos, R. C. H. and Kodde, J. (2012). Seed storage at elevated partial pressure of oxygen, a fast method for analysing seed ageing under dry conditions. *Ann. Bot.* 110(6): 1149–1159.

Han, B., Fernandez, V., Pritchard, H. W. and Colville, L. (2021). Gaseous environment modulates volatile emission and viability loss during seed artificial ageing. *Planta* 253(253): 1–16.

Haslbeck, M. and Vierling, E. (2015). A first line of stress defense: small heat shock proteins and their function in protein homeostasis. *J. Mol. Biol.* 427(7): 1537–1548.

Hay, F. R., Valdez, R., Lee, J. S. and Sta Cruz, P. C. (2018). Seed longevity phenotyping: recommendations on research methodology. *J. Exp. Bot.* 70(2): 425–434.

Hay, F. R., Whitehouse, K. J., Ellis, R. H., Sackville Hamilton, N. R., Lusty, C., Ndjiondjop, M. N., Tia, D., Wenzl, P., Santos, L. G., Yazbek, M., Azevedo, V. C. R., Peerzada, O. H., Abberton, M., Oyatomi, O., de Guzman, F., Capilit, G., Muchugi, A. and Kinyanjui, Z. (2021). CGIAR GeneBank viability data reveal inconsistencies in seed collection management. *Glob. Food Sec.* 30: 100557.

He, H., De Souza Vidigal, D., Basten Snoek, L. B., Schnabel, S., Nijveen, H., Hilhorst, H. and Bentsink, L. (2014). Interaction between parental environment and genotype affects plant and seed performance in Arabidopsis. *J. Exp. Bot.* 65(22): 6603–6615.

Hincha, D. K., Zuther, E. and Popova, A. V. (2021). Stabilization of dry sucrose glasses by four lea_4 proteins from Arabidopsis thaliana. *Biomolecules* 11(5). doi: 10.3390/biom11050615.

Hourston, J. E., Pérez, M., Gawthrop, F., Richards, M., Steinbrecher, T. and Leubner-Metzger, G. (2020). The effects of high oxygen partial pressure on vegetable Allium seeds with a short shelf-life. *Planta* 251(6): 105. doi: 10.1007/s00425-020-03398-y.

Hundertmark, M., Buitink, J., Leprince, O. and Hincha, D. K. (2011). The reduction of seed-specific dehydrins reduces seed longevity in Arabidopsis thaliana. *Seed Sci. Res.* 21(3): 165–173.

Kaur, H., Petla, B. P., Kamble, N. U., Singh, A., Rao, V., Salvi, P., Ghosh, S. and Majee, M. (2015). Differentially expressed seed aging responsive heat shock protein OsHSP18.2 implicates in seed vigor, longevity and improves germination and seedling establishment under abiotic stress. *Front. Plant Sci.* 6: 713.

Kochanek, J., Steadman, K. J., Probert, R. J. and Adkins, S. W. (2011). Parental effects modulate seed longevity: exploring parental and offspring phenotypes to elucidate pre-zygotic environmental influences. *New Phytol.* 191(1): 223–233.

Kranner, I., Birtić, S., Anderson, K. M. and Pritchard, H. W. (2006). Glutathione half-cell reduction potential: a universal stress marker and modulator of programmed cell death? *Free Radic. Biol. Med.* 40(12): 2155–2165.

Laitinen, R. A. E. and Nikoloski, Z. (2019). Genetic basis of plasticity in plants. *J. Exp. Bot.* 70(3): 739–745.

Lalanne, D., Malabarba, J., Ly Vu, J., Hundertmark, M., Delahaie, J., Leprince, O., Buitink, J. and Verdier, J. (2021). Medicago ABI3 splicing isoforms regulate the expression of different gene clusters to orchestrate seed maturation. *Plants (Basel)* 10(8): 1710.

Lee, J. S., Kwak, J. and Hay, F. R. (2020). Genetic markers associated with seed longevity and vitamin E in diverse Aus rice varieties. *Seed Sci. Res.* 30(2): 133–141. doi: 10.1017/S0960258520000173.

Lee, J. S., Velasco-Punzalan, M., Pacleb, M., Valdez, R., Kretzschmar, T., McNally, K. L., Ismail, A. M., Cruz, P. C. S., Sackville Hamilton, N. R. and Hay, F. R. (2019). Variation in seed longevity among diverse Indica rice varieties. *Ann. Bot.* 124(3): 447–460.

Leprince, O. and Hoekstra, F. A. (1998). The responses of cytochrome redox state and energy metabolism to dehydration support a role for cytoplasmic viscosity in desiccation tolerance. *Plant Physiol.* 118(4): 1253–1264.

Leprince, O., Pellizzaro, A., Berriri, S. and Buitink, J. (2017). Late seed maturation: drying without dying. *J. Exp. Bot.* 68(4): 827–841.

Li, D. Z. and Pritchard, H. W. (2009). The science and economics of ex situ plant conservation. *Trends Plant Sci.* 14(11): 614–621.

Long, R. L., Gorecki, M. J., Renton, M., Scott, J. K., Colville, L., Goggin, D. E., Commander, L. E., Westcott, D. A., Cherry, H. and Finch-Savage, W. E. (2015). The ecophysiology of seed persistence: a mechanistic view of the journey to germination or demise. *Biol. Rev. Camb. Philos. Soc.* 90(1): 31–59.

Mène-Saffrané, L., Jones, A. D. and DellaPenna, D. (2010). Plastochromanol-8 and tocopherols are essential lipid-soluble antioxidants during seed desiccation and quiescence in Arabidopsis. *Proc. Natl Acad. Sci. U. S. A.* 107(41): 17815–17820.

Merritt, D. J., Martyn, A. J., Ainsley, P., Young, R. E., Seed, L. U., Thorpe, M., Hay, F. R., Commander, L. E., Shackelford, N., Offord, C. A., Dixon, K. W. and Probert, R. J. (2014). A continental-scale study of seed lifespan in experimental storage examining seed, plant, and environmental traits associated with longevity. *Biodivers. Conserv.* 23(5): 1081–1104.

Michalak, M., Plitta-Michalak, B. P., Nadarajan, J. and Colville, L. (2021). Volatile signature indicates viability of dormant orthodox seeds. *Physiol. Plant.* 173(3): 788–804.

Mira, S., Hill, L. M., González-Benito, M. E., Ibáñez, M. A. and Walters, C. (2016). Volatile emission in dry seeds as a way to probe chemical reactions during initial asymptomatic deterioration. *J. Exp. Bot.* 67(6): 1783–1793.

Mira, S., Nadarajan, J., Liu, U., González-Benito, M. E. and Pritchard, H. W. (2019). Lipid thermal fingerprints of long-term stored seeds of Brassicaceae. *Plants (Basel)* 8(10). doi: 10.3390/plants8100414.

Mira, S., Pirredda, M., Martín-Sánchez, M., Marchessi, J. E. and Martín, C. (2020). DNA methylation and integrity in aged seeds and regenerated plants. *Seed Sci. Res.* 30(2): 92–100.

Miura, K., Lin, Y., Yano, M. and Nagamine, T. (2002). Mapping quantitative trait loci controlling seed longevity in rice (Oryza sativa L.). *Theor. Appl. Genet.* 104(6-7): 981–986.

Mondoni, A., Orsenigo, S., Donà, M., Balestrazzi, A., Probert, R. J., Hay, F. R., Petraglia, A. and Abeli, T. (2014). Environmentally induced transgenerational changes in seed longevity: maternal and genetic influence. *Ann. Bot.* 113(7): 1257–1263.

Mondoni, A., Probert, R. J., Rossi, G., Vegini, E. and Hay, F. R. (2011). Seeds of alpine plants are short lived: implications for long-term conservation. *Ann. Bot.* 107(1): 171–179.

Munz, E., Rolletschek, H., Oeltze-Jafra, S., Fuchs, J., Guendel, A., Neuberger, T., Ortleb, S., Jakob, P. M. and Borisjuk, L. (2017). A functional imaging study of germinating oilseed rape seed. *New Phytol.* 216(4): 1181–1190.

Nagel, M., Holstein, K., Willner, E. and Börner, A. (2018). Machine learning links seed composition, glucosinolates and viability of oilseed rape after 31 years of long-term storage. *Seed Sci. Res.* 28(4): 340–348.

Nagel, M., Kodde, J., Pistrick, S., Mascher, M., Börner, A. and Groot, S. P. C. (2016). Barley seed aging: genetics behind the dry elevated pressure of oxygen aging and moist controlled deterioration. *Front. Plant Sci.* 7: 388.

Nagel, M., Kranner, I., Neumann, K., Rolletschek, H., Seal, C. E., Colville, L., Fernández-Marín, B. and Börner, A. (2015). Genome-wide association mapping and biochemical markers reveal that seed ageing and longevity are intricately affected by genetic background and developmental and environmental conditions in barley. *Plant Cell Environ.* 38(6): 1011–1022.

Narayana, N. K., Vemanna, R. S., Chandrashekar, B. K., Rao, H., Vennapusa, A. R., Narasimaha, A., Makarla, U., Basavaiah, M. R. (2017). Aldo-ketoreductase 1 (AKR1) improves seed longevity in tobacco and rice by detoxifying reactive cytotoxic compounds generated during ageing. *Rice (N Y)* 10(1): 11. doi: 10 1186/s12284-017-0148-3.

Nguyen, T. P., Keizer, P., van Eeuwijk, F., Smeekens, S. and Bentsink, L. (2012). Natural variation for seed longevity and seed dormancy are negatively correlated in Arabidopsis. *Plant Physiol.* 160(4): 2083–2092.

Niedzielski, M., Walters, C., Luczak, W., Hill, L. M., Wheeler, L. J. and Puchalski, J. (2009). Assessment of variation in seed longevity within rye, wheat and the intergeneric hybrid triticale. *Seed Sci. Res.* 19(4): 213–224.

Oenel, A., Fekete, A., Krischke, M., Faul, S. C., Gresser, G., Havaux, M., Mueller, M. J. and Berger, S. (2017). Enzymatic and non-enzymatic mechanisms contribute to lipid oxidation during seed aging. *Plant Cell Physiol.* 58(5): 925–933.

Pellizzaro, A., Neveu, M., Lalanne, D., Vu, B. L., Kanno, Y., Seo, M., Leprince, O. and Buitink, J. (2020). A role for auxin signaling in the acquisition of longevity during seed maturation. *New Phytol.* 225(1): 284–296.

Pence, V. C., Ballesteros, D., Walters, C., Reed, B. M., Philpott, M., Dixon, K. W., Pritchard, H. W., Culley, T. M. and Vanhove, A. C. (2020). Cryobiotechnologies: tools for expanding long-term ex situ conservation to all plant species. *Biol. Conserv.* 250: 108736.

Personat, J. M., Tejedor-Cano, J., Prieto-Dapena, P., Almoguera, C. and Jordano, J. (2014) Co-overexpression of two heat shock factors results in enhanced seed longevity and in synergistic effects on seedling tolerance to severe dehydration and oxidative stress. *BMC Plant Biol.* 14: 56.

Petla, B. P., Kamble, N. U., Kumar, M., Verma, P., Ghosh, S., Singh, A., Rao, V., Salvi, P., Kaur, H., Saxena, S. C. and Majee, M. (2016). Rice PROTEIN L-isoaspartyl methyltransferase isoforms differentially accumulate during seed maturation to restrict deleterious isoAsp and reactive oxygen species accumulation and are implicated in seed vigor and longevity. *New Phytol.* 211(2): 627–645.

Pipatpongpinyo, W., Korkmaz, U., Wu, H., Kena, A., Ye, H., Feng, J. and Gu, X. Y. (2020). Assembling seed dormancy genes into a system identified their effects on seedbank longevity in weedy rice. *Heredity (Edinb)* 124(1): 135–145.

Probert, R., Adams, J., Coneybeer, J., Crawford, A. and Hay, F. (2007). Seed quality for conservation is critically affected by pre-storage factors. *Aust. J. Bot.* 55(3): 326–335.

Probert, R. J., Daws, M. I. and Hay, F. R. (2009). Ecological correlates of ex situ seed longevity: a comparative study on 195 species. *Ann. Bot.* 104(1): 57–69.

Raquid, R., Kohli, A., Reinke, R., Dionisio-Sese, M., Kwak, J., Chebotarov, D., Mo, Y. and Lee, J. S. (2021). Genetic factors enhancing seed longevity in tropical japonica rice. *Curr. Plant Biol.* 26: 100196.

Rehman Arif, M. A. and Börner, A. (2020). An SNP based GWAS analysis of seed longevity in wheat. *Cereal Res. Commun.* 48(2): 149–156.

Rehman Arif, M. A., Nagel, M., Neumann, K., Kobiljski, B., Lohwasser, U. and Börner, A. (2012). Genetic studies of seed longevity in hexaploid wheat using segregation and association mapping approaches. *Euphytica* 186(1): 1–13.

Renard, J., Martínez-Almonacid, I., Castillo, I. Q., Sonntag, A., Hashim, A., Bissoli, G., Campos, L., Muñoz-Bertomeu, J., Niñoles, R., Roach, T., Sánchez-León, S., Ozuna, C.

V., Gadea, J., Lisón, P., Kranner, I., Barro, F., Serrano, R., Molina, I. and Bueso, E. (2021). Apoplastic lipid barriers regulated by conserved homeobox transcription factors extend seed longevity in multiple plant species. *New Phytol.* 231(2): 679–694.

Renard, J., Niñoles, R., Martínez-Almonacid, I., Gayubas, B., Mateos-Fernández, R., Bissoli, G., Bueso, E., Serrano, R. and Gadea, J. (2020). Identification of novel seed longevity genes related to oxidative stress and seed coat by genome-wide association studies and reverse genetics. *Plant Cell Environ.* 43(10): 2523–2539.

Richards, C. M., Lockwood, D. R., Volk, G. M. and Walters, C. (2010). Modeling demographics and genetic diversity in ex situ collections during seed storage and regeneration. *Crop Sci.* 50(6): 2440–2447.

Righetti, K., Vu, J. L., Pelletier, S., Vu, B. L., Glaab, E., Lalanne, D., Pasha, A., Patel, R. V., Provart, N. J., Verdier, J., Leprince, O. and Buitink, J. (2015). Inference of longevity-related genes from a robust coexpression network of seed maturation identifies regulators linking seed storability to biotic defense-related pathways. *Plant Cell* 27(10): 2692–2708.

Roach, T., Nagel, M., Börner, A., Eberle, C. and Kranner, I. (2018). Changes in tocochromanols and glutathione reveal differences in the mechanisms of seed ageing under seedbank conditions and controlled deterioration in barley. *Environ. Exp. Bot.* 156: 8–15.

Roux, F., Mary-Huard, T., Barillot, E., Wenes, E., Botran, L., Durand, S., Villoutreix, R., Martin-Magniette, M.-L., Camilleri, C. and Budar, F. (2016). Cytonuclear interactions affect adaptive traits of the annual plant Arabidopsis thaliana in the field. *Proc. Natl. Acad. Sci.* 113: 3687–3692.

Sano, N., Rajjou, L., North, H. M., Debeaujon, I., Marion-Poll, A. and Seo, M. (2016). Staying alive: molecular aspects of seed longevity. *Plant Cell Physiol.* 57(4): 660–674.

Sasaki, K., Takeuchi, Y., Miura, K., Ando, T., Ebitani, T., Higashitani, A., Yamaya, T., Yano, M. and Sato, T. (2015) Fine mapping of a major quantitative trait locus, qLG-9, that controls seed longevity in rice (Oryza sativa L.). *Theor. Appl. Genet.* 128(4): 769–778.

Schwember, A. R. and Bradford, K. J. (2010). Quantitative trait loci associated with longevity of lettuce seeds under conventional and controlled deterioration storage conditions. *J. Exp. Bot.* 61(15): 4423–4436.

Schwember, A. R. and Bradford, K. J. (2011). Oxygen interacts with priming, moisture content and temperature to affect the longevity of lettuce and onion seeds. *Seed Sci. Res.* 21: 175–185.

Shenzao, F., Guangkun, Y., Xia, X., Shuhua, W., Xinghua, W. and Xinxiong, L. (2018). Levels of crotonaldehyde and 4-hydroxy-(E)-2-nonenal and expression of genes encoding carbonyl-scavenging enzyme at critical node during rice seed aging. *Rice Sci.* 25(3): 152–160.

Solberg, S. Ø., Yndgaard, F., Andreasen, C., von Bothmer, R., Loskutov, I. G. and Asdal, Å. (2020). Long-term storage and longevity of orthodox seeds: a systematic review. *Front. Plant Sci.* 11: 1007.

Springthorpe, V. and Penfield, S. (2015). Flowering time and seed dormancy control use external coincidence to generate life history strategy. *eLife* 4. doi: 10.7554/eLife.05557.

Tejedor-Cano, J., Prieto-Dapena, P., Almogura, C., Carranco, R., Hiratsu, K., Ohme-Takagi, M., Jordano, J. (2010). Loss of function of the HSFA9 seed longevity program. *Plant Cell Environ.* 33(8):1408–1417. doi: 10.1111/j.1365-3040.2010.02159.x.

Thi Hang, N. T., Lin, Q., Liu, L., Liu, X., Liu, S., Wang, W., Li, L., He, N., Liu, Z., Jiang, L. and Wan, J. (2015). Mapping QTLs related to rice seed storability under natural and artificial aging storage conditions. *Euphytica* 203(3): 673–681.

Tognacca, R. S. and Botto, J. F. (2021). Post-transcriptional regulation of seed dormancy and germination: current understanding and future directions. *Plant Commun.* 2(4): 100169.

Walters, C. (2015). Orthodoxy, recalcitrance and in-between: describing variation in seed storage characteristics using threshold responses to water loss. *Planta* 242(2): 397–406.

Walters, C., Ballesteros, D. and Vertucci, V. A. (2010). Structural mechanics of seed deterioration: standing the test of time. *Plant Sci.* 179(6): 565–573.

Walters, C., Wheeler, L. and Stanwood, P. C. (2004). Longevity of cryogenically stored seeds. *Cryobiology* 48(3): 229–244.

Walters, C., Wheeler, L. M. and Grotenhuis, J. M. (2005). Longevity of seeds stored in a GeneBank: species characteristics. *Seed Sci. Res.* 15(1): 1–20.

Waterworth, W. M., Bray, C. M. and West, C. E. (2019). Seeds and the art of genome maintenance. *Front. Plant Sci.* 10: 706.

Whitehouse, K. J., Hay, F. R. and Ellis, R. H. (2015). Increases in the longevity of desiccation-phase developing rice seeds: response to high-temperature drying depends on harvest moisture content. *Ann. Bot.* 116(2): 247–259.

Whitehouse, K. J., Hay, F. R. and Ellis, R. H. (2017). High-temperature stress during drying improves subsequent rice (Oryza sativa L.) seed longevity. *Seed Sci. Res.* 27(4): 281–291.

Whitehouse, K. J., Owoborode, O. F., Adebayo, O. O., Oyatomi, O. A., Olaniyan, A. B., Abberton, M. T. and Hay, F. R. (2018). Further evidence that the GeneBank standards for drying orthodox seeds may not be optimal for subsequent seed longevity. *Biopreserv. Biobank.* 16(5): 327–336.

Wyse, S. V. and Dickie, J. B. (2018). Taxonomic affinity, habitat and seed mass strongly predict seed desiccation response: a boosted regression trees analysis based on 17 539 species. *Ann. Bot.* 121(1): 71–83.

Yuan, Z., Fan, K., Wang, Y., Tian, L., Zhang, C., Sun, W., He, H. and Yu, S. (2021). OsGRETCHENHAGEN3-2 modulates rice seed storability via accumulation of abscisic acid and protective substances. *Plant Physiol.* 186(1): 469–482.

Zani, D. and Müller, J. V. (2017). Climatic control of seed longevity of Silene during the post-zygotic phase: do seeds from warm, dry climates possess higher maturity and desiccation tolerance than seeds from cold, wet climates? *Ecol. Res.* 32(6): 983–994.

Zhang, X., Hina, A., Song, S., Kong, J., Bhat, J. A. and Zhao, T. (2019). Whole-genome mapping identified novel "QTL hotspots regions" for seed storability in soybean (Glycine max L.). *BMC Genomics* 20(1): 499.

Zhou, W., Chen, F., Luo, X., Dai, Y., Yang, Y., Zheng, C., Yang, W. and Shu, K. (2020). A matter of life and death: molecular, physiological, and environmental regulation of seed longevity. *Plant Cell Environ.* 43(2): 293–302.

Zinsmeister, J., Lalanne, D., Terrasson, E., Chatelain, E., Vandecasteele, C., Vu, B. L., Dubois-Laurent, C., Geoffriau, E., Signor, C. L., Dalmais, M., Gutbrod, K., Dörmann, P., Gallardo, K., Bendahmane, A., Buitink, J. and Leprince, O. (2016). ABI5 is a regulator of seed maturation and longevity in legumes. *Plant Cell* 28(11): 2735–2754.

Zinsmeister, J., Berriri, S., Basso, D. P., Ly-Vu, B., Dang, T. T., Lalanne, D., da Silva, E. A. A., Leprince, O. and Buitink, J. (2020a). The seed-specific heat shock factor A9 regulates the depth of dormancy in Medicago truncatula seeds via ABA signalling. *Plant Cell Environ.* 43(10): 2508–2522.

Zinsmeister, J., Leprince, O. and Buitink, J. (2020b). Molecular and environmental factors regulating seed longevity. *Biochem. J.* 477(2): 305–323.

Zuo, J., Liu, J., Gao, F., Yin, G., Wang, Z., Chen, F., Li, X., Xu, J., Chen, T., Li, L., Li, Y., Xia, X., Cao, H. and Liu, Y. (2018). Genome-wide linkage mapping reveals QTls for seed vigor-related traits under artificial aging in common wheat (Triticum aestivum). *Front. Plant Sci.* 9: 1101. doi: 10.3389/fpls.2018.01101.

Zuo, J. H., Chen, F. Y., Li, X. Y., Xia, X. C., Cao, H., Liu, J. D. and Liu, Y. X. (2020). Genome-wide association study reveals loci associated with seed longevity in common wheat (Triticum aestivum L.). *Plant Breed.* 139(2): 295–303.

Part 2

Seed quality control and treatment

Chapter 6

Advances in seed phenotyping using X-ray imaging

Sherif Hamdy, Aurélie Charrier and Laurence Le Corre, GEVES, France; Pejman Rasti, Université d'Angers and École d'ingénieur Informatique et Environnement (ESAIP), France; and David Rousseau, Université d'Angers, France

1 Introduction

Seed phenotyping based on imaging techniques has received considerable attention over the past two decades (see [1-12] for detailed large-scale reviews). In a nutshell, X-ray imaging, fluorescence, spectral imaging, laser speckle, Lidar imaging, and thermal imaging have demonstrated their value in this domain. The first proofs of feasibility have extended to high-throughput imaging systems which are on the way to standardization as illustrated in Fig. 1. These techniques have been shown useful for the phenotyping of various stages of development of seed and seedling. As summarized in Table 1, this includes the morphogenesis of seed, the characterization of the internal anatomy of dry seeds, the imbibition process of the seed, the germination and elongation process, or the seedling growth (emergence, opening of cotyledon, and appearance of first leaves). As visible in Table 1, the two most versatile techniques are spectral reflectance imaging (RGB imaging and multi-hyperspectral imaging) and X-ray imaging. While spectral reflectance imaging has been extensively reviewed (see [6-8,11,12] for detailed and very recent overviews), X-ray imaging has not benefited so far from a synthetic review. We

http://dx.doi.org/10.19103/AS.2022.0105.08

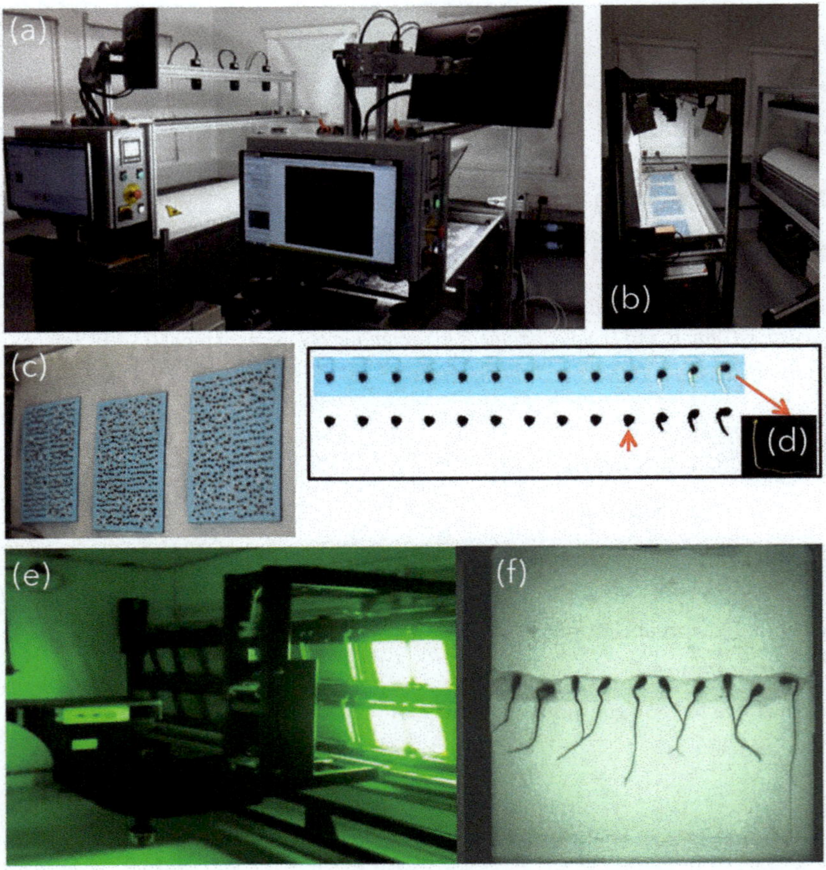

Figure 1 Examples of standard spectral reflectance imaging seed imbibition (panel a–d) and seed elongation (e–f). Post-acquisition images (d or f) are processed to compute the targeted trait. Each seed or seedling is segmented from the background and tracked over time to produce time-lapse as in panel d (typically one image every 2–4 h) during the whole process (4–30 days for seed germination and elongation depending on the species and the temperature). In the system of panel (a), cameras are replicated, while in system (e), the camera is moved across each petri dish. In system a, seeds are positioned on a wet blotter, while in the system shown in (e), seeds are placed between two papers sheets to have seed watered during the process. Systems developed at GEVES France, PHENOTIC Platform.

therefore propose a focus on X-ray imaging applied to seed phenotyping in this chapter.

By discovering the X-ray in 1895, the German physicist Wilhelm Roentgen offered the scientific community a revolutionary tool that contributed to reaching unprecedented achievements and outstanding applications in various research areas. Soon afterward, the first seed analysis (testing) trial using X-ray was carried out in 1903 by Lundstróm. Since then, this powerful tool has been

Table 1 Global view of the potentiality of the main imaging modalities for seed/seedling phenotyping

	Morphogenesis	Internal dry	Imbibition	Germination/ elongation	Seedling
X-ray	X	X	X	X	
Fluorescence		X			X
Spectral reflectance		X	X	X	X
Laser			X		
Lidar					X
Thermal			X	X	

integrated into the suite of technologies dedicated to bypassing the obstacles placed in the path of seed production and crop breeding development. X-ray has extensively contributed to taking seed quality assessment and phenotyping to a distinctly advanced stage of precision, accuracy, and rapidity [13–34]. Radiography and tomography are distinguished from other imaging techniques by giving access to perform a qualitative analysis and observation of the condition of internal characteristics of objects. This is specially adapted to seeds that have internal structures of interest not directly visible without being destructive. Also, the relatively recent reduction of the cost of X-ray imaging systems has democratized its use for seed phenotyping. While review articles have been proposed for several imaging methods for seed (including hyperspectral or thermal imaging), no review with X-ray has yet been proposed.

In this chapter, we review the various applications of X-ray imaging applied to seed phenotyping. As illustrated in Fig. 2, X-ray imaging follows a workflow that starts with sample preparation, projection acquisition, image reconstruction, and information extraction with image processing algorithms.

We analyze the currently existing literature along this process also describe the panel of biological traits which can be measured on seeds. Finally, we discuss the current challenges and bottlenecks for the wider uses of X-ray for seed phenotyping.

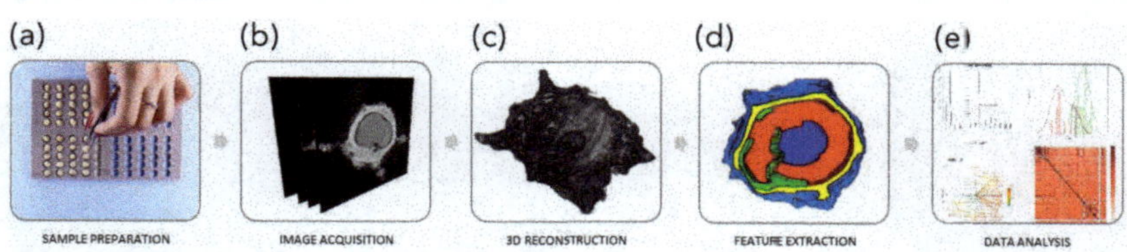

Figure 2 Illustration of the workflow used for seed phenotyping using computerized tomography (CT) imaging. (a) Sample preparation and setup, (b) acquisition of X-ray CT projection images, (c) 3D reconstruction and generation of a volumetric image, (d) extraction and measurement of seed's internal and external features, and (e) data analysis.

2 Sample preparation

X-ray imaging provides contrast related to the variation of attenuation of X-ray photons. These particles have an energy range of 5–10 keV. Attenuation depends on the material and its width according to Beer–Lambert law. The higher the atomic number the higher the attenuation. Depending on the species, seeds are constituted of cell walls, seed storage reserves such as proteins, starch or lipids, water, and air. The main contrast to be expected is therefore between these materials. However, seeds must be placed in the field of view of the X-ray beam and should not move during the acquisition protocol. It is therefore necessary to hold the sample. Some examples of sample holders are displayed in Fig. 3.

Special care should be paid to the selection of the material to be used for this sample holder and to avoid any metallic items.

When the focus is put on a single seed, glue can be used carefully in the field of view. For seed phenotyping, higher throughputs are often targeted, and multiple seeds can be imaged at the same time on a regular grid to allow individual analysis. The stability of the samples is tremendously important for some acquisition protocols (especially for those where the sample is moving during acquisition). Dry foam can be used to place multiple seeds. Higher throughputs can even be obtained when a global analysis of an ensemble of seeds is to be done without the need for individual sorting afterward. In this case, the seed can just be poured into a cylindrical sample holder. The fact that, in this case, the seeds are touching each other can be solved numerically with image post-processing [35]. The preparation of the sample can also be automated via a conveyer belt for even higher throughputs. When applied to imbibed tissues, the X-ray contrast inside the seeds vanishes (air is replaced by water) [7,36].

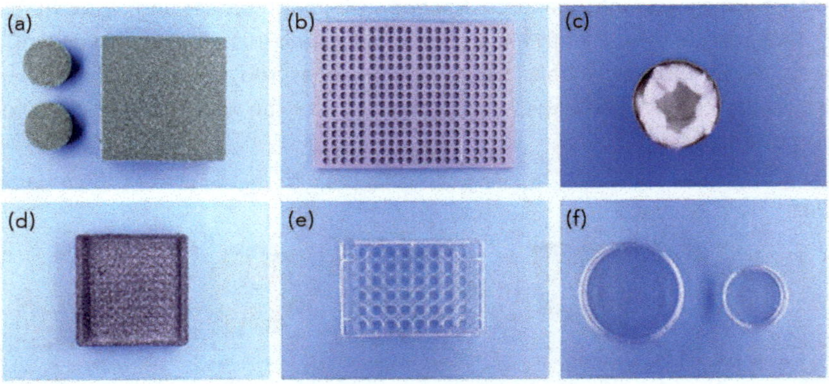

Figure 3 A collection of X-ray sample holders. (a) Floral foam, (b) mesh-based plastic, (c) cardboard container internally coated with cotton, (d) 3D-printed polyamide (nylon plastic), (e) mesh-based crystal plastic, and (f) crystal plastic petri dish.

Table 2 Summary of applications using X-ray imaging for phenotyping and seed testing

Ref	Species	2D /3D	Voltage	Time	Current	Sample size	Distance	Resolution	Method	Traits	Substrate	State
[13]	Brachiaria ruziziensis	2D	23 kV	10 sec	-	50	Focal distance of 27.8 cm	image contrast calibrated at 13915 (width) × 5374 (center)	Multilevel thresholding followed by particle analysis of ROIs	Height, width, area, perimeter, circularity, solidity, relative density, inetegrated density, and seed filling	Contact adhesive paper	Dry
[14]	Corylopsis coreana and C. sinensis	2D	20 kV	15 sec	-	50	-	-	-	Classification into (full, mostly full, partially empty, and empty)	Index cards and Elmer's glue	Immersed
[15]	Crotalaria juncea L.	2D	Auto	Auto	-	25	Source to sample distance of 19.1 cm	-	-	Evaluation of damages (cracks in the cotyledons, cracks in diverse parts, stink bug damage in the cotyledons, stink bug damage in embryo and cotyledon)	Transparent acetate and double-sided adhesive tape	Dry
[16]	Cucumis melo L.	2D	23 kV	10 sec	-	20	Focal distance of 41.6 cm	contrast of the calibrated image in 16383 (width) × 3124 (center)	Multilevel Thresholding followed by particle analysis of ROIs	Area, perimeter, circularity, solidity, relative density, integrated density, and seed filling	Adhesive paper and tape	Dry
[17]	Platypodium elegans Vog.	2D	26 kV	1.2 sec	-	200	Source to sample distance of 28 cm	-	Visual observation	Classified into partially full, full, and empty seeds	Transparent double-face sellotape	Dry

(Continued)

Table 2 (*Continued*)

Ref	Species	2D /3D	Voltage	Time	Current	Sample size	Distance	Resolution	Method	Traits	Substrate	State
[18]	Quinoa	3D	50 kV	10 min	200 µA	30	-	20 µm voxel size	Multilevel thresholding followed by particle analysis of ROIs	Volume, maximum feret diameter, minimum feret diameter, feret diameter ratio of the whole seed, and volume of the embryo	3 PMMA Cylinder (2.5 cm diameter and 2.8 cm height) whose internal surface was coated with a polystyrene	Dry
[18]	Quinoa	3D	50 kV	20 min	200 µA	1	-	2 µm voxel size	Multilevel thresholding followed by particle analysis of ROIs	Volumes of the solid phase of the whole seed and of each seed part (embryo, pericarp, and perisperm)	A PMMA Cylinder (2.5 cm diameter and 2.8 cm height) whose internal surface was coated with a polystyrene	Dry
[19]	Tomato	3D	50 kV	54 min	200 µA	18	Source to sample distance of 75 mm	6.5 µm voxel size	Visual analysis - Thresholding followed by particle analysis of ROIs	Total volume, solid-phase volume, free space, seed surface, structure model index, structure thickness, equivalent rod length, major diameter, volume equivalent sphere diameter, surface-equivalent sphere diameter, sauter diameter, sphericity, and surface convexity index	Polystyrene cylinder	Dry
[20]	Walnut	3D	120 kV	14 min	300 µA	5 to 16	-	992 × 992 × 2991	Multilevel thresholding followed by particle analysis of ROIs	Volume, area, rugosity, sphericity and shape indexes, and filling ratio	Foral foam	Dry

[21]	Sugar beet	3D	45 kV	-	178 μA	25	-	976 × 976 × 976	Multilevel thresholding followed by particle analysis of ROIs	Volume, surface area, shape factor, filling factor, and the spherical diameter	Floral foam	Dry
[22]	Muskmelon	3D	40 kV	35 ms	250 μA	1	Sample to source distance of 70.310 mm and detector to source distance of 210.666 mm	11.77μm voxel size	Segmentation using balanced histogram thresholding method then feature extraction process (intensity, contrast, and texture)	High dimension features including height, width, major axis length, minor axis length, centre of gravity, and orientation of the seed components were extracted	Waxed over a sample holder	Dry
[23]	Pansy	3D	8 kV	1.5 sec	1.54 A	1	Sample-to-detector distance of 7 mm and the source-to-sample distance of 260 mm	1.25 μm voxel size	Manual segmentation	-	Waxed on a cylindrical plastic pellet	Dry
[24]	Bell pepper	2D	10 kV	260 sec	-	50	Sample to source distance of 58 cm	-	Visual analysis	Classified into multiple categories according to the proportion between the area occupied by the embryo and endosperm in relation to the total cavity area in the internal seed structure as well as abnormalities and malformation	Acrylic plate	Dry
[25]	Rice	2D	14 kV	360 sec	-	100	Sample to source distance of 35 cm	-	-	Classified from 1 to 3, where 1 = without fissures, 2= non-severe fissures, and 3= severe fissures	Acrylic plate with a transparent sticky tape in the bottom	Dry

(*Continued*)

Table 2 (*Continued*)

Ref	Species	2D /3D	Voltage	Time	Current	Sample size	Distance	Resolution	Method	Traits	Substrate	State
[26]	Watermelon	2D	50 kV	0.05 sec	100 μA	90	5 μm focal spot size	2304 × 1300	Machine and deep learning	Classifying seeds into viable and non-viable seeds	Plastic transparent plate	Dry
[27]	Salvia spinosa, Salvia aegyptiaca, Ochradenus baccatus, chradenus arabicus, Suaeda aegyptiaca, Suaeda vermiculata, Prosopis farcta, and Panicum turgidum	2D	18 kV	10 sec	-	50 / type	-	-	-	Evaluation based on the presence and morphology of the embryo and endosperm as well as the percentage of seeds with a whole embryo, damaged embryo, or no embryo	-	Dry
[28]	Sweet Corn	2D	15 kV	5 min	-	50	Sample to source distance of 57.2 cm	-	Visual analysis	Evaluation according to the severity and location of the damage	Acrylic plate with transparent tape	Dry
[29]	Tomato	2D	20 kV	20 sec	-	20	-	-	Using opening operation, then manual selection of seeds, after that, range selection and image calibration followed by measurement	Area, perimeter, and assessment of embryo and endosperm	Fixed on a transparent plastic sheet	Dry
[30]	Sugar beet	3D	130 kV	250 ms	100 μA	-	-	20 μm	Using morphological operation	Volume, soil volume, surface area, and filling	Polypropylene cylinders(25 mm diameter × 70 mm length) filled with sandy loam soil	Dry

[31]	Castor bean	2D	20 to 50 kV	15 to 75 sec	-	400	-	-	Visual analysis	Classified into seven categories according to their internal morphology and seed filling	Styrofoam sheets	Dry
[32]	Rice	3D	90 kV	~2 min / panicle	3.2 mA	One Panicle	-	512 × 512 × 450	Binarization using global adaptive threshold segmentation followed by successive morphological operations, PCA, triangle pitching, and trait extraction	Number of traits relevant to shape, size, and density	Plastic holder	Dry
[33]	Wheat	3D	100 kV	5 sec	20 μA	50	Focal length of the scanner 5 μm	Final resolution of a slice (0.1 mm, 1024 × 1024)	Image intensity equalization and thresholding followed by PCA	Length, width, thickness, radius, surface area, volume, compactness, and sphericity	Plastic container with multiple sub-boxes (70 mm)	Dry
[33]	Soybean	3D	100 kV	5 sec	20 μA	30-36	Focal length of the scanner 5 μm	Final resolution of a slice (0.1 mm, 1024 × 1024)	Image intensity equalization and thresholding followed by PCA	Length, width, thickness, radius, surface area, volume, compactness, and sphericity	Plastic container with multiple sub-boxes (80 - 90 mm)	Dry
[33]	Peanut, Pine nuts, and Pistachio	3D	100 kV	5 sec	20 μA	21	Focal length of the scanner 5 μm	Final resolution of a slice (0.1 mm, 1024 × 1024)	Image intensity equalization and thresholding followed by PCA	Length, width, thickness, radius, surface area, volume, compactness, and sphericity	Plastic container with multiple sub-boxes (90 mm)	Dry

In this case, an additional step in sample preparation can be necessary with the use of contrast agents. Some trials have been carried out using solutions brought from the medical domain some of which have a high atomic number (metals) (see refs [37,38] with details on solutions and concentration). While temporal information on the diffusion kinetics of such contrast agents in human tissues is well documented, information on plant science, in general, and seed, in particular, is still missing [39]. In Table 2, a summary of multiple applications is provided where radiography and micro-computed tomography (CT) imaging techniques were exploited to develop various phenotyping methods. An extensive range of materials was introduced as sample holders in the diverse experiments such as index papers [13,16], floral foam [20,21], PMMA cylinders [18], acrylic plates [24,28], polystyrene cylinders [19], or even sandy loam soil [30] that were used indirectly as a filler inside polypropylene cylinders while studying the seed–soil interactions. Additionally, with the rise of the usage of 3-dimensional (D) printing, some sample holders are recently designed and printed using certain types of materials such as polyamide as shown in Fig. 3d.

3 Projection acquisition

Once the sample is positioned, it is ready to be acquired. There are two main acquisition configurations: radiography (for 2D imaging) and tomography (for 3D imaging). Radiography is obtained from a single projection of the X-ray beam through the sample. In tomography, the sample is rotated along one axis iteratively and a projection is acquired for each angle. Several parameters can be adjusted at this stage including the distance of the sample to the camera and sample to source and the emitting parameter of the source [40]. Shortly, the emitting current of the source determines the intensity of the X-ray beam, while the voltage controls its energy and thus the contrast in the image. The distance of the sample to the camera is related to the field of view and thus also the spatial resolution. The influence of these parameters is illustrated qualitatively in Figs 4 and 5.

The typical values for these parameters are provided in Table 2 for the main literature on seed phenotyping with X-ray. Exposure time is one of the key image acquisition settings that influences the image quality and has a strong relationship with the quantity of X-rays released. Typically, image acquisition using radiography takes a significantly shorter time than in micro-CT while in radiography, it took around 10 s to scan 50 seeds of *Brachiaria ruziziensis* [13], 15 s for a sample of 50 seeds of *Corylopsis coreana* and *C. Sinensis* [14], and as small as 0.05 s which was found to be sufficient to acquire images of watermelon seeds [26]. On the other hand, a micro-CT scan took approximately 14 min for obtaining an image of approximately 5–16 Walnuts [20], 20 min for 1 Quinoa seed to obtain higher resolution and extra details

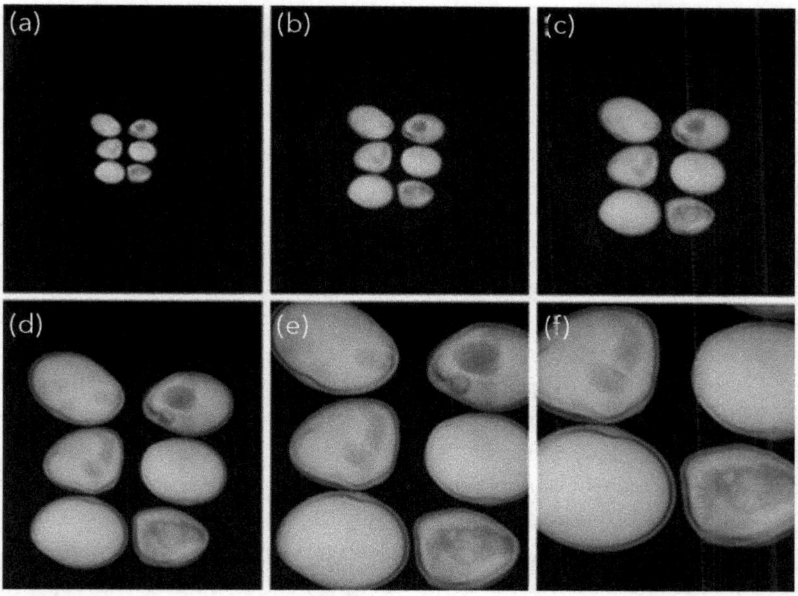

Figure 4 A sample of 5 Faba bean seeds acquired at various magnification levels. (a) mag × 1, (b) mag × 1.5, (c) mag × 2, (d) mag × 3, (e) mag × 4, and (f) mag × 5.

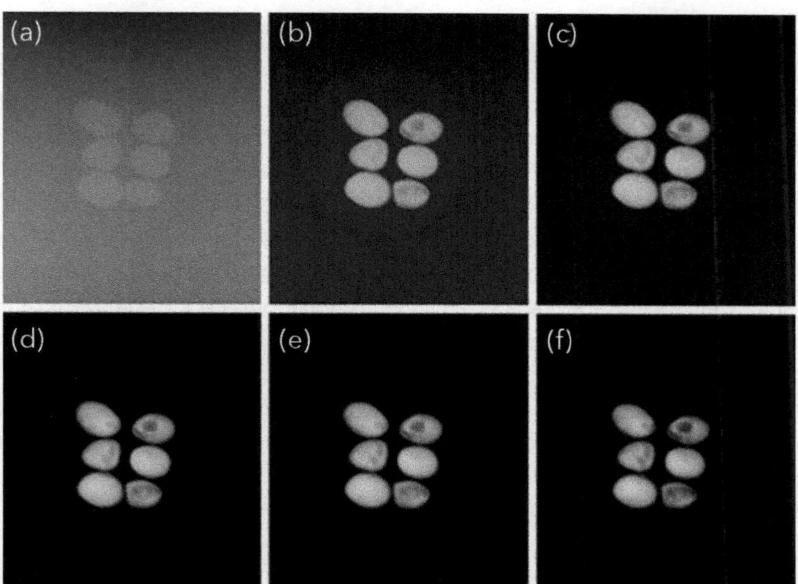

Figure 5 X-ray radiographs of a sample of 6 Faba bean seeds acquired at various tube voltages while fixing sample-to-source distance, sample-to-detector distance, and exposure time. (a) 10 kV, (b) 15 kV, (c) 20 kV, (d) 25 kV, (e) 30 kV, and (f) 35 kV.

[18], and up to 54 min for 18 tomato seeds [19]. Equally, sample-to-source distance (Fig. 4) plays a significant role to control the visibility of the samples by setting a reasonable magnification factor to allow observability and to avoid acquiring poor radiographs where the images are arbitrarily cropped, which in return could lead to potential image processing fails and severe loss of information. Similarly, the optimal voltage must be correctly estimated to avoid noisy images, loss of details, overexposure, and saturation (Fig. 5). Due to the lack of optimized and standard imaging protocols, choices of some of the scanning parameters such as tube voltage, exposure time, current, and sample-to-source distance are adapted experimentally in the function of the high variance between the seed species in density, size, the number of seeds per scan, in addition to the purpose of the analysis. It is to be noticed that other types of X-ray imaging also exist when coherent sources of X-ray such as the one provided by the synchrotron radiation facility [37,38,41]. Such sources of X-ray are similar to laser beams but in the range of wavelength corresponding to 10–100 nm [40]. This opens the possibility of interferometric contrast which is sensitive to a fraction of wavelength. Such radiations have been applied to the imaging of fine details of seeds including dry and imbibed samples [7]. The increase of contrast is very useful to detect fine details. However, the currently limited access to these coherent sources of X-ray constitutes a limit to the widespread of these experiments.

4 Image reconstruction

In tomographic mode, projections serve as input to produce a 3D digital reconstruction of the sample. Various methods exist to perform this reconstruction. Analytic methods based on filtered back-projection algorithm or radon transform assume that the X-ray beam is parallel. They are fast but depend on the geometry of the apparatus and require a massive number of projections [40]. Algebraic methods solve the inverse problem of image reconstruction iteratively and are therefore less sensitive to noise but are independent of the acquisition geometry [40]. They can be improved by regularization with the addition of prior information. For seed, the sample is steady and relatively small compared to the size of the beam. Consequently, contrarily to the biomedical domain, the choice of the reconstruction algorithm is not critical in seed X-ray imaging. Examples of 2D radiographic acquisition and 3D reconstruction of walnut and sugar beet seeds are provided for illustration in Figs 6 and 7.

5 Image processing

Visual assessment is currently strongly present in seed analysis. In this way, phenotyping heavily depends on the analyst's competence and knowledge to

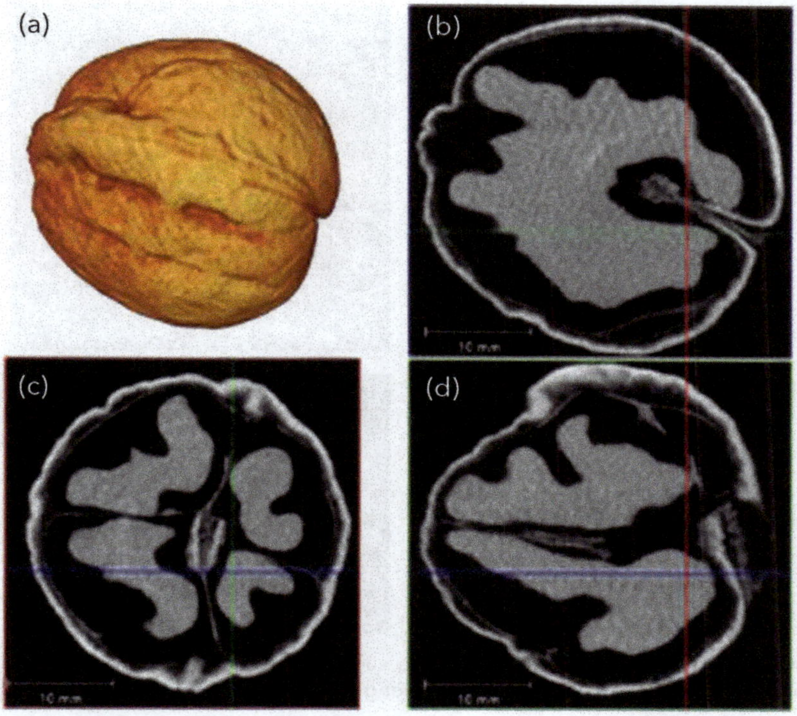

Figure 6 Micro-CT volume rendering of a walnut (a). Cross-section views showing the internal structure of the seed along the *Z*-axis (b), *X*-axis (c), and *Y*-axis (d).

classify the seeds based on their conditions and the quality of the microstructures of the seeds. Various seed quality indicators can be observed as described in Fig. 8. The quality assessment process may therefore get influenced by subjectivity bias on the part of the analysts. A way to reduce this bias is to head towards the automation of the assessment via image processing. Automation of the processing of the images has been undertaken for a variety of traits related to morphology, such as parameters related to seed size and form, structural damage such as fissures or holes or malformation, completeness of organ formation, and ratio between organs and density (Table 2, Fig. 8). Examples are the presence of internal organs in bell pepper [24], the severity of fissures in sweet corn and rice [25,28], infestation level in the embryo and cotyledon of *Crotalaria L.* seeds [15], and malformation of castor bean seeds [21]. This approach was also used for pre-screening as a preliminary selection step before further analysis [21]. Several fully automated or semi-automated pipelines were introduced as rapid and high-throughput approaches that mainly include conventional image processing tasks such as thresholding, morphological operations (opening, closing, etc.), and labelling followed by particle analysis as presented in sugar beet and *Cucumis melo L.* seeds [21,16].

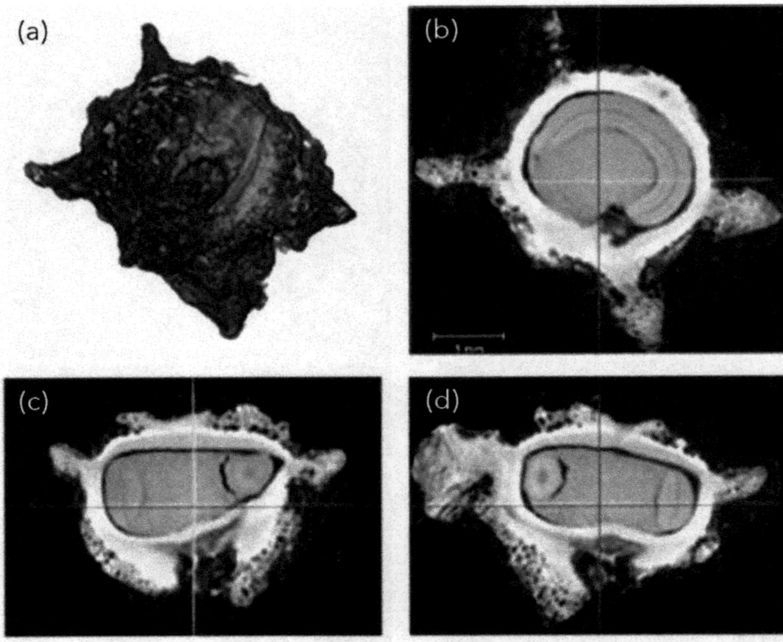

Figure 7 Micro-CT volume rendering of a sugar beet seed (a). Cross-section views showing the internal structure of the seed along the *Z*-axis (b), *X*-axis (c), and *Y*-axis (d).

While artificial intelligence-based approaches now constitute the state-of-the-art in most applied computer vision problems, till the moment, deep learning methods have received insufficient attention for seed X-ray imaging. Surprisingly, only a few successful deep learning-based applications were reported in [34] for classifying the quality of one of the important oil sources *Crambe abyssinica* and in watermelon seeds [26]. So far, pure hand-crafted shallow learning-based approaches have mainly been used. Several reasons can explain this situation. First, deep learning requires annotated data set. At the moment, very few annotated datasets are available for seed X-ray imaging. Second, there exists a large diversity in the shape of seeds in nature. Therefore, the transferability of one species to another is likely to fail, and in consequence, it is by no means yet clear whether a universal model could indeed be obtained like in X-ray imaging for biomedical application. Also, as stated before and contrarily to biomedical imaging, the acquisition protocols are not fully standardized. This diversity may also constitute a specific difficulty to reach some generic models. Finally, while 2D X-ray images are easy to annotate, 3D X-ray images of seeds are more challenging to annotate. While some specific protocols for visualization, annotation, and pre-segmentation exist in medical imaging, such tools are still missing for X-ray imaging of seeds. In general, reported results were either qualitative scores given to the seeds with regards

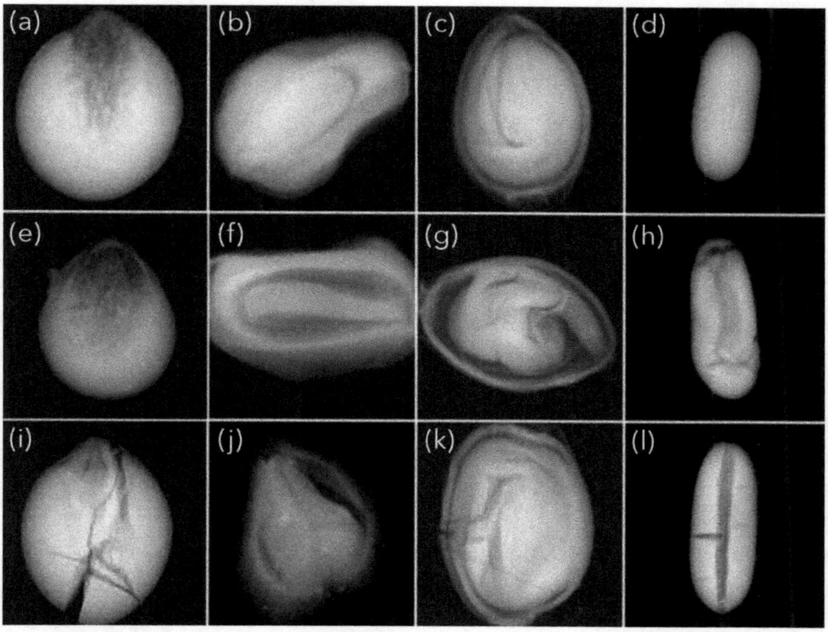

Figure 8 Evaluation of seed internal quality using X-ray. Full seeds (a, b, c, d), with malformation (e, f, g, h), with mechanical damage (i, j, k, l) of Sorghum, rose achene, Cannabis, and wheat from left to right, respectively.

to their condition or quantitative analysis that mostly requires 3D imaging and was performed on wheat, soybean, peanut, pine nuts, pistachio [33], sugar beet [21], pansy seeds [23], and muskmelon [22]. Typically, the quantitative analysis provides precise high-dimensional measures of the various morphometric internal and external traits that include, but are not limited to, volume (mm^3), surface area (mm^2), length (mm), width (mm), filling ratio (%), shape descriptors, as well as surface indices such as sphericity and rugosity. Urgent research questions are rising out of the notably extended research that aims to study the relationship between seed traits, germination and seed structure, genotype-phenotype correlation, resistance, and tolerance to extreme temperatures. Thus, to find reliable responses, large amounts of rich phenotype datasets are analysed with the aid of modern data analysis techniques and modelling algorithms.

6 Conclusion and outlook

In this chapter, we have reviewed the different steps of X-ray imaging when applied to seed phenotyping. X-rays have been used extensively in research programmes. In this case, the protocols are often adapted and adjusted according to the characteristics to be observed. However, X-rays also have great potential for application in seed industries, especially for seed quality testing. Seed quality is based on purity, germination, and sanitary evaluation of seed lots. The current tests, as described in the International Seed Testing Association (ISTA) Rules, are regularly reviewed within the ISTA technical committees with the main objective of their development. At present, these analyses are still mainly carried out manually and visually and rely totally on the expertise of the analysts. The development of new techniques, such as X-ray coupled with automated imaging analysis, could contribute to improving processes in terms of both efficiency and accuracy of observations. However, while this imaging modality has been demonstrated of high generic value, its wider dissemination still requires some effort from the research community before they can be considered as standardized methods for routine use, especially in terms of reproducibility, robustness, increasing the range of species, ensuring non-destructive conditions, and reducing costs. Indeed, a more reproducible imaging protocol requires the acquisition parameters to be provided more systematically. More robust algorithms would require the production of larger public data sets. Another step will be the potential of the application of X-ray analysis in a wide range of plant species. Indeed, seed testing methods must be relevant and applicable to many plant species that are to be used in seed quality testing. Finally, another important direction to make X-ray imaging definitely mature would lay in the systematic study of the invasiveness of X-ray on the normal development of the seeds. Indeed, although some studies have shown that the effect of X-rays does not seem to influence plant physiology [42–45], it will be necessary to confirm that 2D X-rays and CT scans are always non-destructive techniques or to determine the association between species/X-ray exposure conditions to ensure the non-destructive aspect of the method. The prospects for X-ray imaging are very challenging with the aim of integrating new automated methods to assist in seed quality evaluation.

7 Future trends

As advocated in the previous sections on X-ray, but also applicable for the other imaging techniques listed in the introductive section, acquisition protocols for seed imaging are heading towards standardization. Therefore, one can envision that the future bottleneck in seed phenotyping will soon shift from image acquisition to image processing and data analysis. The trend in image processing for a decade is the massive use of data-based algorithms rather than model-based ones. This prodigious revolution benefits from artificial intelligence innovation and specifically from neural network models. Indeed, in a model-based era, the development of algorithms took time to translate, in geometrical terms, the expertise of a biologist into a programmer. Each individual question required this interdisciplinary crosstalk to be refueled. Undoubtedly, this is time consuming and tends to produce a jungle of algorithms with limited applicability yet requiring heavy computational developments. On the contrary, in the data-driven era, very few algorithms are needed since generic universal algorithms are now available to perform classification, object recognition, and segmentation [46]. Classification takes an image as input and produces a class as output. It can for instance be used to recognize species or types of defaults in images with a single seed and in cases where the localization is not required. Object recognition enables object classification within an image. It can for instance be used to recognize species or types of defaults in images with multiple seeds and in cases where the delineation of the targeted objects is not required. Segmentation produces pixel classification that can for instance be used to measure the volume of a given subpart of a seed. In the data-driven era, biologists just need to identify the type of task (classification, object recognition, and segmentation) they are facing and feed the identified algorithm with annotated data. These algorithms are publicly available in open source. Some first attempts have started to appear in this deep learning trend applied to seed imaging [47–49] and one can bet that it is only the beginning. Consequently, to contribute to this shift from model to data-driven algorithms, it becomes crucial to encourage the release of publicly available annotated datasets to reduce the time of annotation. Ergonomic interfaces to use deep learning methods start to be available [50]. Also, it is important to empower the next generation of biologists with programming skills so that they can use the available set of image processing algorithms and have a clear understanding of the principal advantages and pitfalls of these promising algorithms for seed phenotyping.

8 Where to look for further information

As complementary information to the literature pointed out in this chapter, the reader can refer to the Youtube channel of the authors which provides courses and tutorials on machine learning applied to plant and seed imaging: https://www.youtube.com/channel/UCsd9Dt6N7O-fydynsWEfkww. Biologists can start with machine learning and deep learning by reading the book [51]. Publicly available datasets and algorithms for seed imaging can be found at: https://www.quantitative-plant.org/.

9 References

1. Dell'Aquila, A. (2006). Computerised seed imaging: a new tool to evaluate germination quality. *Communications in Biometry and Crop Science* 1(1), 20–31.

2. Dell'Aquila, A. (2007). Towards new computer imaging techniques applied to seed quality testing and sorting. *Seed Science and Technology* 35(3), 519–538.
3. Dell'Aquila, A. (2009). Digital imaging information technology applied to seed germination testing: a review. *Agronomy for Sustainable Development* 29(1), 213–221.
4. Huang, M., Wang, Q. G., Zhu, Q. B., Qin, J. W. and Huang, G. (2015). Review of seed quality and safety tests using optical sensing technologies. *Seed Science and Technology* 43(3), 337–366.
5. Rahman, A. and Cho, B. K. (2016). Assessment of seed quality using non-destructive measurement techniques: a review. *Seed Science Research* 26(4), 285–305.
6. Belin, E., Douarre, C., Gillard, N., Franconi, F., Rojas-Varela, J., Chapeau-Blondeau, F., Demilly, D., Adrien, J., Maire, E. and Rousseau, D. (2018). Evaluation of 3D/2D imaging and image processing techniques for the monitoring of seed imbibition. *Journal of Imaging*, 4(7), 83.
7. Boelt, B., Shrestha, S., Salimi, Z., Jørgensen, J. R., Nicolaisen, M. and Carstensen, J. M. (2018). Multispectral imaging–a new tool in seed quality assessment? *Seed Science Research* 28(3), 222–228.
8. Feng, L., Zhu, S., Liu, F., He, Y., Bao, Y. and Zhang, C. (2019). Hyperspectral imaging for seed quality and safety inspection: a review. *Plant Methods* 15(1), 91.
9. Xia, Y., Xu, Y., Li, J., Zhang, C. and Fan, S. (2019). Recent advances in emerging techniques for non-destructive detection of seed viability: a review. *Artificial Intelligence in Agriculture* 1, 35–47.
10. ElMasry, G., ElGamal, R., Mandour, N., Gou, P., Al-Rejaie, S., Belin, E. and Rousseau, D. (2020). Emerging thermal imaging techniques for seed quality evaluation: principles and applications. *Food Research International* 131, 109025.
11. ElMasry, G., Mandour, N., Al-Rejaie, S., Belin, E. and Rousseau, D. (2019). Recent applications of multispectral imaging in seed phenotyping and quality monitoring–an overview. *Sensors* 19(5), 1090.
12. Long, W., Jin, S., Lu, Y., Jia, L., Xu, H. and Jiang, L. (2021). A review of artificial intelligence methods for seed quality inspection based on spectral imaging and analysis. *Journal of Physics: Conference Series* 1769(1), 012013.
13. Medeiros, A. D. D., Silva, L. J. D., Pereira, M. D., Oliveira, A. M. S. and Dias, D. C. F. S. (2020). High-throughput phenotyping of brachiaria grass seeds using free access tool for analyzing X-ray images. *An Anais da Academia Brasileira de Ciencias* 92(Suppl. 1), e20190209, doi: 10.1590/0001-3765202020190209.
14. Ji Hee, K., Kyung Ok, K., Ae Kyung, L., Mark S, R. and Jeung Keun, S. (2017). Germination of Corylopsis seeds evaluated by X-ray imaging and cold stratification. *Horticultural Science* 44(2), 105–111, doi: 10.17221/194/2015-HORTSCI.
15. Arruda, N., Cicero, S. M. and Guilhien Gomes-Junior, F. (2016). Radiographic analysis to assess the seed structure of Crotalaria juncea L. *Journal of Seed Science* 38(2), 161–168, doi: 10.1590/2317-1545v38n2155116.
16. Medeiros, A. Dd, Martins, M. S., Silva, L. Jd, Pereira, M. D., León, M. J. Z. and Dias, D. C. FdS. (2020). X-ray imaging and digital processing application in non-destructive assessing of melon seed quality. *Journal of Seed Science* 42, e202042005, doi: 10.1590/2317-1545v42229761.
17. Paradelo, K., Martins, J., Soares, I. and de Carvalho, R. (2016). X-ray test to evaluate the physiological potencial of Platypodium elegans Vog. Seeds (Fabaceae). *Scientia Agropecuaria* 7, 305–3111, doi: 10.17268/sci.agropecu.2016.03.19.

18. Gargiulo, L., Grimberg, Å., Repo-Carrasco-Valencia, R., Carlsson, A. S. and Mele, G. (2019). Morpho-densitometric traits for quinoa (Chenopodium quinoa Willd.) seed phenotyping by two X-ray micro-CT scanning approaches. *Journal of Cereal Science* 90, 102829, doi: 10.1016/j.jcs.2019.102829.
19. Gargiulo, L., Leonarduzzi, C. and Mele, G. (2020). Micro-CT imaging of tomato seeds: predictive potential of 3D morphometry on germination. *Biosystems Engineering* 200, 112–122, doi: 10.1016/j.biosystemseng.2020.09.003.
20. Bernard, A., Hamdy, S., Corre, L. L., Dirlewanger, E. and Lheureux, F. (2020). 3D characterization of walnut morphological traits using X-ray computed tomography, preprint. *Plant Methods* 16, 115. doi: 10.1186/s13007-020-00657-7.
21. Ducournau, S., Charrier, A., Demilly, D., Wagner, M. H., Trigui, G., Dupont, A., Hamdy, S., Boudehri-Giresse, K., Le Corre, L., Landais, L., Delanoue, A., Charruaud, D., Henry, K., Henry, N., Ledroit, L. and Dürr, C. (2020). High throughput phenotyping dataset related to seed and seedling traits of sugar beet genotypes. *Data in Brief* 29, 105201, doi: 10.1016/j.dib.2020.105201.
22. Ahmed, M. R., Yasmin, J., Collins, W. and Cho, B.-K. (2018). X-ray CT image analysis for morphology of muskmelon seed in relation to germination. *Biosystems Engineering* 175, 183–193, doi: 10.1016/j.biosystemseng.2018.09.015.
23. Kunishima, N., Takeda, Y., Hirose, R., Kalasová, D., Šalplachta, J. and Omote, K. (2020). Visualization of internal 3D structure of small live seed on germination by laboratory-based X-ray microscopy with phase contrast computed tomography. *Plant Methods* 16(1), 7, doi: 10.1186/s13007-020-0557-y.
24. Gagliardi, B. and Marcos-Filho, J. (2011). Relationship between germination and bell pepper seed structure assessed by the X-ray test. *Scientia Agricola* 68(4), 411–416, doi: 10.1590/S0103-90162011000400004.
25. de Menezes, N. Ld, Cicero, S. M., Villela, F. A. and Bortolotto, R. P. (2012). Using X-rays to evaluate fissures in rice seeds dried artificially. *Revista Brasileira de Sementes* 34(1), 70–77.
26. Ahmed, M. R., Yasmin, J., Park, E., Kim, G., Kim, M. S., Wakholi, C., Mo, C. and Cho, B. K. (2020). Classification of watermelon seeds using morphological patterns of X-ray imaging: a comparison of conventional machine learning and deep learning. *Sensors* 20(23), 15.
27. Al-Turki, T. A. and Baskin, C. C. (2017). Determination of seed viability of eight wild Saudi Arabian species by germination and X-ray tests. *Saudi J Biol Sci.*, 24(4), 822–829, doi: 10.1016/j.sjbs.2016.06.009.
28. Junior, F. G. G. and Cicero, S. M. (2012). X-ray analysis to assess mechanical damage in sweet corn seeds. *Revista Brasileira de Sementes* 34, 8.
29. Borges, S. RdS., Silva, P. Pd, Araújo, F. S., Souza, F. FdJ. and Nascimento, W. M. (2019). Tomato seed image analysis during the maturation. *Journal of Seed Science* 41(1), 22–31, doi: 10.1590/2317-1545v41n1191888.
30. Blunk, S., Malik, A. H., de Heer, M. I., Ekblad, T., Bussell, J., Sparkes, D., Fredlund, K., Sturrock, C. J. and Mooney, S. J. (2017). Quantification of seed–soil contact of sugar beet (Beta vulgaris) using X-ray computed tomography. *Plant Methods* 13, 71.
31. Carvalho, M. L. Md, Alves, R. A. and Oliveira, L. M. D. (2010). Radiographic analysis in castor bean seeds (Ricinus communis L.). *Revista Brasileira de Sementes* 32(1), 170–175, doi: 10.1590/S0101-31222010000100019.
32. Hu, W., Zhang, C., Jiang, Y., Huang, C., Liu, Q., Xiong, L., Yang, W. and Chen, F. (2020). Nondestructive 3D image analysis pipeline to extract rice grain traits

using X-ray computed tomography. *Plant Phenomics* 2020, 3414926, doi: 10.34133/2020/3414926.
33. Liu, W., Liu, C., Jin, J., Li, D., Fu, Y. and Yuan, X. (2020). High-throughput phenotyping of morphological seed and fruit characteristics using X-ray computed tomography. *Frontiers in Plant Science* 11, 601475, doi: 10.3389/fpls.2020.601475.
34. de Medeiros, A. D., Bernardes, R. C., da Silva, L. J., de Freitas, B. A. L., Dias, D. C. FdS. and da Silva, C. B. (2021). Deep learning-based approach using X-ray images for classifying Crambe abyssinica seed quality. *Industrial Crops and Products* 164, 113378, doi: 10.1016/j.indcrop.2021.113378.
35. Ross, L. and Russ, J. C. (2011). The image processing handbook, Sixth Edition, John C. Russ. CRC Press, Boca Raton FL, 2011, 972 pages. ISBN 1-4398-4045-0(Hardcover). *Microscopy and Microanalysis* 17(5), 843.
36. Staedler, Y. M., Masson, D. and Schönenberger, J. (2013). Plant tissues in 3D via X-ray tomography: simple contrasting methods allow high resolution imaging. *PLoS ONE* 8(9), e75295.
37. Rousseau, D., Widiez, T., Di Tommaso, S., Rositi, H., Adrien, J., Maire, E., Langer, M., Olivier, C., Peyrin, F. and Rogowsky, P. (2015). Fast virtual histology using X-ray in-line phase tomography: application to the 3D anatomy of maize developing seeds. *Plant Methods* 11(1), 55.
38. Wang, Z., Verboven, P. and Nicolai, B. (2017). Contrast-enhanced 3D micro-CT of plant tissues using different impregnation techniques. *Plant Methods* 13(1), 105.
39. Koç, M. M., Aslan, N., Kao, A. P. and Barber, A. H. (2019). Evaluation of X-ray tomography contrast agents: a review of production, protocols, and biological applications. *Microscopy Research and Technique* 82(6), 812–848.
40. Russo, P. (Ed.) (2017). *Handbook of X-ray Imaging: Physics and Technology*. CRC Press, Boca Raton.
41. Friis, E. M., Marone, F., Pedersen, K. R., Crane, P. R. and Stampanoni, M. (2014). Three-dimensional visualization of fossil flowers, fruits, seeds, and other plant remains using synchrotron radiation X-ray tomographic microscopy (SRXTM): new insights into Cretaceous plant diversity. *Journal of Paleontology* 88(4), 684–701.
42. Gudkov, S. V., Grinberg, M. A., Sukhov, V. and Vodeneev, V. (2019). Effect of ionizing radiation on physiological and molecular processes in plants. *Journal of Environmental Radioactivity* 202, 8–24.
43. Blaser, S. R. G. A., Schlüter, S. and Vetterlein, D. (2018). How much is too much?– influence of X-ray dose on root growth of faba bean (Vicia faba) and barley (Hordeum vulgare). *PLoS ONE* 13(3), e0193669.
44. Zappala, S., Helliwell, J. R., Tracy, S. R., Mairhofer, S., Sturrock, C. J., Pridmore, T., Bennett, M. and Mooney, S. J. (2013). Effects of X-ray dose on rhizosphere studies using X-ray computed tomography. *PLoS ONE* 8(6), e67250.
45. Kehinde, D. O., Ogunwenmo, K. O., Ajeniya, B., Ogunowo, A. A. and Onigbinde, A. O. (2017). Effects of X-ray irradiation on growth physiology of Arachis hypogaea (Var. Kampala). *Chemistry International* 3(3), 296–300.
46. Jiang, Y. and Li, C. (2020). Convolutional neural networks for image-based high-throughput plant phenotyping: a review. *Plant Phenomics* 2020, 4152816.
47. Samiei, S., Rasti, P., Vu, J. L., Buitink, J. and Rousseau, D. (2020). Deep learning-based detection of seedling development. *Plant Methods* 16(1), 103.
48. Colmer, J., O'Neill, C. M., Wells, R., Bostrom, A., Reynolds, D., Websdale, D., Shiralagi, G., Lu, W., Lou, Q., Le Cornu, T., Ball, J., Renema, J., Flores Andaluz, G., Benjamins, R.,

Penfield, S. and Zhou, J. (2020). SeedGerm: a cost-effective phenotyping platform for automated seed imaging and machine-learning based phenotypic analysis of crop seed germination. *New Phytologist* 228(2), 778–793.
49. Yang, S., Zheng, L., He, P., Wu, T., Sun, S. and Wang, M. (2021). High-throughput soybean seeds phenotyping with convolutional neural networks and transfer learning. *Plant Methods* 17(1), 50.
50. Gómez-de-Mariscal, E., García-López-de-Haro, C., Ouyang, W., Donati, L., Lundberg, E., Unser, M., Muñoz-Barrutia, A. and Sage, D. (2021). DeepImageJ: a user-friendly environment to run deep learning models in ImageJ. *Nature Methods* 18(10), 1192–1195.
51. Ramsundar, B., Eastman, P., Walters, P. and Pande, V. (2019). *Deep Learning for the Life Sciences: Applying Deep Learning to Genomics, Microscopy, Drug Discovery, and More*. O'Reilly Media, Inc., Boca Raton.

Chapter 7

Advances in testing seed health

Peter Bonants, Wageningen University & Research, The Netherlands

1 Introduction

We need plants to feed the world's growing population. Plants must be healthy to meet this need. Plant pests destroy up to 40% of global crops annually with losses estimated at $220 billion. Plant pests are becoming more widespread due to climate change, posing a threat to food security (Delgado-Baquerizo et al., 2020). Warming increases the abundance of soil-borne potential fungal plant pathogens.

Many plants are grown from seeds, and these seeds are transported all over the world for breeding and production. Seed-borne pests pose a serious threat to cropping systems as they can be disseminated to many regions around the world (Chalam et al., 2020). With increasing global seed production and trade, seed-health testing is an important quality control measure to prevent the introduction and spread of these harmful pests (Kumar et al., 2020).

Several fungi, bacteria, viruses, viroids and even nematodes are recognized as seed transmittable. Seed-borne pathogenic fungi can cause serious damage to crops by reducing the germination, vigor and emergence of the seeds. An example of world-wide dissemination of a seed-borne pathogen is the outbreak of potato spindle tuber viroid in tomato (Van Brunschot et al., 2014b). There is continuing research on whether some pathogens are seed-transmissible, for example, seed transmission of Zucchini yellow mosaic virus has been

http://dx.doi.org/10.19103/AS.2022.0105.10

experimentally verified in the last 10 years or so (Simmons et al., 2011). A recent review is 'Seed-Borne Diseases of Agricultural Crops: Detection, Diagnosis & Management' (Eds. Kumar et al., 2020) which lists many fungi, bacteria, viruses and viroids as seed-transmissible. It also summarises several methods for detection.

A seed health test has historically consisted of three steps:

- isolation of the pathogen from seeds;
- detection and identification of the pathogen;
- confirmation of the viability of the isolate and its pathogenicity by inoculating it into assay plants.

This type of test identifies the presence of a pathogen directly and is only possible for bacteria but not for other pests like viruses. More recently, indirect tests have been developed to complement direct tests, including enzyme-linked immunosorbent assay (ELISA) and polymerase chain reaction (PCR) tests, which identify the presence of a protein or a piece of the genome of the pathogen. Indirect tests used for pathogen detection do not yield any information about pathogen viability and are not sufficient to prove pathogenicity.

Traditional methods for detection are:

Fungi:	Blotter or media test combined with morphological examination.
Bacteria:	Dilution plating, (semi-)selective media, confirmation by PCR and/or immunofluorescence (IF), ELISA. PCR has now been for more than 25 years.
Viruses:	antibody-based detection, including ELISA and electron microscopy.

Morphological observation is a lengthy and laborious process. Poly- or monoclonal antibodies have been used extensively since their development in the 1980s. Since their introduction, molecular methods such as PCR, qPCR (quantitative polymerase chain reaction) and isothermal methods have been developed and tested on infected seed lots. The last decade has also seen the use of high-throughput sequencing (HTS) methods in seed-health testing as well as phenotyping methods using spectral camera systems.

Internationally agreements have been concluded to secure the quality of seeds. International organizations such as ISF (International Seed Federation), ISHI-Veg (International Seed Health Initiative for Vegetable Crops), ISTA (International Seed Testing Association), EPPO (European Plant Protection Organization) and IPPC (International Plant Protection Convention) have developed rules for controlling seed quality:

- ISF is a non-governmental, non-profit organization which has represented its members' interests since 1924 and is widely regarded as the voice of the global seed industry.

- ISHI-Veg came into being in 1993 to provide a forum for exchanging information, sharing expertise on seed-health testing and facilitating the development of new seed-health tests. Today, their members account for 70%–75% of vegetable seed that is traded internationally.
- ISTA is an independent organization supported by experienced seed scientists and analysts and has been developing seed testing methods since 1924. It is regarded as a leading authority in seed science and technology.
- EPPO is an intergovernmental organization responsible for cooperation in plant health within the Euro-Mediterranean region. Founded in 1951 by 15 European countries, EPPO now has 52 members. It develops standards on diagnostics methods together with expert panels.
- IPPC is an intergovernmental treaty signed by over 180 countries aiming to protecting the world's plant resources from the spread and introduction of pests and promoting safe trade in plants and seeds for breeding and other purposes. The convention introduced International Standards for Phytosanitary Measures (ISPMs) to govern the movement of plants and seeds, making it the sole global standard setting organization for plant health.

ISF, ISHI-Veg, ISTA and EPPO have developed, validated and published testing procedures using dedicated seed-health committees and diagnostic panels. Global standards in seed-health testing have been outlined (Aveling, 2014). All these methods are based on detecting a single pathogen. Some methods have not been validated. Method validation is a critical aspect of method development in ISHI-Veg, ISTA and EPPO. Performance criteria include:

- specificity;
- sensitivity;
- selectivity;
- repeatability;
- reproducibility.

Validation involves assessing a method against these criteria to demonstrate that the method is fit for purpose. Certification of disease-free seed, seed potatoes, bulbs and cuttings is essential to allow safe trade of these materials between countries. Phytosanitary certification of seed lots is supported by validated test methods.

Pests can be present externally on the surface of the seed or internally in the seed. This may have a significant impact on detection and makes the development of an efficient detection method. A test method must also account for factors such as mode of infection and the number of seeds to be tested.

2 Traditional detection methods

Conventional methods for detecting different seed pathogens (fungi, bacteria, viruses and viroids) include visual and/or microscopic examination, deep freezing blotter methods, NaOH seed soak methods, use of selective growth media, electron microscopy and bioassays. Disadvantages of these methods include poor reliability, specificity, sensitivity, as well as being time-consuming to perform.

ISHI-Veg has conducted a comparative test on four semi-selective agar media for the detection of *Xanthomonas campestris* pv. *campestris* in brassica seeds to establish the media to use (Koenraadt et al., 2005). A seed-soaking method has been developed to detect the seed-borne fungi *Sclerotinia sclerotiorum* and *Phomopsis* spp. in soybean seeds (Ramiro et al., 2019). Blotter and plating assays have been compared with PCR based methods for the detection of *Alternaria* species in carrot seeds (Konstantinova et al., 2002).

Because the traditional deep-freezing blotter method and plating on a selective medium are time-consuming and laborious, a PCR-based assay has been developed. A positive correlation was found between the percentage of seed infection identified by the blotter method and the PCR assay. The PCR-assay based on specific primers seems to be a good alternative to the deep-freezing blotter method and plating on *Alternaria radicina* selective agar, especially when speed of detection is an important issue. A selective medium has been developed for the specific detection of *Alternaria dauci* and *A. radicina* in carrot seeds (Strandberg, 2002). The advantage of using (semi-) selective media is the ability to detect living organisms although time is needed for growth.

3 Serological detection techniques

Serology is the scientific study of serum and other body fluids. In practice, the term usually refers to the diagnostic identification of antibodies in blood serum. Such antibodies are typically formed in response to an infection. By injecting specific proteins or whole plant pathogens (viruses and bacteria) into rabbits, one can, after a certain period, isolate antibodies from blood serum.

An antibody (Ab), also known as an immunoglobulin (Ig), is a large, Y-shaped protein used by the immune system to identify and neutralize foreign objects such as pathogenic bacteria and viruses. Abs are heavy (~150 kDa) proteins of about 10 nm. Abs can either be polyclonal or monoclonal, depending on the way they are produced. Polyclonal Abs (pAbs) are Abs that are secreted by different B cell lineages within the body, whereas monoclonal antibodies come from a single B cell lineage.

pAbs have been developed for many plant pathogenic viruses and bacteria and used in several immunological methods. The most well-known method is DAS-ELISA (double antibody sandwich ELISA). An ELISA method has for example, been developed to detect *Corynebacterium sepedonicum* in seed potato tubers (Corbière et al., 1987). Three immuno-sera were needed, at least two from different animals. An important aspect of antibody tests is specificity, especially when pAbs are used. An example is a study of the specificity of two antisera to whole cells of *Erwinia chrysanthemi* (Ech) serogroup O1Ha in a DAS-ELISA (Van der Wolf and Gussenhoven, 1992). The antisera were tested against 100 strains of different plant pathogenic bacteria, including 39 Ech strains. Two saprophytes (A254 and A256), both identified as pectinolytic *Pseudomonas fluorescens* species, cross-reacted strongly with pAbs against Ech. Non-specific reactions were found in DAS-ELISA with 16 saprophytes due to cross reactivity of the pAbs used.

A DAS-ELISA in combination with PCR has been developed and tested for the detection of *Tilletia caries*, causal agent of common bunt of wheat, using mycelial homogenates of cultures of *T. caries* for immunization (Eibel et al., 2005). To develop a practical detection programme for Bean pod mottle virus (BPMV), a cocktail ELISA nested reverse transcriptase (RT)-PCR using a combination of serological and molecular methods was designed for soya bean seeds (Jin et al., 2016).

Many serological assays developed so far for the detection of plant pathogens are able to identify the presence or absence of the target pathogen in soil and plant materials. However, they are not able to discriminate between virulent and avirulent strains of the pathogen and were not specific to strains and races. Using virulent bacterial cells of *Ralstonia solanacearum* (encapsulated with mucin) from tomato seeds as an antigen for immunization, polyclonal antisera were developed in rabbit using a classical immunization protocol (Rajeshwari et al., 1998). The resulting ELISA has been found to be sensitive, specific and rapid, therefore suitable for the detection of *R. solanacearum* isolates from tomato seeds during routine assays. However, the ELISA is also virulent strain-specific as it recognizes the virulent exopolysaccharide component as an antigenic determinant.

ELISA has traditionally used ELISA micro-titre plates (96 wells). The last decade has also seen the development of microfluidics ELISA techniques for the detection of plant pathogens. An example is a microfluidic-based sandwich ELISA for *Acidovorax citrulli*, watermelon silver mottle virus and melon yellow spot virus (Thaitrong et al., 2013). The immunoassay occurs on the surface of a reaction chamber within a microfluidic channel. The capillary force within the microchannel draws a reagent into the reaction chamber and facilitates assay incubation, resulting in more rapid analysis.

Fluorescently labelled antibodies can also be used in 'fluorescent in situ hybridization'. An example of this technique is detection of the seed-borne virus Cucumber green mottle mosaic virus (CGMMV) in plant tissue (Shargil et al., 2015).

Antibodies can also be used in lateral flow devices (LFDs). LFDs are simple devices intended to detect the presence of a target substance in a liquid sample. LFDs operate on the same principles as ELISAs. In essence, these tests run the liquid sample along the surface of a pad with reactive molecules that show a visual positive or negative result. Different LFD assays have been developed for Potato virus Y (PVY) (Razo et al., 2018), Potato leafroll virus (PLRV) (Panferov et al., 2018), Potato virus X (PVX) (Drygin et al., 2012) and more generally for plant pathogens (Danks and Barker, 2000). An example for detection of seed pathogens is a LFD assay for soybean mosaic virus (Zhu et al., 2016).

Many different antibodies can also be spotted on an immuno-array which combines a number of immunological detectors. An example is simple, rapid, visual antibody array for the simultaneous detection of multiple plant pathogens in seeds (Xiong et al., 2013). This array system was used to screen 24 seed samples (for 10 different plant pathogens) and has shown a high consistency (240/240) compared with results from ELISA tests, indicating that this antibody array assay system can be used as a rapid method to simultaneously analyse at least ten pathogens, not only for lab research but also for practical applications.

4 Nucleic acid-based detection methods

4.1 PCR

Since the discovery of the structure of DNA and RNA, new methods have been developed for detection of pathogens. In 1993, the Nobel Prize in Chemistry was awarded to Mullis and Smith for their invention of PCR and oligonucleotide-based, site-directed mutagenesis. Many researchers have subsequently used PCR in developing detection methods for RNA and DNA targets. Specific sequences of plant pathogen RNA and DNA have been used to develop a broad range of methods to detect pathogens in all kind of substrates such as plants, seeds, soil, water and even in air.

Specific RNA and DNA sequences have been used in the seed sector for PCR-based detection of many seed-borne pathogens, including viruses (López et al., 2006.), bacteria (Audy et al., 1996; Koenraadt et al., 2009) and fungi (Pryor and Gilbertson, 2001). Different variations using PCR amplification methodology have also been developed. One example is BIO-PCR in which the sample is incubated on a (semi-)selective medium, and, after colony growth, colonies are checked for the correct identification by colony PCR or the whole plate is checked by PCR for the target organism. This approach has been used in BIO-PCR detection of potato blackleg pathogens from stem, tuber and soil samples with perfect sensitivity (Aono et al., 2021). BIO-PCR could both detect that an organism was alive and identify all four blackleg pathogens from symptomatic stems, progeny tubers and the soil around the stem of diseased plants. This is a promising technique for efficient detection of potato blackleg pathogens not only for seed inspection but also for epidemiological research and disease surveillance.

Another example of BIO-PCR has been developed for the detection of *Xanthomonas oryzae* pv. *oryzae* using hrp gene sequences (Singh et al., 2015). The bacterium was detected from seeds after 8 months of storage. Magnetic capture hybridization-PCR has been performed for the simultaneous detection of *Acidovorax avenae* subsp. *citrulli* and *Didymella bryoniae* in cucurbit seed lots (Ha et al., 2009). Single-stranded DNA hybridization capture probes have been used to target DNA of *A. avenae* subsp. *citrulli* (causal agent of bacterial fruit blotch) and *D. bryoniae* (causal agent of gummy stem blight) from cucurbit seed samples. DNA was concentrated by being covalently attached to magnetic particles. Sequestered template DNA was subsequently amplified by multiplex real-time PCR using pathogen-specific TaqMan PCR assays.

Many of the methods for detection of seed-borne pests have been published on websites by the relevant authorities, for example:

- ISHI-Veg protocols (https://worldseed.org/our-work/phytosanitary-matters/seed-health/ishi-veg-protocols/);
- ISTA rules (https://www.seedtest.org/en/home.html);
- EPPO standards (https://www.eppo.int/RESOURCES/eppo_standards).

In the development of PCR-based methods for detection, it is very important to know not just the sequences of isolates of the target species but also of their closest relatives which need to be negative in the PCR reaction. The 3′-region of the forward and reverse primers of isolates of pathogens needs to be identified. Therefore reference collections are indispensable. The Microbial Resource Research Infrastructure (MIRRI) is the pan-European distributed research infrastructure for the preservation, systematic investigation, provision and valorization of microbial resources and biodiversity. Many national reference collections are part of MIRRI (https://www.mirri.org/), such as Insitute Pasteur, Belgian Coordinated Collections of Microorganisms, Westerdijk Fungal Biodiversity Institute and many more.

An important aspect of using molecular methods is that the DNA/RNA must be extracted from the seeds being analysed. During this extraction, substances that can inhibit the PCR reaction, so-called inhibitors, may also be introduced. Examples of inhibitors are humic acids, phenolic components, fulvic acids and heavy metals. Substances such as polyvinylpyrodon and bovine serum albumin can inactivate these inhibitors during the PCR reaction. Because removing those inhibitors is important, diagnostic companies are developing numerous

DNA/RNA extraction kits to address this problem. Adding an internal extraction and amplification control reveals inhibition in a sample.

Combinations of serological and molecular-based methods have also been developed. A cocktail ELISA and nested PCR has been developed to detect BPMV in the large quantities of soya bean seed imported into China (Jin et al., 2016). The single-vessel detection assay was performed in a 96-well ELISA plate, which also served as a carrier for the subsequent nested RT-PCR assay. This method was over 100 times more sensitive than immunocapture-RT-PCR. This assay resulted in successful micro-extraction from soya bean seeds and combined the advantages of each individual technique.

4.2 Quantitative PCR (q-PCR) and multiplex q-PCR

Quantitative or qPCR was developed in the 1990a by Kary Mullis. It monitors the amplification of a targeted DNA molecule during PCR (i.e. in real time) rather than at the end, as in conventional PCR. Primers used for normal PCR are added to a target specific probe including a fluorescent dye at the 5′ end and a second quencher molecule at the 3′ end of the probe. Using a specific Taq polymerase with exonuclease activity, the fluorescent dye is no longer quenched and fluorescence is increased during PCR amplification. By monitoring the increase of fluorescence during each cycle, the quantity of target DNA can be calculated using a dilution series of target DNA.

There are many examples of the use of qPCR for seed-borne pathogens. Detection of pospiviroids in pepper and tomato seeds has been performed using (real-time) RT-PCR (Botermans et al., 2020). The test consists of two multiplex reactions running in parallel which can detect seven viroids (Citrus Exocortis Viroid, Columnea Latent Viroid, Pepper chat fruit viroid (PCFVd), Potato Spindle Tuber Viroid, Tomato Apical Stunt Viroid, Tomato Chlorotic Dwarf Viroid and Tomato Planta Macho). Dahlia latent viroid was used as an internal extraction and amplification control. Many RNA viruses have been amplified by quantitative RT-PCR (Çevik et al., 2019). A qPCR protocol has been developed for the bacterial pathogen *Xanthomonas citri* pv. *malvacearum* in cotton seeds (Wang et al., 2019) or *Clavibacter michiganensis* subsp. *michiganensis* in tomato seeds (Han et al., 2018). qPCR methods have also been described for seed transmittable fungi, For example, *Pyricularia oryzae* in wheat (Thierry et al., 2020) and *Colletotrichum truncatum* in soybean (Júnior et al., 2020).

Labelling a TaqMan probe with different dyes makes multiplex PCR possible. Depending on PCR machine and the dyes used, a multiplex of six is possible. There are many different dyes in the wavelength range from 450 to 740 nm. The most used dyes are: 6-6-Carboxyfluorescein (6-FAM), VIC, TET, hexachlorofluorescein (HEX), JOE, Yakima Yellow, ROX, Texas Red, ATTO and Cy5. Using different dyes for a TaqMan probe, a multiplex pPCR assay has been developed for different pathogens in soybean seeds (Ciampi-Guillardi et al., 2020). Internal controls labelled with different dyes have been used in multiplex detection of *C. michiganensis (*Ramachandran et al., 2021). A fully synthetic universal internal control has been designed to detect PCR inhibitors and false-negative results in qPCRs. Multiplex PCR detection can also be used by targeting multiple genes of the same organism. In this way, extra specificity and confirmation can be included in the assay. A triplex TaqMan PCR has been used to target two genes of the bacterium *Xylella fastidiosa* (Bonants et al., 2019). Dyes used were 6-FAM, HEX and Texas Red.

The sequence of the TaqMan probe needs to be highly specific for the target species to be detected. Inclusion of locked nucleic acid nucleotides and/or the use of minor groove binder probes will increase specificity when needed. This option of an extra TaqMan probe makes it possible to increase specificity even more.

Since PCR and qPCR can amplify DNA from both dead and living organisms, there is a need to be able to distinguish viable organisms using PCR. One elegant solution has been to use propidium monoazide (PMA) and its improved structure PMAxx for the detection of *C. michiganensis* subsp. *michiganensis* (Han et al., 2018). This chemical substance can penetrate membranes of dead cells but not membranes of living cells. Once penetrated, it can be covalently bound to the DNA and therefore block DNA amplification by PCR. Both chemicals were used to pretreat *Cmm* prior to DNA extraction, followed by qPCR. Both PMA and PMAxx could bind to the chromosomal DNA of dead bacterial cells and therefore block DNA amplification by PCR. A clear distinction between viable and dead cells of *Cmm* could therefore be made. The viable but non-culturable state of *Cmm* cells was demonstrated in tomato seeds, but false results for alive/dead cells can occur with this technique. *Pseudomonas syringae* pv. viable cells were detected in tomato seed using PMA and a real-time PCR assay (Chai et al., 2020). The detection of RNA instead of DNA can also be used for the distinction between dead and living organisms. RNA is associated with living and DNA with both dead and living organisms. A RNA-based PCR assay has been evaluated to detect viable *Candidatus Liberibacter solanacearum* (Lso) in Lso-contaminated carrot seeds using different disinfection methods (Fujikawa et al., 2020).

4.3 Digital PCR (dPCR)

Digital PCR (dPCR) is another variant of PCR. The PCR reaction is not performed in traditional 10–25 μL but instead in nanoliter reaction vesicles which include all the PCR reaction ingredients and the sample to be analysed. Applications focus on absolute nucleic acid quantification, mutation detection, copy number variation, gene expression studies, gene-editing analysis and next-generation sequencing (NGS) sample quantification. Different platforms are available with different capabilities for multiplexing:

- BioRad: QX ONE Droplet Digital PCR (ddPCR) system 4-plex;
- Qiagen: QIAcuity One 5-plex in 2 h;
- Thermo Fisher Scientific: QuantStudio™ 12K Flex Real-time PCR System, 6 channels;
- Stilla: Naica system, 6-Color, easy to use multiplex digital PCR system with reliable measurements in 3 h.

Droplet digital PCR has been performed for the detection of teliospores of *Tilletia controversa* Kühn in soil to increase sensitivity (Liu et al., 2020). A comparison has been made between PCR, real-time PCR and ddPCR for the qualitative and quantitative measurement of teliospores of *T. controversa* in soil. ddPCR was found to have the most sensitive detection at 2.1 copies/μL.

4.4 Micro-arrays

Micro-arrays can be used to detect multiple pathogens simultaneously. Target-specific probes are bound to a membrane. After hybridization of labelled sample RNA with the membrane and scanning of the membrane, results can be analysed. Microarray detection provides the greatest capability for parallel yet specific testing; the technique can be used to detect individual viruses and combinations of viruses with a sensitivity comparable to ELISA. Various steps in microarray technology (including probe development, array fabrication, assay target preparation, hybridization, washing, scanning and interpretation) are needed for identification of plant viruses (Boonham et al., 2008).

The use of polyprobes or riboprobes, carrying partial sequences of different plant viruses or viroïds fused in tandem, has permitted the detection of up to ten main viruses and viroïds affecting tomato crops (Sánchez-Navarro et al., 2019). The analysis of tomato seeds revealed that both single micro-arrays and polyprobes were able to detect an infected seed in a pool of 250 healthy seeds.

4.5 Padlock probes

Padlock probes (PLPs) are long oligonucleotides with ends that are complementary to adjacent target sequences. Upon hybridization to the target, the two ends are brought into contact, allowing PLP circularization by ligation. Multiplexing is also possible using this approach. PLPs have been evaluated for ten different plant pathogens (Szemes et al., 2005). The use of PLPs has also been demonstrated for simultaneous detection of *X. oryzae* pv. *oryzae* and *X. oryzae* pv. *oryzicola* in rice seed (Tian et al., 2014). It has also been shown that PLPs (P-Xoo and P-Xoc), combined with reverse dot-blot hybridization, can be used to simultaneously detect multiple pathogens in naturally infested commercial rice seed lots.

4.6 LAMP: Loop-mediated isothermal amplification (LAMP)

Point-of-care technology is a term used to describe technology used at the 'site-of-need,' i.e. at the point analysis is needed such as in field conditions. It is difficult to use PCR in the field given the need for thermal cycling equipment across three temperatures. This limitation has led to the development of diagnostic methods which can be used in the field, greenhouses, etc (Cassedy et al., 2020).

Isothermal methods typically employ unique DNA polymerases for separating duplex DNA. DNA polymerases with this ability include:

- Klenow exo-, Bsu large fragment and phi29 for moderate temperature reactions (25–40°C);
- large fragment of Bst DNA polymerase for higher temperature (50–65°C) reactions.

One of the first examples was LAMP (loop-mediated amplification) using Bst enzyme to amplify DNA at 65°C (Notomi et al., 2000) and now commercialized by Eiken Japan (https://www.eiken.co.jp/). Reverse transcription loop-mediated isothermal amplification (RT-LAMP) combines LAMP with a reverse transcription step to allow the detection of RNA. A LAMP application for the on-site detection of Pepper chat fruit viroid (PCFVd) in leaves has been evaluated (Tangkanchanapas et al., 2018). Using sequence-characterized amplified regions, a rapid, visual and sensitive LAMP assay has been developed for *D. bryoniae* (Yao et al., 2016). The sensitivity of the LAMP assay was 1000-fold higher than that of conventional PCR with a detection limit of 0.1 fg μL^{-1} of targeted DNA.

Detection of seed-borne viruses and virus-like pathogens using LAMP or RT-LAMP has also been described (Kaur et al., 2020). PCR and LAMP-based assays have been developed for the detection of *Fusarium fujikuroi* based upon unique NRPS31 gene sequences (Sunani et al., 2019). The diagnostics developed are efficient, rapid and sensitive for the detection of *F. fujikuroi* which is useful for testing seed health, validating varietal resistance and monitoring the pathogen inoculum.

Amplifying multiple pathogens using LAMP simultaneously (multiplex) is challenging. A multiplex method for the detection of soybean root diseases caused by several fungal and oomycete pathogens has been evaluated (Ye et al., 2020). Thirteen LAMP assays were developed and tested on 159 samples.

Most samples (86%) contained two to six pathogen species. The LAMP detection results were confirmed by traditional isolation methods. The isolated strains were found to exhibit different degrees of virulence.

Other enzymes for isothermal amplification have been found and several applications have been developed. These include an ultrasensitive isothermal detection method for a plant pathogen (*Leptosphaeria maculans*) by using recombinase polymerase amplification (RPA) and a gold nanoparticle-enhanced microcantilever sensor (Lei et al., 2021). Considering its high sensitivity and specificity, it was concluded that the proposed RPA-microcantilever assay could have wide applications in nucleic acid diagnostics for on-site plant pathogen detection and species identification.

Transcription-mediated amplification is another isothermal nucleic acid amplification system utilizing two enzymes, RNA polymerase and reverse transcriptase. nucleic acid sequence-based amplification and reverse transcription-free exponential amplification reaction are also isothermal amplification systems.

A combination of LAMP and LFD has been developed for the detection of *Rhizoctonia solani* (Patel et al., 2015). The use of PMA in combination with LAMP has also been explored to detect viable cells of *Xanthomonas hortorum* pv. *carotae* in carrot seeds (Temple et al., 2013).

4.7 Luminex xMAP

An excellent platform for multiplex detection is Luminex xMAP (multi-analyte profiling) technology. The xMAP technology uses different coloured magnetic beads. By mixing red and infrared fluorophores dyes in different proportions, 100 different beads are obtained. The conjugation of a distinct antibody to a distinct bead allows for analysis of multiple analytes in a single well. Multiple analyte-specific beads can then be combined in a single well of a 96- or 384-well microplate to detect and quantify multiple targets simultaneously using one of the Luminex instruments for analysis.

A mixture of beads are incubated with the sample and analysed in the Luminex xMAP detection system. Lasers can measure the incubated mixture of beads passing the two different laser beams, one measuring the colour of the bead and the other measuring if this bead is loaded with the target. Multiplex detection of three different potato viruses (PVX, PVY and PLRV) in seed potatoes has been described (Bergervoet et al., 2008). The Luminex xMAP platform was able to screen for three different viruses simultaneously. The multiplex system has now been developed to detect six potato viruses (Priegnitz et al., 2019).

A true multiplexed diagnostic method for the detection and identification of all nine currently recognized pospiviroid species in one assay using Luminex bead-based suspension array technology has been developed (Van Brunschot et al., 2014a). All assays of the multiplexed array were shown to be 100% specific, sensitive and reproducible.

4.8 DNA barcoding

DNA barcoding involves the production of PCR amplicons from particular regions to allow sequencing. These sequence data are then used to identify or 'barcode' an organism so it can be distinguished from other species. Specific genes are selected for different organisms. Those genes should have low intraspecies variation and high interspecies variation. For insects, the CO-I gene has been selected as the optimal barcode gene. The 16S rRNA gene has been used mainly for bacteria and phytoplasmas. However, resolution is sometimes not high enough and other genes must be used. the internal transcribed spacer (ITS) regions of the rDNA gene cluster have been used for barcoding fungi but, again, these do not always discriminate between species.

DNA barcoding protocols for selected European Union (EU)-regulated arthropods, bacteria, fungi, nematodes and phytoplasmas have been developed within the Quarantine Organisms Barcoding of Life Project financed by EU 7th framework research programme (Bonants et al., 2010). Specific barcodes were selected to be used to discriminate between the target organism and the most closely related organism. Barcode sequences of relevant quarantine organisms and their relatives have been deposited in a database Qbank (https://www.eppo.int/RESOURCES/eppo_databases/eppo_q_bank), which is now hosted by EPPO (Bonants, 2014). It is also important to test DNA barcoding techniques in different laboratories to validate performance. This has been undertaken for quarantine organisms (Vossenberg van de et al., 2013).

4.9 Next-generation sequencing (NGS)

In the last decade, traditional Sanger sequencing technology has been replaced by NGS platforms. NGS, also known as HTS, is a catch-all term used to describe a number of different modern sequencing technologies, each with their own characteristics. These include:

- Illumina (Solexa) sequencing;
- Roche 454 sequencing;
- Ion torrent Proton/PGM sequencing;
- SOLiD sequencing.

These recent technologies allow HTS of DNA and RNA more rapidly and cheaply than Sanger sequencing. NGS has thus revolutionized the study of genomics and molecular biology. The four main advantages of NGS over classical Sanger sequencing are speed, cost, sample size and accuracy. However, third-generation sequencing techniques, such as the single-molecule real-time technique from PacBio and the MinION technique from Oxford Nanopore, can generate long, error-prone sequencing reads which pose new challenges for fragment assembly algorithms.

The possibility of using the Oxford Nanopore single-molecule sequencing platform as a general method for diagnosis of plant diseases has been examined (Chalupowicz et al., 2019). It was tested both on symptomatic plant samples inoculated with known pathogens as well as seeds. In all the inoculated plants/seeds, pathogens were identified within 1–2 h of running the Nanopore sequencer and were classified into the species or genus level. This study indicates that adoption of the Nanopore platform will significantly improve routine laboratory diagnosis.

Comparative genomics has been used to develop a multiplex PCR assay for detection of the seedborn pathogen *C. michiganensis* (Thapa et al., 2020). Genome-wide phylogenic analyses revealed additional diversity within the genus *Clavibacter*. Pathogenic *C. michiganensis* strains varied in plasmid composition, highlighting the need for detection methods based on chromosomal targets. Two chromosomal genes (rhuM and tomA) were used together with an internal control to develop a triplex assay. This method represents a robust, specific tool for detection of *C. michiganensis* in tomato seed and infected plants. Comparative genomics and phylogenetic analyses have suggested several novel species within the genus *Clavibacter*, including non-pathogenic tomato-associated strains (Osdaghi et al., 2020). The results of phylogenetic analyses revealed the existence of nine hypothetical new species among the investigated strains. Detection and diagnosis of seed-borne viruses and virus-like pathogens using NGS has been reviewed by Kaur et al. (2020).

5 Spectroscopy-based methods

A recently developed tool called a Videometer, which is able to distinguish infected from healthy seeds, is a multispectral vision system which can also measure the colour, texture and chemical composition of seed surfaces (Boelt et al., 2018). Results from multispectral imaging were compared with reference methods with a high correlation. Since there is a need for non-destructive detection methods, multispectral imaging has potential for measuring seed quality in the future.

A hand-held Raman Spectrometer has been used for the non-destructive detection of the insect *Callosobruchus maculatus* within cowpea seeds (Sanchez et al., 2019). Raman spectroscopy can also detect chemical signatures of bruchid larvae as well as their excrement inside intact seeds. This method may prove useful as a rapid means of screening crops for internal pests.

Multispectral imaging using support vector machine has been used to discriminate varieties in rice seeds (Liu et al., 2014). This used a Videometer (Videometer A/S, Hørsholm, Denmark) able to acquire multispectral images at 19 different wavelengths (from the visual to the lower wavelengths of the near-infrared (NIR) region). Multispectral imaging combined with an appropriate algorithm has proved a powerful way for classifying varieties of rice seeds because it is non-destructive, simple and rapid with no pre-treatments required. Non-destructive hyperspectral imaging technology, combined with machine vision and NIR spectroscopy, has been used successfully applied to identify barley seed varieties (Sun et al., 2021).

Hyperspectral image data spanning the visible/NIr wavelength region between 400 nm and 1000 nm have been collected from 500 healthy and diseased rice seeds (Baek et al., 2019). Only two to three wavelengths are needed to differentiate between discoloured, diseased and healthy rice seeds, instead of using the entire hyperspectral imaging wavelength regions. This demonstrates the feasibility of developing a low-cost multispectral imaging technology based on these selected wavelengths for non-destructive and HTS of diseased rice seed.

6 Case study: detecting pathogens in tomato seeds

In this chapter we have discussed several methods for seed-health testing and its applications to different seeds. In this section, we evaluate the use of NGS sequencing to detect the most important seed-transmitted pathogens in tomato and compare them with different available TaqMan PCR methods.

Different DNA and RNA extraction kits were evaluated in combination with the KingFisher platform and TaqMan PCR assays. The method was validated against different performance characteristics:

- sensitivity;
- specificity;
- repeatability;
- reproducibility;
- selectivity;
- robustness.

Tests used tomato seed lots spiked with different tomato pathogens and naturally contaminated tomato seed lots. DNA and RNA extracts were also analysed by NGS sequencing using a CLC genomics-based automated workflow. This showed that NGS can be used for detection of pathogens with similar qualitative and quantitative detection specifications.

6.1 Plant pathogens used in the NGS/TaqMan PCR study

C. michiganensis subsp. *michiganensis* (Cmm), a seedborn bacterial pathogen, is the causal agent of bacterial canker of tomato. Dilution plating on a semi-selective medium has been used for initial screening (GSPP, 2010). Given its specificity and sensitivity, amplified fragment length polymorphism (AFLP) analysis (Rijlaarsdam et al., 2004) was conducted on a large set of Cmm and so-called look-alikes, resulting in a specific AFLP fragment unique to all pathogenic strains of Cmm. Sequence analysis was performed and a specific TaqMan PCR was developed with 100% specificity (Berendsen et al., 2011).

Pepino mosaic virus (PepMV), a positive, single-stranded RNA virus, is a member of the genus Potexvirus, family Flexiviridae (Adams et al., 2004) and is pathogenic to tomato (*Solanum lycopersicum* L.) (van der Vlugt et al., 2000). The genome of PepMV is approximately 6500 bp in length and is composed of five open reading frames. A TaqMan PCR has been described for the detection of this virus (Kai-Shu Ling et al., 2007).

A complex of four *Xanthomonas* species is responsible for bacterial spot disease on tomato and pepper: *X. euvesicatoria* (*Xe*); *X. perforans* (Xp); *X. vesicatoria* (Xv) and *X. gardneri* (Xg) (Jones et al., 2004). Several studies have been published which detail diagnostic primers for bacterial spot pathogens. Most notably four primer pairs were developed from the results of an AFLP study, which allow the identification of the four Bacterial spot-causing xanthomonads (BSX) species (Koenraadt et al., 2009). These primers were validated on a collection of 70 strains. In addition, further validation of the specificity of these primers was published by a research group which tested a collection of 160 BSX strains in Brazil. In these studies, the primer pairs were 100% specific. Baldwin et al. (see Table 1) developed TaqMan PCRs based on

Table 1 Contaminated seed lots used in the study

1 Cmm inoc. Fera
2 Cmm crop Canada 2
3 Cmm crop Canada 3
4 Cmm crop Canada 4
5 X. sp. crop Canada
6 X. ves. inoc. Fera
7 ZZB-652
8 ZZB-533

the four specific PCRs, and these primers and probes have been used in the current study.

P. syringae pv. *tomato* (Pst) is the causal agent of bacterial speck of tomato plants. This bacterium produces angular necrotic spots on leaves and fruits, followed by the appearance of a chlorotic halo. The disease causes serious economic losses especially on fruits of susceptible genotypes (Zaccardelli et al., 2002). Zaccardelli et al. (2005) developed a rapid detection method based on PCR amplification of Pst. Based upon these PCR primers, a specific TaqMan PCR method for Pst has been developed by H. Koenraadt (Naktuinbouw) and used in the study described in this section.

6.2 Isolates used in the study

C. michiganensis michiganensis IPO542 (Cmm), *X. perforans* IPO1067 (Xp), *X. euvesicatoria* IPO1088 (Xeuv), *X. vesicatoria* IPO1059 (Xves), *X. gardneri* IPO1073 (Xg), *P. syringae tomato* IPO919 (Pst), *Acidovorax cattleyae*, isolate number NBC 430, IPO3677 (Acat).

6.3 DNA-extraction method for tomato seeds

Ten thousand tomato seeds were mixed with 150 mL SEB (seed extraction buffer) (pH 7.4): 7.75 g Na_2HPO_4 + 1.65 g KH_2PO_4 + 0.2 mL Tween 20 in 1 L) in a stomacher bag (universal long; art.# 470100 Bioreba) on one side of the gaze. This bag was incubated for 16 h at 4°C. The bag was treated in a stomacher (Bagmixer 400 CC (catalog# 024 230 InterScience) for at least 4 min at setting 4. Fifty millilitres of liquid supernatant (looking milky and with seeds partly broken) were taken from the other side of the gaze. The bacterial spike was added to the supernatant. The 50 mL tubes were centrifuged at 1000 × g for 1 min. Forty-five millilitres of the supernatant were transferred to a new 50 mL tube and centrifuged at 5000 × g for 5 min. After removing the supernatant, the pellet was adjusted to 3 mL and divided over three Eppendorf tubes and used for DNA extraction.

Ninety milligram of glass beads (106 μm and finer; catalog# G-4649 Sigma) and one stainless steel bead of 3.2 mm (catalog# 11079132c, BioSpec Products) were added to eight-strip collection microtubes (catalog# 19560 Qiagen) with caps (catalog# 19566 Qiagen). Bead beating was performed in a TissueLyser II Qiagen with adapter set (catalog# 85300 and 69984 Qiagen). Further DNA extraction was performed with the BioSprint 96 One-For-All Vet kit (catalog# 947057 Qiagen Benelux B.V.) according to the manufacturer's instruction on the KingFisher platform (type and Company). DNA was eluted in 100 μL.

6.4 RNA-extraction method tomato seeds

One thousand tomato seeds were mixed with 20 mL GH + buffer (6M Guanidine hydrochloride, 0.2 M sodium acetate, pH 5.2, 25 mM EDTA, 2.5% PVP-10) in a stomacher bag (universal long; art. # 470100 Bioreba) on one side of the gaze. The bag was incubated for 1 h at room temperature. Stomacher (Bagmixer 400 CC (catalog# 024 230 InterScience) was performed for 4 min at setting 4. Ten millilitres of the liquid (milky look and with seeds partly broken) were taken from the other side of the gaze. The liquid was pipetted into a 50 mL tube, which was subsequently centrifuged at 5000 × g for 5 min. The supernatant was used for RNA extraction using the BioSprint 96 One-For-All Vet kit (catalog# 947057 Qiagen Benelux B.V.) on the KingFisher platform. Finally, the RNA was eluted in 100 µL.

6.5 TaqMan PCR

The TaqMan PCR reactions for bacteria were performed in a final volume of 25 µL, containing 1 × Takara Premix ExTaq (cat. No. RR 039), 0.5 × Rox Dye II, 300 nM both forward and reverse primer, 100 nM probe and 2 µL DNA extract. Thermal cycling conditions were: initial denaturation 2 min 95°C, 40 cycles for 15 s at 95°C and 1 min at 60°C.

The TaqMan PCR reactions for PepMV were performed in 25 µL, containing 1 × TaqMan RT-PCR mix, TaqMan RT-Enzyme (Life Technology TaqMan RNA to Ct 1-Step kit), 300 nM both forward (2) and reverse (2) primers, 100 nM probe and 2 µL total RNA extract. Thermal cycling conditions were incubation for 30 min at 48°C, denaturation for 10 min at 95°C, 40 cycles for 15 sec at 95°C and 1 min at 60°C.

6.6 NGS

The Illumina HiSeq 2500 platform has been used in our study to investigate detection of tomato pathogens in the different tomato seed lots.

6.7 Results

NGS analysis was performed on DNA and RNA extracts of eight different tomato seed lots (Table 1). In the EU project 'TESTA,' these DNA extracts were analysed following this NGS pipeline (Fig. 1). Simultaneously, the DNA extracts were also analysed with the TaqMan PCRs available for the different pathogens (Table 2). The results of the sequencing analysis and the TaqMan PCR for the different seed lots are shown in Table 3.

When analysing eight tomato seed lots for the presence of *C. michiganensis michiganensis*, we found a very high number of reads for

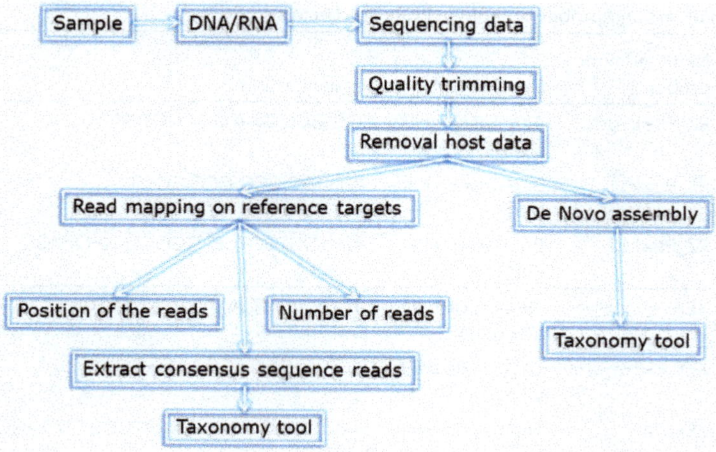

Figure 1 Sequence analysis pipeline used in this study.

tomato. A good correlation between the TaqMan PCR results and the NGS results for Cmm was observed for seed lot 1–4. Seed lots 1, 2, 3, 4, 7 and 8 were clearly contaminated with Cmm. The number of Cmm reads for lot 7 and 8 were low, while TaqMan results were clearly positive. Seed lot number 5 showed a relatively high number of Cmm reads compared to 7 and 8 but a low Ct value from the TaqMan PCR. A possible explanation is the reference sequences used for mapping reads for the different seed lots. Single-nucleotide polymorphisms could be found between the Cmm contaminated tomato seed lots and the Cmm542 reference genome used for the mapping showing that NGS can be used for 'track and trace' studies (Fig. 2). RNA extracts of the seed lots were also analysed for the presence of the RNA virus PepMV. Results (Fig. 3) show clearly that in this seed lot the Chili-2 strain of PepMV was present.

7 Summary and future trends

In summary, we can state that the development of detection methods has made enormous progress over recent decades. From traditional plating, microscopic analysis, via immunological methods such as ELISA to molecular tools, sensitivity and specificity have improved. Since the development of PCR in 1993, molecular methods have been used increasingly for sensitivity and specificity detection and identification of many seed-borne plant pathogens. DNA and RNA sequencing has made great progress since the introduction of the third-generation of sequence technology. The price of genome sequencing has also dropped significantly and made these methods affordable for many seed-testing laboratories.

Table 2 Primers and probes of TaqMan PCRs used in this study

Target	Name primers/ probes	Developed by	Sequence (5′-3′)
Cmm	RZ_Ptssk10	Berendsen et al. (2011)	GGG GCC GAA GGT GCT GGTG
	RZ_Ptssk11	Berendsen et al. (2011)	CGT CGC CCG CCC GCTG
	RZ_Ptssk12	Berendsen et al. (2011)	FAM-TGGTCGTCCTCGGCG-NFQ-MGB
Xe	XEF	Thomas Baldwin, Vilmourin	CTCGCTCATCAAAGTGATAACGCC
	XER	Thomas Baldwin, Vilmourin	GGGCTTGGCAGGAACGGC
	XEFAM	Thomas Baldwin, Vilmourin	FAM-TCCGGCGAGGCAATGCGCTATAGCT-BHQ
Xv	XVF	Thomas Baldwin, Vilmourin	GTGCCGTTGAAATACTTGCTAGCAG
	XVR	Thomas Baldwin, Vilmourin	CACGCTACGGGCCGCAA
	XVFAM	Thomas Baldwin, Vilmourin	FAM-TTCGCACCGCGGGCCCTGTTCT-BHQ
Xp	XPF	Thomas Baldwin, Vilmourin	GTCGTGTTGATGGAGCGTTCCC
	XPR	Thomas Baldwin, Vilmourin	CCGTCTGCTACACGACTTCCGA
	XPFAM	Thomas Baldwin, Vilmourin	FAM-TCTCCCACACCGCGATAGGATTGACAGTAGA-BHQ
Xg	XGF	Thomas Baldwin, Vilmourin	ACCTGCTCCACAACGCGCTC
	XGR	Thomas Baldwin, Vilmourin	GCTTGAATCTGTTTTTTCATTGGGATG
	XGFAM	Thomas Baldwin, Vilmourin	FAM-TCCCATCAATAGTTGCTGCGCTATAG CTTTTCT-BHQ
Pst	HrpZ-F	Harrie Koenraadt, Naktuinbouw	ACAGCTCCCTGGGTAATCCT
	HrpZ-R	Harrie Koenraadt, Naktuinbouw	AAACCACGGTCGATGAGTTG
	HrpZ-VIC	Harrie Koenraadt, Naktuinbouw	VIC-CGCGGCCAATGGCAATGC-BHQ1

(*Continued*)

Table 2 (Continued)

Target	Name primers/ probes	Developed by	Sequence (5′-3′)
PepMV	KL05-48F1	Ling et al. (2007)	ACTCCTAGAGCTGACCTCAC
	KL05-49F2	Ling et al. (2007)	ACTCCTAGAGCTGATCTTAC
	KL05-51R1	Ling et al. (2007)	TCTCCAGCAACAGGTTGGTA
	KL05-52R2	Ling et al. (2007)	TCACCTGCAACTGGTTGATA
	KL05-50	Ling et al. (2007)	FAM-TGTCAGCTTGCATTTACTTCCAAAA-BHQ
PSTVd	PSTV-231F	Boonham et al. (2004)	GCCCCCTTTGCGCTGT
	PSTV-296R	Boonham et al. (2004)	AAGCGGTTCTCGGGAGCTT
	PSTV-251T	Boonham et al. (2004)	FAM-CAGTTGTTTCCACCGGGTAGTAGCCGA-BHQ
Acat	Acat_NPC290d	Harrie Koenraadt, Naktuinbouw	TGTAGCGATCCTTCACAAG
	Acat_NPC290e	Harrie Koenraadt, Naktuinbouw	TGTCGATAGATGCTCACAAT
	Acat_NPC290f-P	Harrie Koenraadt, Naktuinbouw	Texas red-CTTGCTCTGCTTCTCTATCACG-BHQ
Nad5	Nad5-F	Menzel et al. (2002)	GATGCTTCTTGGGGCTTCTTGTT
	Nad5-R	Menzel et al. (2002)	CTCCAGTCACCAACATTGGCATAA
	Nad5-P	Botermans et al. (2013)	VIC- AGGATCCGCATAGCCCTCGATT TATGTG-BHQ

All of the different methods described have their pros and cons. Antibody-based methods rely on the specificity of the antibodies used. Monoclonal antibodies are preferred above pAbs. Molecular methods are often selected to increase sensitivity and specificity. PCR and TaqMan PCR use specific sequences from a particular region of pathogen DNA or RNA. Often barcode regions are used because of genetic information held in databases. It is often discussed which gene should be used to have optimal sequence homology within a species and optimal sequence difference between species. This is known as the barcoding gap. PCR-based methods use only a very small part of the genome, sometimes less than 100 nucleotides. Small broken fragments of the genome present in the sample can be amplified and give positive results even when the whole genome of the organism is not present in the chosen sample. Making a selection for other genes which are more involved in infection,

Table 3 NGS results obtained for eight seed lots for the presence different bacterial pathogens of tomato and correlation with the corresponding TaqMan PCRs

Name	Total read count	%Reads	RPKM	Taqman CT	Total read count	%Reads	RPKM	Taqman CT	Total read count	%Reads	RPKM	Taqman CT	Total read count	%Reads	RPKM	Taqman CT
	1				2				3				4			
	Cmm inoc. Fera				Cmm crop Canada 2				Cmm crop Canada 3				Cmm crop Canada 4			
Total tomato chromosome	1480338	95.81	1.2		2189528	97.37	1.21		4730329	97.63	1.2		8514232	98.15	1.19	
Total Xeu	1970	0.13	0.25	ND	1358	0.06	0.12	ND	1408	0.03	0.06	ND	3259	0.04	0.07	ND
Total Xv	2506	0.16	0.32	ND	1594	0.07	0.14	ND	1721	0.04	0.07	ND	4766	0.05	0.1	ND
Total Xg	2712	0.18	0.34	ND	10878	0.48	0.92	24.56	2873	0.06	0.11	31.28	45036	0.52	0.97	25.18
Total Xp	2025	0.13	0.27	ND	1483	0.07	0.13	36.02	1416	0.03	0.06	ND	3606	0.04	0.08	35.22
Total Pst	20775	1.34	2.21	ND	11432	0.51	0.83	32.48	16427	0.34	0.55	36.22	16657	0.19	0.31	31.95
Total Cmm	12039	0.78	2.39	18.47	9725	0.43	1.32	21.9	8561	0.18	0.53	23.6	20014	0.23	0.69	24
Total mitochondrion	4376	0.28	10.78		4975	0.22	8.36		17489	0.36	13.46		15833	0.18	6.73	
	5				6				7				8			
	X. sp. crop Canada				X. ves. inoc. Fera				ZZB-652				ZZB-533			
Total tomato chromosome	8783304	98.13	1.18		3293425	91.84	1.11		5407763	94.93	1.15		8607946	97.31	1.19	
Total Xeu	1251	0.01	0.03	ND	1411	0.04	0.07	ND	1276	0.02	0.04	ND	310	0	0.01	ND
Total Xv	1623	0.02	0.03	36.25	131894	3.68	6.94	16.92	95312	1.67	3.17	35.64	26229	0.3	0.56	ND
Total Xg	5154	0.06	0.11	32.13	4590	0.13	0.24	ND	3826	0.07	0.13	ND	1173	0.01	0.02	ND
Total Xp	1345	0.02	0.03	ND	1173	0.03	0.07	ND	1080	0.02	0.04	ND	331	0	0.01	ND
Total Pst	12451	0.14	0.22	34.96	87405	2.44	3.88	36.98	72835	1.28	2.04	31.83	20989	0.24	0.38	34.46
Total Cmm	5194	0.06	0.17	36.5	304	0.01	0.03	ND	2607	0.05	0.14	26.1	694	0.01	0.02	26.8
Total mitochondrion	31720	0.35	13.02		11019	0.31	11.31		21501	0.38	13.94		34901	0.39	14.66	

RPKM, number of reads per Kilobase mapped reads.

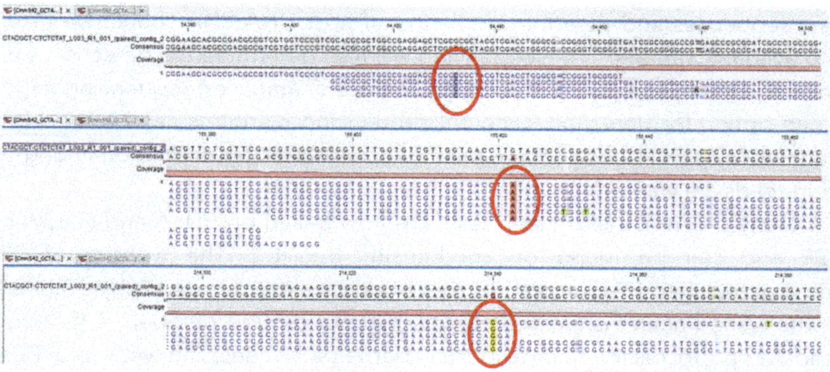

Figure 2 Sequence analysis of DNA extract of a tomato seed lot for the presence of Cmm. Genome sequence of Cmm isolate 542 was used as reference. SNP positions are indicated with red circles.

so-called pathogenicity genes, would constitute a big step forward in functional detection. Because PCR can also detect DNA of dead organisms, the detection of living infectious organisms would increase the understanding of risk.

Multiplex assays, including Luminex and micro-arrays, are more difficult to design and optimization takes more time. On-site methods such as RPA and LAMP are generally less sensitive than TaqMan PCR assays, but their advantage lies in the on-site application.

NGS analysis provides a new way of analysing the whole sample in one. Sequencing the complete DNA/RNA extract of a seed sample will lead to many

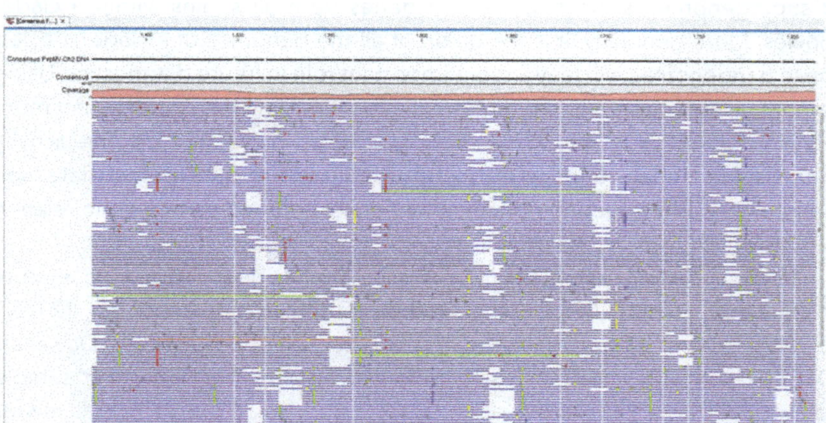

Figure 3 Sequence analysis of RNA extract of a tomato seed lot for the presence of PepMV. Genome sequence of PepMV Chili-2 strain was used as reference.

host sequences and therefore depletion of host DNA/RNA should be taken into account. The depth of sequencing will also determine the detection limit one can obtain for a certain pathogen. The use of amplicon sequencing might be an option for detection. Different sequencing platforms can be used but Illumina, MinION and PacBio are most often selected depending on if long or short reads are wanted.

Many different seed health tests are now based on ELISA and (TaqMan) PCR and can be found on seed-health websites. The website of ISF (International Seed Federation: https://worldseed.org/our-work/phytosanitary-matters/seed-health/ishi-veg-protocols/) shows a huge number of ISHI-Veg validated protocols for a range of seed-borne pathogens. The website of ISTA (International Seed Testing Agency: https://www.seedtest.org/en/home.html) and EPPO (European Plant Protection Organisation: https://www.eppo.int/) can help in finding seed-health tests.

New trends are focussing on on-site test methods such as LAMP, RPA and others. These technologies can be performed by non-specialists, do not need specific lab equipment and can be performed at any site. Mostly they are used on plant material (leaves and stems) with symptoms and not directly on seeds.

What is very important in all kinds of detection and identification methods is the availability of reference isolates, reference genomes and reference databases. But who will decide what is the reference isolate of a certain plant pathogen? And what is the reference genome sequence? Sequence variation within a certain species should be known and therefore more genome sequences of a certain species should be included as a reference. In the fields of molecular biology and genetics, a pan-genome (pangenome or supragenome) is the entire set of genes from all strains within a clade/species. More generally, it is the union of all the genomes of a clade/species. These pangenomes will play a more important role in future diagnostics. Using comparative genomics and bioinformatic tools, we will be able to select more relevant functional genes to be used as targets in diagnostics. The activity of these functional genes will also be used in more detail. Most detection methods are now targeting ribosomal or housekeeping genes with no relation to function of the selected organism.

In the analysis of seed lots, much discussion has been around the amount of seeds to be tested. In analysing seed lots of tomato and capsicum, 12 000–40 000 seeds per lot were analysed for the presence of pospiviroïds (Dall et al., 2019). Statistical calculations using binomial distributions show that sample sizes of 3000 and 9400 seeds allow detection of viroid prevalence as low as 0.1% and 0.032%, respectively, with 95% confidence.

Another growing area is the study of the microbiome of seeds: the community of micro-organisms living in close proximity of seeds. What micro-organisms are there and what is their function in seed health? In a recent study, the seed fungal microbiome (mycobiome) of six commercial sorghum cultivars from South Africa was investigated using a deep amplicon NGS approach based on the ITS region of the ribosomal operon (Kinge et al., 2019). It was shown that a number of molecular operational taxonomic units could be identified at the species level and that certain known pathogens were not present in the tested seeds, for instance, in the Aspergillus group. Other groups could not be identified, not even to genus level. Although ITS data might not be the optimal region for analysis, it was shown that amplicon sequencing is a valuable diagnostic tool for seed disease control and prevention in some cases.

8 Where to look for further information

- ISTA https://www.seedtest.org/en/home.html: Founded in 1924, with the aim to develop and publish standard procedures in the field of seed testing, ISTA (Int. Seed Testing Association) is inextricably linked with the history of seed testing. It has member laboratories in over 80 countries/ distinct economies worldwide.
- ISF International Seed Federation https://www.worldseed.org/: The seed industry has a twofold responsibility for seed health: to deliver sufficiently healthy seed to its customers and to respect national phytosanitary regulations.
- ISHI Veg Int Seed Health Initiative Vegetables: ISHI-Veg's methods for seed health testing are internationally recognized as reference methods and accepted as industry standards https://www.worldseed.org/resources/ishi-veg-protocols/.
- EPPO https://www.eppo.int/: EPPO is an intergovernmental organization responsible for cooperation in plant health within the Euro-Mediterranean region. EPPO standards has been written by experts for the diagnostics of a huge number of pathogens: https://www.eppo.int/RESOURCES/eppo_standards/pm7_diagnostics.
- GSPP https://www.gspp.eu/index.php Good Seed and Plant Practices (GSPP) is an international, transparent business chain system. The purpose of Good Seed and Plant Practices (GSPP) is to prevent tomato seed and plant lots from being infected by *Clavibacter michiganensis* subsp. *michiganensis* (Cmm).
- NSHS https://seedhealth.org/ The National Seed Health System (NSHS) is a program authorized by USDA-APHIS (USA) and administered by the Iowa State University Seed Science Center to accredit both private and public entities to perform certain activities needed to support the issuance of Federal phytosanitary certificates for the international movement of seed.

Detection of seed and propagating material-borne bacterial diseases of economically important crops. 2020. Book: see pdf Seed-Borne Diseases of Agricultural Crops: Detection, Diagnosis & Management. Ravindra Kumar Anuja Gupta Editors (Springer) https://doi.org/10.1007/978-981-32-9046-4.

9 References

Adams, M. J., Antoniw, J. F., Bar-Joseph, M., Brunt, A. A., Candresse, T., Foster, G. D., Martelli, G. P., Milne, R. G., Zavriev, S. K. and Fauquet, C. M. 2004. The new plant virus family *Flexiviridae* and assessment of molecular criteria for species demarcation. *Archives of Virology* 149(5), 1045–1060.

Aono, Y., Nakayama, T., Ozawa, T., Ushio, Y., Yasuoka, S., Fujimoto, T., Ohki, T., Oka, N. and Maoka, T. 2021. Simple and sensitive BIO-PCR detection of potato blackleg pathogens from stem, tuber, and soil samples. *Journal of General Plant Pathology* 87(4), 209–218.

Audy, P., Braat, C. E., Saindon, G., Huang, H. C. and Laroche, A. 1996. A rapid and sensitive PCR-based assay for concurrent detection of bacteria causing common and halo blights in bean seed. *Phytopathology* 86(4), 361–366.

Aveling, T. A. S. 2014. Global standards in seed health testing. *Global Perspectives on the Health of Seeds and Plant Propagation Material* 6(17), 28.

Baek, I., Kim, M. S., Cho, B.-K., Mo, C., Barnaby, J. Y., McClung, A. M. and Oh, M. 2019. Selection of optimal hyperspectral wavebands for detection of discoloured, diseased rice seeds. *Applied Sciences (Switzerland)* 9(5), 1027.

Berendsen, S. M. H., Koenraadt, H., Woudt, B. and Oosterhof, J. 2011. The development of a specific Real-Time TaqMan for the detection of *Clavibacter michiganensis* subsp. *michiganensis*. Poster August 6–10, 2011 APS-IPPC Meeting, Hawaii.

Bergervoet, J. H., Peters, J., van Beckhoven, J. R., van den Bovenkamp, G. W., Jacobson, J. W. and van der Wolf, J. M. 2008. Multiplex microsphere immuno-detection of potato virus Y, X and PLRV. *Journal of Virological Methods* 149(1), 63–68.

Boelt, B., Shrestha, S., Salimi, Z., Jørgensen, J. R., Nicolaisen, M. and Carstensen, J. M. 2018. Multispectral imaging - a new tool in seed quality assessment? *Seed Science Research* 28(3), 222–228.

Bonants, P., Groenewald, E., Rasplus, J. Y., Maes, M., de Vos, P., Frey, J., Boonham, N., Nicolaisen, M., Bertacini, A., Robert, V., Barker, I., Kox, L., Ravnikar, M., Tomankova, K., Caffier, D., Li, M., Armstrong, K., Freitas-Astúa, J., Stefani, E., Cubero, J. and Mostert, L. 2010. QBOL: a new EU project focusing on DNA barcoding of quarantine organisms. *EPPO Bulletin* 40(1), 30–33.

Bonants, P. J. M. 2014. Results of the EU project QBOL, focusing on DNA barcoding of quarantine organisms, added to an international database (Q-bank) on identification of plant quarantine pathogens and relatives (book chapter). In: *Detection and Diagnostics of Plant Pathogens*, 119-133 Gullino, M. L. and Bonants, P. J. M. (Eds), Springer, Dordrecht.

Bonants, P., Griekspoor, Y., Houwers, I., Krijger, M., van der Zouwen, P., van der Lee, T. A. J. and van der Wolf, J. 2019. Development and evaluation of a triplex TaqMan assay and next-generation sequence analysis for improved detection of Xylella in plant material. *Plant Disease* 103(4), 645–655.

Boonham, N., Tomlinson, J. and Mumford, R. 2008. Microarrays for rapid identification of plant viruses. *Annual Review of Phytopathology* 45, 307–328.

Botermans, M., Roenhorst, J. W., Hooftman, M., Verhoeven, J. Th. J., Metz, E., van Veen, E. J., Geraats, B. P. J., Kemper, M., Beugelsdijk, D. C. M., Koenraadt, H., Jodlowska, A. and Westenberg, M. 2020. Development and validation of a real-time RT-PCR test for screening pepper and tomato seed lots for the presence of pospiviroids. *PLoS ONE* 15(9), e0232502.

Cassedy, A., Mullins, E. and O'Kennedy, R. 2020. Sowing seeds for the future: the need for on-site plant diagnostics. *Biotechnology Advances* 39, 107358.

Çevik, B., Kıvrak, H. and Şahin-Çevik, M. 2019. Development of a graft inoculation method and a real-time RT-PCR assay for monitoring tomato chlorosis virus infection in tomato. *Journal of Virological Methods* 265, 1–8.

Chai, A. L., Ben, H. Y., Guo, W. T., Shi, Y. X., Xie, X. W., Li, L. and Li, B. J. 2020. Quantification of Viable Cells of *Pseudomonas syringae* pv. *tomato* in tomato seed using propidium monoazide and a real-time PCR assay. *Plant Disease* 104(8), 2225–2232.

Chalam, V. C., Deepika, D. D., Abhishek, G. J. and Maurya, A. K. 2020. Major seed-borne diseases of agricultural crops: international trade of agricultural products and role of quarantine (book chapter). In: *Seed-Borne Diseases of Agricultural Crops: Detection, Diagnosis & Management*, pp. 25–64 Kumar, R. and Gupta, A. (Eds), Springer.

Chalupowicz, L., Dombrovsky, A., Gaba, V., Luria, N., Reuven, M., Beerman, A., Lachman, O., Dror, O., Nissan, G. and Manulis-Sasson, S. 2019. Diagnosis of plant diseases using the nanopore sequencing platform. *Plant Pathology* 68(2), 229–238.

Ciampi-Guillardi, M., Ramiro, J., De Moraes, M. H. D., Barbieri, M. C. G. and Massola, N. S. 2020. Multiplex qPCR assay for direct detection and quantification of Colletotrichum truncatum, Corynespora cassiicola, and Sclerotinia sclerotiorum in soybean seeds. *Plant Disease* 104(11), 3002–3009.

Corbière, R., Hingand, L. and Jouan, B. 1987. Application des méthodes ELISA et immunofluorescence pour la detection de *Corynebacterium sepedonicum*: réponses variétales de la pomme de terre au flétrissement bactérien. *Potato Research* 30(3), 539–549.

Dall, D., Penrose, L., Daly, A., Constable, F. and Gibbs, M. 2019. Prevalences of pospiviroid contamination in large seed lots of tomato and capsicum, and related seed testing considerations. *Viruses* 11(11), 1034.

Danks, C. and Barker, I. 2000. On-site detection of plant pathogens using lateral-flow devices. *EPPO Bulletin* 30(3–4), 421–426.

Delgado-Baquerizo, M., Guerra, C. A., Cano-Díaz, C., Egidi, E., Wang, J.-T., Eisenhauer, N., Singh, B. K. and Maestre, F. T. 2020. The proportion of soil-borne pathogens increases with warming at the global scale. *Nature Climate Change* 10(6), 550–554.

Drygin, Y. F., Blintsov, A. N., Grigorenko, V. G., Andreeva, I. P., Osipov, A. P., Varitzev, Y. A., Uskov, A. I., Kravchenko, D. V. and Atabekov, J. G. 2012. Highly sensitive field test lateral flow immunodiagnostics of PVX infection. *Applied Microbiology and Biotechnology* 93(1), 179–189.

Eibel, P., Wolf, G. A. and Koch, E. 2005. Detection of *Tilletia* caries, causal agent of common bunt of wheat, by ELISA and PCR. *Journal of Phytopathology* 153(5), 297-306.

Fujikawa, T., Taguchi, M., Kido, K., Kusano, S., Enya, J., Mihara, T., Kaku, H. and Sato, M. 2020. Evaluation of an RNA-based PCR assay to detect viable *Candidatus Liberibacter solanacearum* (Lso) in Lso-contaminated carrot seeds using different disinfection methods. *Journal of Plant Pathology* 102(1), 205–211.

GSPP 2010. Good seed and plant practice. Gouda, The Netherlands. Available at: http://www.gspp.eu/index.php Accessed: 18 August 2021.

Ha, Y., Fessehaie, A., Ling, K. S., Wechter, W. P., Keinath, A. P. and Walcott, R. R. 2009. Simultaneous detection of *Acidovorax avenae* subsp. *citrulli* and *Didymella bryoniae* in cucurbit seed lots using magnetic capture hybridization and real-time polymerase chain reaction. *Phytopathology* 99(6), 666–678.

Han, S., Jiang, N., Lv, Q., Kan, Y., Hao, J., Li, J. and Luo, L. 2018. Detection of *Clavibacter michiganensis* subsp. *michiganensis* in viable but nonculturable state from tomato seed using improved qPCR. *PLoS ONE* 13(5), e0196525.

Jin, J., Shen, J., Cai, W., Liao, F., Gao, F., Chen, X. and Wu, Z. 2016. A cocktail ELISA semi-nested RT-PCR assay to detect Bean pod mottle virus in soya bean seeds. *Journal of Phytopathology* 164(11-12), 904–912.

Jones, J. B., Lacy, G. H., Bouzar, H., Stall, R. E. and Schaad, N. W. 2004. Reclassification of the *Xanthomonas* associated with bacterial spot disease of tomato and pepper. *Systematic and Applied Microbiology* 27(6), 755–762.

Júnior, M. B. S., Resende, M. L. V., Pozza, E. A., Botelho, D. M. S., Cardoso, A. M. S., Siqueira, C. S., Machado, J. C., Resende, A. R. M., Silveira, G. C. D. and Guimarães, S. S. C. 2020. qPCR-based detection of Colletotrichum truncatum in soybean seeds. *Tropical Plant Pathology* 45(5), 550–555.

Kaur, S. I., Kashyap, P. L., Kang, S. S. and Sharma, A. 2020. Detection and diagnosis of seed-borne viruses and virus-like pathogens (book chapter). In: *Seed-Borne Diseases of Agricultural Crops: Detection, Diagnosis & Management*, pp. 169–199 Kumar, R. and Gupta, A. (Eds), Springer.

Kinge, T. R., Cason, E. D., Valverde, A., Nyaga, M. and Gryzenhout, M. 2019. Endophytic seed mycobiome of six sorghum (Sorghum bicolor) cultivars from commercial seedlots using an Illumina sequencing approach. *Mycosphere* 10(1), 739–756.

Koenraadt, H., Van Bilsen, J. G. P. M. and Roberts, S. J. 2005. Comparative test of four semi-selective agar media for the detection of *Xanthomonas campestris* pv. *campestris* in brassica seeds. *Seed Science and Technology* 33(1), 115–125.

Koenraadt, H., van Betteray, B., Germain, R., Hiddink, G., Jones, J. B., Oosterhof, J., Rijlaarsdam, A., Roorda, P. and Woudt, B. 2009. Development of specific primers for the molecular detection of bacterial spot of pepper and tomato. In: *Proceedings of the IInd International Symposiumosium on Tomato Diseases. Acta Horticulturae*, pp. 99–102 Saygili, H. (Eds), Illinois State Historical Society.

Konstantinova, P., Bonants, P. J. M., van Gent-Pelzer, M. P. E., van der Zouwen, P. and van den Bulk, R. 2002. Development of specific primers for detection and identification of *Alternaria* spp. in carrot material by PCR and comparison with blotter and plating assays. *Mycological Research* 106(1), 23–33.

Kumar, R., Gupta, A., Srivastava, S., Devi, G., Singh, V. K., Goswami, S. K., Gurjar, M. S., Aggarwal, R. 2020. Diagnosis and detection of seed-borne fungal phytopathogens. In: *Seed-Borne Diseases of Agricultural Crops: Detection, Diagnosis & Management*, pp 107–142.

Lei, R., Wu, P., Li, L., Huang, Q., Wang, J., Zhang, D., Li, M., Chen, N. and Wang, X. 2021. Ultrasensitive isothermal detection of a plant pathogen by using a gold nanoparticle-enhanced microcantilever sensor. *Sensors and Actuators, Part B,: Chemical* 338, 129874.

Ling, K.-S. W., Patrick, W. P. and Jordan, R. 2007. Development of a one-step immunocapture real-time TaqMan RT-PCR assay for the broad spectrum detection of *Pepino mosaic virus*. *Journal of Virological Methods* 144(1-2), 65-72.

Liu, J., Li, C., Muhae-Ud-Din, G., Liu, T., Chen, W., Zhang, J. and Gao, L. 2020. Development of the droplet digital PCR to detect the teliospores of *Tilletia controversa* Kühn in the soil With greatly enhanced sensitivity. *Frontiers in Microbiology* 11(4), 1-9.

Liu, W., Liu, C. and Zheng, L. 2014. Discrimination in varieties of rice seeds with multi-spectral imaging using support vector machine. *Nongye Gongcheng Xuebao/ Transactions of the Chinese Society of Agricultural Engineering* 30(10), 145-151.

López, R., Asensio, C., Guzman, M. M. and Boonham, N. 2006. Development of real-time and conventional RT-PCR assays for the detection of potato yellow vein virus (PYVV). *Journal of Virological Methods* 136(1-2), 24-29.

Notomi, T., Okayama, H., Masubuchi, H., Yonekawa, T., Watanabe, K., Amino, N. and Hase, T. 2000. Loop-mediated isothermal amplification of DNA. *Nucleic Acids Research* 28(12), E63.

Osdaghi, E., Rahimi, T., Taghavi, S. M., Ansari, M., Zarei, S., Portier, P., Briand, M. and Jacques, M. A. 2020. Comparative genomics and phylogenetic analyses suggest several novel species within the genus *Clavibacter*, including nonpathogenic tomato-associated strains. *Applied and Environmental Microbiology* 86(6), e02873-19.

Panferov, V. G., Safenkova, I. V., Byzova, N. A., Varitsev, Y. A., Zherdev, A. V. and Dzantiev, B. B. 2018. Silver-enhanced lateral flow immunoassay for highly-sensitive detection of potato leafroll virus. *Food and Agricultural Immunology* 29(1), 445-457.

Patel, J. S., Brennan, M. S., Khan, A. and Ali, G. S. 2015. Implementation of loop-mediated isothermal amplification methods in lateral flow devices for the detection of *Rhizoctonia solani*. *Canadian Journal of Plant Pathology* 37(1), 118-129.

Priegnitz, U., Lommen, W. J. M., van der Vlugt, R. A. A. and Struik, P. C. 2019. Impact of positive selection on incidence of different viruses during multiple generations of potato seed tubers in Uganda. *Potato Research* 62(1), 1-30.

Pryor, B. M. and Gilbertson, R. L. 2001. A PCR-based assay for detection of Alternaria radicina on carrot seed. *Plant Disease* 85(1), 18-23.

Rajeshwari, N., Shylaja, M. D., Krishnappa, M., Shetty, H. S., Mortensen, C. N. and Mathur, S. B. 1998. Development of ELISA for the detection of *Ralstonia solanacearum* in tomato: its application in seed health testing. *World Journal of Microbiology and Biotechnology* 14(5), 697-704.

Ramachandran, S., Dobhal, S., Alvarez, A. M. and Arif, M. 2021. Improved multiplex TaqMan qPCR assay with universal internal control offers reliable and accurate detection of *Clavibacter michiganensis*. *Journal of Applied Microbiology* 131(3), 1405-1416. Accessed: 23 January 2021, Available at: https://doi.org/10.1111/jam.15017

Ramiro, J., Ciampi-Guillardi, M., Caldas, D. G. G., de Moraes, M. H. D., Barbieri, M. C. G., Pereira, W. V. and Massola, N. S. 2019. Quick and accurate detection of *Sclerotinia sclerotiorum* and *Phomopsis* spp. in soybean seeds using qPCR and seed-soaking method. *Journal of Phytopathology* 167(5), 273-282.

Razo, S. C., Panferov, V. G., Safenkova, I. V., Varitsev, Y. A., Zherdev, A. V., Pakina, E. N. and Dzantiev, B. B. 2018. How to improve sensitivity of sandwich lateral flow immunoassay for corpuscular antigens on the example of potato virus Y? *Sensors* 18(11), 3975.

Rijlaarsdam, A., Woudt, B., Simons, G., Koenraadt, H., Oosterhof, J., Asma, M. 2004. Development of specific primers for the molecular detection of *Clavibacter michiganensis* subsp. *michiganensis* EPPO Conference on quality of diagnosis and new diagnostic methods for plant pests. Noordwijkerhout, The Netherlands, 19-22 April 2004.

Sanchez, L., Farber, C., Lei, J., Zhu-Salzman, K. and Kurouski, D. 2019. Noninvasive and nondestructive detection of cowpea bruchid within cowpea seeds with a Hand-Held Raman spectrometer. *Analytical Chemistry* 91(3), 1733–1737.

Sánchez-Navarro, J. Á., Corachán, L., Font, I., Alfaro-Fernández, A. and Pallás, V. 2019. Polyvalent detection of twelve viruses and four viroids affecting tomato by using a unique polyprobe. *European Journal of Plant Pathology* 155(1), 361–368.

Shargil, D., Zemach, H., Belausov, E., Lachman, O., Kamenetsky, R. and Dombrovsky, A. 2015. Development of a fluorescent in situ hybridization (FISH) technique for visualizing *CGMMV* in plant tissues. *Journal of Virological Methods* 223, 55–60.

Simmons, H. E., Holmes, E. C., Gildow, F. E., Bothe-Goralczyk, M. A. and Stephenson, A. G. 2011. Experimental verification of seed transmission of Zucchini yellow mosaic virus. *Plant Disease* 95(6), 751–754.

Singh, D., Sinha, S. and Singh, R. P. 2015. Detection of Xanthomonas oryzae pv oryzae from seeds and leaves of rice (Oryza sativa) using hrp gene based BIO-PCR marker. *Indian Journal of Agricultural Sciences* 85(4), 519–524.

Strandberg, J. O. 2002. A selective medium for the detection of *Alternaria dauci* and *Alternaria radicina*. *Phytoparasitica* 30(3), 269–284.

Sun, H., Zhang, L., Li, H., Rao, Z. and Ji, H. 2021. Nondestructive identification of barley seeds varieties using hyperspectral data from two sides of barley seeds. *Journal of Food Process Engineering* 44(8), e13769.

Sunani, S. K., Bashyal, B. M., Rawat, K., Manjunatha, C., Sharma, S., Prakash, G., Krishnan, S. G., Singh, A. K. and Aggarwal, R. 2019. Development of PCR and loop mediated isothermal amplification assay for the detection of bakanae pathogen *Fusarium fujikuroi*. *European Journal of Plant Pathology* 154(3), 715–725.

Szemes, M., Bonants, P., de Weerdt, M., Baner, J., Landegren, U. and Schoen, C. D. 2005. Diagnostic application of padlock probes–multiplex detection of plant pathogens using universal microarrays. *Nucleic Acids Research* 33(8), e70.

Tangkanchanapas, P., Höfte, M. and De Jonghe, K. 2018. Reverse transcription loop-mediated isothermal amplification (RT-LAMP) designed for fast and sensitive on-site detection of Pepper chat fruit viroid (PCFVd). *Journal of Virological Methods* 259, 81–91

Temple, T. N., du Toit, L. J., Derie, M. L. and Johnson, K. B. 2013. Quantitative molecular detection of Xanthomonas hortorum pv. carotae in carrot seed before and after hot-water treatment. *Plant Disease* 97(12), 1585–1592.

Thaitrong, N., Charlermroj, R., Himananto, O., Seepiban, C. and Karoonuthaisiri, N. 2013. Implementation of microfluidic sandwich ELISA for superior detection of plant pathogens. *PLoS ONE* 8(12), e83231.

Thapa, S. P., O'Leary, M., Jacques, M. A., Gilbertson, R. L. and Coaker, G. 2020. Comparative genomics to develop a specific multiplex PCR assay for detection of *Clavibacter michiganensis*. *Phytopathology* 110(3), 556–566.

Thierry, M., Gladieux, P., Fournier, E., Tharreau, D. and Ioos, R. 2020. A genomic approach to develop a new qPCR test enabling detection of the Pyricularia oryzae lineage causing wheat blast. *Plant Disease* 104(1), 60–70.

Tian, Y., Zhao, Y., Xu, R., Liu, F., Hu, B. and Walcott, R. R. 2014. Simultaneous detection of Xanthomonas oryzae pv. oryzae and X. oryzae pv. oryzicola in rice seed using a padlock probe-based assay. *Phytopathology* 104(10), 1130–1137.

Van Brunschot, S. L., Bergervoet, J. H. W., Pagendam, D. E., De Weerdt, M., Geering, A. D. W., Drenth, A. and Van Der Vlugt, R. A. A. 2014a. Development of a multiplexed bead-based suspension array for the detection and discrimination of Pospiviroid plant pathogens. *PLoS ONE* 9(1), e84743.

van Brunschot, S. L., Verhoeven, J. T. J., Persley, D. M., Geering, A. D. W., Drenth, A., Thomas, J. E. 2014b. An outbreak of Potato spindle tuber viroid in tomato is linked to imported seed. *European Journal of Plant Pathology* 139(1), 1–7.

Van der Vlugt, R. A. A., Stijger, C. C. M. M., Verhoeven, J. Th. J. and Lesemann, D. E. 2000. First report of *Pepino mosaic virus* on tomato. *Plant Disease* 84(1), 103.

Van der Wolf, J. M. and Gussenhoven, G. C. 1992. Reaction of saprophytic bacteria from potato peel extracts and plant pathogenic bacteria in ELISA with antisera to *Erwinia chrysanthemi* (serogroup O1Ha). *Netherlands Journal of Plant Pathology* 98(1), 33–44.

Vossenberg van de , B. T. L. H., Westenberg, M. and Bonants, P. J. M. 2013. DNA barcoding as an identification tool for selected EU-regulated plant pests: an international collaborative test performance study among 14 laboratories. *EPPO Bulletin* 43(2), 216–228.

Wang, X. Q., Allen, T. W., Wang, H., Peterson, D. G., Nichols, R. L., Liu, A., Li, X. D., Deng, P., Jia, D. and Lu, S. E. 2019. Development of a qPCR protocol to detect the cotton bacterial blight pathogen, *Xanthomonas citri* pv. *malvacearum*, from cotton leaves and seeds. *Plant Disease* 103(3), 422–429.

Xiong, L., Gao, L., Liu, Q., Xia, J., Han, X. and Liu, Y. 2013. A simple, rapid and visual antibody array for the simultaneous detection of multiple plant pathogens. *Analytical Methods* 5(9), 2413–2418.

Yao, X., Li, P., Xu, J., Zhang, M., Ren, R., Liu, G. and Yang, X. 2016. Rapid and sensitive detection of *Didymella bryoniae* by visual loop-mediated isothermal amplification assay. *Frontiers in Microbiology* 7, 1372.

Ye, W.-W., Zeng, D.-D., Xu, M., Yang, J., Ma, J., Wang, Y.-C. and Zheng, X.-B. 2020. A LAMP-assay-based specific microbiota analysis reveals community dynamics and potential interactions of 13 major soybean root pathogens. *Journal of Integrative Agriculture* 19(8), 2056–2063.

Zaccardelli, M. Parisi, M. and Giordano, I. 2002. Susceptibility of tomato genotypes to *Pseudomonas syringae* pv. *tomato* in the field conditions. *Journal of Plant Pathology* 84, 200.

Zaccardelli, M., Spasiano, A., Bazzi, C. and Merighi, M. 2005. Identification and in planta detection of *Pseudomonas syringae* pv. *tomato* using PCR amplification of hrpZPst. *European Journal of Plant Pathology* 111(1), 85–90.

Zhu, M., Zhang, W. N., Tian, J. Y., Zhao, W. Y., Chen, Z. Q., Sun, L. H., Xue, F., Liu, Y., Tan, X. Q., Wang, L. M., Liu, F. Q. and Tao, X. R. 2016. Development of a lateral-flow assay (LFA) for rapid detection of soybean mosaic virus. *Journal of Virological Methods* 235, 51–57.

Chapter 8

Advances in preservation of seed vigour during storage

Fiona R. Hay, Aarhus University, Denmark

1 Introduction

Seed longevity in storage – how long seeds retain the ability to germinate upon removal from storage if germination requirements are met (water, temperature, light) – is an inherent aspect of seed physiological quality. It is also key to the maintenance of physiological quality. If a sample of seeds taken from a seed lot germinates quickly and uniformly under both ideal laboratory conditions and field or other stressed conditions, it is described as having high vigour. Indeed, evaluation of this ability to germinate under non-ideal conditions is the basis of many of the tests that are currently used for assessing seed lot vigour, such as the cold test (Table 1).

Since a high-vigour seed lot can be considered 'young' in terms of physiological age, it is also likely to maintain high levels of viability and vigour for some period of time, i.e. have high storage potential (Fig. 1), depending on the conditions under which the seed lot is stored. This is the basis of other seed vigour tests, for example, the controlled deterioration or accelerated ageing tests (Table 1). Seed lots that have higher vigour (i.e. that are further to the left in Fig. 1B) will have higher germination after a standard storage treatment (with duration, temperature, and moisture content controlled) than seed lots

http://dx.doi.org/10.19103/AS.2022.0105.11

Table 1 Examples of seed lot vigour tests that have been prescribed by official seed testing organizations (the International Seed Testing Association and the Association of Official Seed Analysts) or which have been proposed in the scientific literature. In the case of the latter, the list shown does not include publications that have applied an already official method (e.g. conductivity test), to species not covered by publications of the official testing agencies

Method	Principle	Species	Correlated with	Reference
Seed ageing-based methods				
Accelerated ageing	Seeds are exposed to high humidity (over water) and high temperature for a standard period of time and then tested for germination. Seeds that are older and therefore lower in vigour will have lower germination after the ageing treatment than high vigour seeds.	*Glycine max.*	Field emergence, storage potential.	ISTA (2021)
		Glycine max, *Solanum lycopersicum*, *Sorghum bicolor.*	Field emergence, storage potential.	AOSA (2009)
Saturated salt accelerated ageing	Similar to AA but saturated salt solutions are used (KCl, NaCl, NaBr, KNO_3). Suitable for small-seeded species (flowers and vegetables). Salt solutions reduce the RH of the ageing environment, thus preventing fungal growth.	*Impatiens walleriana*, *Allium cepa.*	Field emergence, storage potential.	AOSA (2009)
Controlled deterioration	Seed moisture content is raised to a specified moisture content and then the seeds are sealed inside aluminium foil laminate packets and placed at a high temperature for a standard period of time. This is followed by either a germination test or a conductivity test, the results of which will be respectively lower or higher for seed lots with reduced vigour.	*Brassica* spp.	Field emergence, storage potential.	ISTA (2021)
		Medicago sativa, *Trifolium pratense*, *Festuca arundinacea*, *Pisum sativum.*	Field emergence, storage potential.	AOSA (2009)

Germination performance-based methods				
Cold test	Assesses ability of seeds to germinate under stress conditions.	*Zea mays*, *Glycine max*, *Gossypium hirsutum*, *Allium cepa*.	Field emergence.	AOSA (2009)
Cool germination test		*Gossypium hirsutum*.	Field emergence.	AOSA (2009)
Saturated cold test		*Zea mays*, *Oryza sativa*, *Solanum lycopersicum*.	Field emergence under very wet, cold soils.	AOSA (2009)
Speed of germination test	Calculations based on time and percentage of germinated seeds. Implemented as a standard germination test with stricter control of test conditions (temperature, moisture availability, and light).	*Zea mays*, *Oryza sativa*, *Solanum lycopersicum*, *Allium cepa*, *Cucumis sativus*, *Triticum aestivum*, *Medicago sativa*.	Field emergence.	AOSA (2009)
Computer imaging tests	Images of seedlings are processed with computer software to produce a seed vigour index to augment traditional seed quality analyses. Can be used with other tests.	*Zea mays*, *Glycine max*, *Lactuca sativa*.	Standard and other germination tests.	AOSA (2009)
Radicle emergence	Since vigour is related to speed of emergence, a single early count of the number of seeds with radicle emergence in a standard germination test will be indicative of seed lot vigour.	*Brassica napus*, *Triticum aestivum* subsp. *aestivum*, *Raphanus sativus*, *Zea mays*.	Field emergence.	ISTA (2021)
Determination of parameters related to germination performance	Various parameters that measure the speed and uniformity of germination of a sample of seeds in a germination test.	Reported in various publications.	-	See e.g. Ranal and de Santana (2006) and Joosen et al. (2010)
Seedling growth rate test	Conducted as part of a standard germination test with data of dry weight gain or seedling length determined at the completion of the germination test. High growth rate is a reflection of high seed vigour.	Many food and forage species.	Field emergence.	AOSA (2009)

(*Continued*)

Table 1 (*Continued*)

Method	Principle	Species	Correlated with	Reference
Metabolism-related methods				
Tetrazolium	The test uses a colourless solution of 2,3,5-triphenyl tetrazolium chloride, which is reduced by living cells to a red non-diffusible compound formazan. Variation in the location and intensity of the red staining reflects variation in respiration activity (the deeper the staining, the greater the vigour).	*Glycine max.*	Field emergence.	AOSA (2009)
		Zea mays, Arachis hypogaea, Gossypium hirsutum, Phaseolus vulgaris.		AOSA (2009)
Seedling reserve mobilisation	Measurement of conversion of seed storage reserves to seedling.	*Phaseolus vulgaris.*	Accelerated ageing.	Padilha et al. (2020)
Single seed respiration	Respiration is a measure of metabolic activity which is required for germination. Respiration is measured by the depletion of oxygen in a sealed vial containing a single seed and the germination medium (e.g. water agar gel).	*Brassica, Allium cepa.*		Bradford et al. (2013)
Ethanol production	Damage to mitochondrial membranes in the cells of aged seeds results in anaerobic respiration and release of ethanol that can be detected with an alcohol breath analyzer.	*Brassica napus.*	Radicle protrusion after 2 days in germination test, seedling growth.	Buckley and Huang (2011)
		Raphanus.	Controlled deterioration.	Kucukhuseyin et al. (2021)
		Citrullus lanatus.	Seedling emergence (soil laboratory test).	

		Capsicum annuum.	Controlled deterioration, seedling emergence.	
		Solanum melongena, *Allium porrum*, *A. cepa.*	None of those tested.	
Other				
Conductivity test	Membranes are damaged during seed ageing such that when seeds are placed in water, electrolytes are leaked. The extent of leakage can be assessed by measuring the electrical conductivity of the water after the seeds have been soaked.	*Cicer arietinum*, *Glycine max*, *Phaseolus vulgaris*, *Pisum sativum* (garden peas only), *Raphanus sativus.*	Field emergence.	ISTA (2021)
		Grain, legume and vegetable crops.	Field emergence.	AOSA (2009)
Single seed conductivity test	Individual seeds are tested as individual replicates to eliminate the possibility of a false reading in samples composed of many seeds in which one seed exhibits severe damage with excessive leakage.	*Zea mays*, *Glycine max*, *Gossypium hirsutum*, *Pisum sativum*, small-seeded crops.	Field emergence.	AOSA (2009)
Potassium leakage test	Conductivity test specific for potassium ion leakage.	*Zea mays*, *Pisum sativum.*	General seed vigour, effects of storage.	AOSA (2009)
X-ray image analysis	Measures density of storage reserves as well as structural integrity.	*Brassica oleraceae* var. *italica.*	Seedling length.	Abud et al. (2018)
		Jatropha curcas.	Germination speed.	De Medeiros et al. (2020)

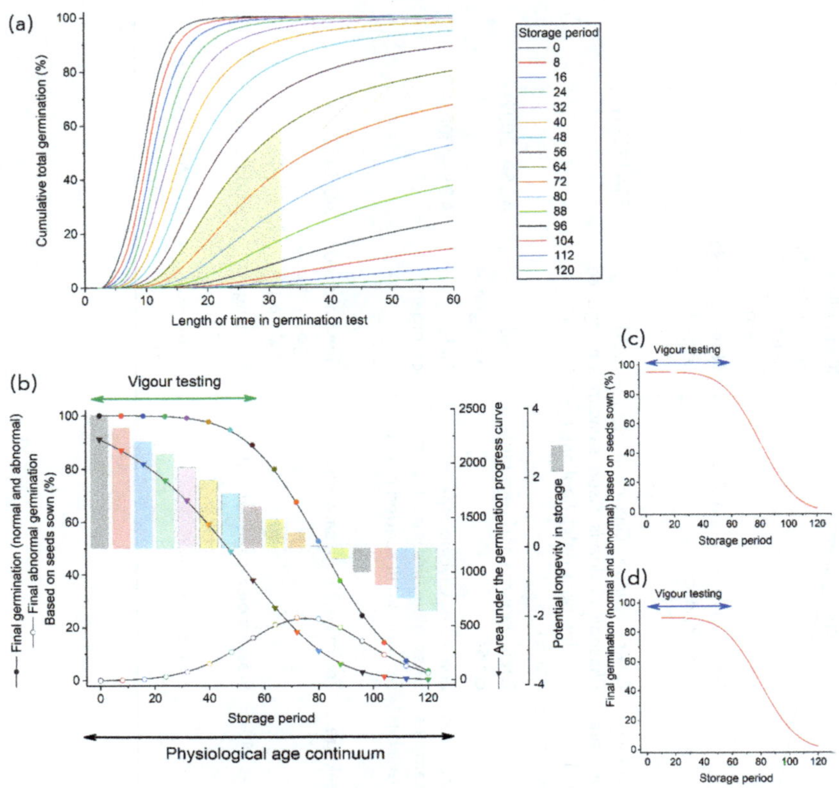

Figure 1 The relationships between seed germination, longevity in storage and vigour. Conceptualization based on the modelling presented in Bradford et al. (1993). (a) As seeds age, the speed and uniformity of germination of a sample of those seeds decrease. The final germination that is reached is also reduced. The area under the germination progress curve (AUC) is a parameter which incorporates information about speed, uniformity and the germination percentage reached by a certain time in the germination test (shown as the filled area for seeds aged for 64 (unspecified) units of time) (Joosen et al., 2010). After a certain length of time in the germination test (indicated by the vertical line), any further seeds that germinate (independent of storage period) give rise to abnormal seedlings (shown as the slashed area under the curve). (b) Parameters extracted from the germination progress curves shown in (a) for seeds aged for different lengths of time (storage period). For a seed lot with high initial physiological quality, high total final germination is maintained, but vigour (indicated by the area under the progress curve) declines immediately. The storage potential of the seeds (expressed in probits, where a value of 0 indicates that half the population of seeds still retains the ability to germinate and the other half has died; K_i in eqn. 1) also starts to decline immediately (columns). With further storage, not only are final germination, vigour and storage potential reduced, but the proportion of seeds that produce abnormal seeds increases. For seeds stored for the longest period of time, most or all of the seeds that germinate are abnormal. These parameters together can be taken as an indicator of the physiological age of a seed lot, which is not necessarily directly related to chronological age, as it will depend on the conditions the seeds experience after harvest and indeed, the maturity of the seeds at harvest. In (c) and (d), the plateau before viability decline is at<100% viability (shown for 95% and 90% viability, respectively); vigour tests would still discriminate between seed lots at different points along that plateau phase.

that have already aged (further to the right in Fig. 1B), despite similar high percentage germination (ISTA, 2021). The result of such a vigour test is usually just the standard germination percentage after the storage treatment; more precisely, the potential longevity in storage (right-most axis in Fig. 1B), as a measure of the vigour/physiological age of a seed lot, could be determined, by sampling seeds for germination testing on a number of different occasions during storage (again, under controlled conditions) and using, for example, probit regression analysis to model the data (see, for example, Yadav and Ellis, 2018; Hay et al., 2014, 2019). Because seed vigour is directly related to the establishment of a crop in the field, where conditions may be variable and certainly less than optimal, ensuring the availability of seed lots with high physiological quality is of fundamental importance for global food security (Finch-Savage and Bassel, 2016).

Seed vigour tests are normally only applied to seed lots which have high germination in a standard germination test. 'High' germination typically means seed lots that meet the minimum standard for commercial seed lots, which varies between countries/regions and, within a country, among different crop species. 'High' germination could mean germination as low as 75%, which in theory (Fig. 1B), would mean that a seed lot has already aged considerably. However, because of random sampling error and/or perhaps suboptimal germination conditions even in a standard germination test, and because seed lots may show the maintenance of high viability at lower germination levels than 100% (Fig. 1C,D), vigour tests, due to their greater sensitivity compared with standard germination tests, are still valid for comparing and ranking seed lots with such germination levels to predict field performance.

Achieving and then retaining a seed lot with high vigour depends on the conditions the seeds experience before harvest, the conditions at and timing of seed harvest, and the conditions and treatments after harvest. The conditions developing seeds experience in the field are perhaps difficult to control, certainly when producing seeds at scale, though of course, good agronomic practices will help ensure a good yield of healthy, high-quality seeds. How far seeds have progressed through seed development by the time they are harvested will also have a profound effect on the quality of the harvested seeds (Hay and Smith, 2003; Finch-Savage and Bassel, 2016; Ellis, 2019). However, without harvesting individual seeds one by one, it may be inevitable that seed lots contain seeds with a range of maturities. Furthermore, for species that readily disperse their seeds, including, for example, some *Brassica* taxa and many crop wild relatives, it may be necessary to harvest seeds before they shatter, or at least bag the infructescences, so that the seeds are not lost. In contrast, for shatter-resistant species (e.g. cultivated cereals), the seeds may remain on the plant long after physiological quality peaks (Ellis, 2019). While this means that seeds should not be lost, it does mean that physiological quality could decline, due to ageing in the field, if there are delays in harvesting. After harvest, it may be possible to treat the seeds in such a way that physiological quality can be enhanced, for example, by controlling the conditions the seeds experience immediately after harvest or through some sort of priming treatment. Otherwise, the seeds will be ageing and losing vigour and viability, at rates that depend primarily on the moisture content of the seeds and the temperature and gaseous composition of the storage environment, but which also vary within and among species (Walters et al., 2005; Probert et al., 2009; Solberg et al., 2020). It is possible to predict these rates of viability loss for some species, though not the decline in vigour; otherwise, if it is important that viability and vigour are known at any point in time, it will be necessary to carry out monitoring tests. This review will discuss these various aspects of achieving and preserving vigour during seed storage.

2 How can conditions during seed production affect seed quality?

When choosing a location for seed production, factors such as local climate, properties of the soil, potential disease pressures, and accessibility should be taken into account, and best agronomic practices followed, including in relation to planting time (Finch et al., 2014; Kameswara Rao et al., 2017; Cruz et al., 2020; Oliveira et al., 2021). However, even at seemingly good locations, freak weather conditions or, for example, failure in irrigation equipment, may affect seed development. The impact of such stresses will depend on when they occur during seed development. Seed development following fertilization, cell division, and histodifferentiation can be divided into two stages, occurring respectively, before and after mass maturity: the seed filling stage during which dry matter is accumulated, and the maturation drying stage during which the seeds dry to equilibrium with ambient environmental conditions (Fig. 2; Ellis, 2019). By the time a developing cohort of seeds has reached mass maturity, most of the seeds are able to germinate upon removal from the parent plant (i.e. without drying). The timing of the acquisition of desiccation tolerance, the ability of the seeds to germinate after drying, relative to the acquisition of germinability varies between species (Probert et al., 2007), and improvements in storage potential and other vigour parameters can only occur after desiccation tolerance has been acquired. Indeed, vigour as a concept is generally not applied to seeds of recalcitrant species, which do not acquire desiccation tolerance and which, since they are usually non-dormant, typically germinate very readily.

The improvements in storage potential and vigour during maturation drying can only be realized after seeds have been harvested and hence it is difficult to know when to harvest seeds for maximum physiological quality.

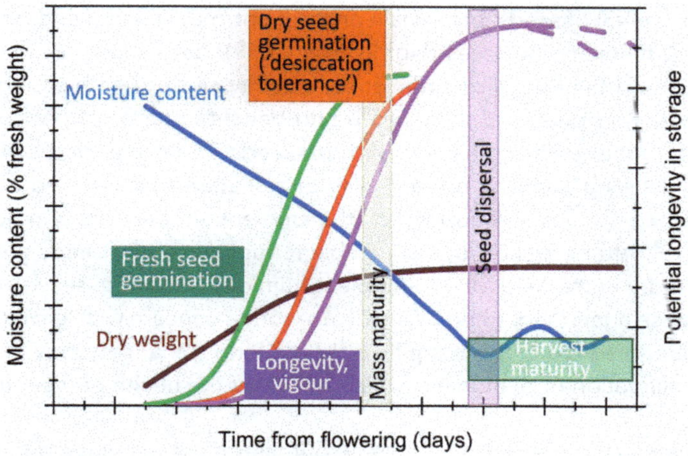

Figure 2 The progress of seed development may be followed by monitoring the moisture content and weight of the seeds and changes in seed physiological quality. Seed moisture content is initially very high, but then declines as dry matter accumulates. After seed dry weight plateaus, at mass maturity, moisture content declines more rapidly (the seeds no longer receive water from the parent plant) to equilibriate with the ambient conditions. If seeds are not naturally dispersed during this period, moisture content may fluctuate as ambient conditions fluctuate. Physiological quality reflects the ability of the seeds to germinate fresh, upon removal from the plant, and after drying, and by the vigour and longevity of the seeds (see Fig. 1). The relative timings of the acquisition of the ability to germinate and of desiccation tolerance varies among species, but for many species 100% germinability before and after drying is achieved at or soon after mass maturity. In contrast, seed longevity and vigour continue to improve after mass maturity, when the seeds are drying to equilibrium with ambient conditions. If seeds are not naturally dispersed, physiological quality may be maintained for some time or decline, depending on the ambient conditions. For such shatter resistant species, seeds will typically be harvested when moisture content is reduced (seeds reach harvest maturity).

For seeds of wild species and crops that normally fall from the maternal plant, the best time to harvest is as close as possible to that point, and there may be natural indicators, such as changes in seed or fruit colour or signs of fruit dehiscence, which can help decide on timing (Hay and Smith, 2003). The use of chlorophyll fluorescence to assess seed maturity, based on the extent of breakdown of chlorophyll in the covering structures of a seed, is also possible for some species (Yadav et al., 2015; Galletti et al., 2020; Bello and Bradford, 2021), though it may be hard to identify a unique value across diverse varieties (Hay et al., 2015). Otherwise, the best way to decide whether seeds are ready for harvesting is to evaluate how close seeds are to reaching equilibrium with the ambient conditions (Hay and Probert, 2011); this will indicate whether the seeds have progressed through natural on-plant maturation drying. The seed moisture status is also key to understanding whether seeds of shatter-resistant species have reached harvest maturity. After the initial maturation drying, the moisture content of seeds that remain on the plant, may subsequently increase, if ambient humidity increases (Fig. 2; Ellis, 2019), but as long as the seeds are air-dry and their equilibrium relative humidity close to ambient humidity, longevity should peak and there will be no benefit in terms of physiological quality, to further delaying harvest of these seeds. The extent of the negative effect of delaying harvest on the quality of a shatter-resistant seed crop will vary between species and depend on the ambient temperature and humidity (Ellis, 2019). After a period of high ambient humidity, if the seeds are again able to re-dry in the field, due to ambient humidity dropping, seed longevity may increase again (Yadav and Ellis, 2016). Another concern with leaving seeds from shatter-resistant crops in the field for too long is that they may be prone to the proliferation of pathogens, which will affect the health of the harvested seed lot.

If seeds are harvested prematurely (but after mass maturity), perhaps for logistical reasons, it may be possible for the seed maturation drying programme to proceed off-plant. This might be by keeping seeds at high humidity rather than immediately exposing them to low humidity (Hay and Probert, 1995, 2011; Probert et al., 2007) or by drying them at a relatively high temperature (see case study). The use of more advanced technologies, such as remote sensing and hyperspectral imaging, to predict optimal harvest times for crops is gaining traction (e.g. Thenkabail et al., 2000; Xu et al., 2019; Singh et al., 2021) and will likely also be widely adopted for seed crops, in which case complementary studies of the development of seed physiological quality will also be required.

Some studies have considered the impact of maternal environment on seed quality and/or of various abiotic stresses on the development of seed physiological quality, and depending on when they occur during seed development. For example, Li et al. (2017) reported that it was better to produce *Vicia sativa* seeds in an environment where the temperature was relatively low during seed development; and Nguyen et al. (2021) found that seed quality was higher in *Petunia* x *Hybrida* seeds produced under controlled conditions (23°C) rather than under sub-optimal greenhouse conditions (22°C to 31°C). In wheat, high-temperature stress early in seed development, when seeds were still accumulating dry matter, negatively impacted seed quality, while exposure to the same high temperature later in seed development improved subsequent longevity (Nasehzadeh and Ellis, 2017). It is perhaps, in particular, in relation to rising temperatures that concerns about the effect of climate change on the production of high-quality seeds has been expressed. However, higher CO_2 levels may also impact seed quality. As reviewed by Lamichaney and Maity (2021), different studies involving different species have shown negative, positive or no impact of growing plants under elevated CO_2 conditions, on the germination of harvested seeds. For example, Lamichaney et al. (2019) found that elevated CO_2 (610 ppm and 720 ppm cf. 420 ppm as current ambient CO_2

level) reduced the germination and hence vigour of rice seeds. in contrast, elevated CO_2 (600 ppm) did not seem to have such a negative effect on the germination and vigour of chickpea seeds (Lamichaney et al., 2021). Overall however, Lamichaney and Maity (2021) noted in their review, that there were not many studies that considered the impact of elevated CO_2 on the vigour of harvested seeds, especially over multiple generations.

Droughts caused by changes in rainfall patterns attributed to global climate change are also already apparent and affecting crop production (e.g. Lesk et al., 2016). In the case of seed production, drought imposed during the seed filling stage was found to reduce final seed weight but was beneficial to the subsequent longevity of rapid-cycling Brassica seeds, attributed to greater accumulation of heat-stable proteins (Sinniah et al., 1998a,b). Abdul Rahman and Ellis (2019) considered the effects of both drought and high temperature during seed development on the subsequent quality of rice seeds. In contrast to the results for Brassica (Sinniah et al., 1998a, cited above), they found that seeds subjected to drought at different stages were not able to attain the same storage potential as seeds from plants that were irrigated throughout and that high-temperature stress was more damaging the earlier it occurred during seed development. Combining the two stresses was even more damaging for subsequent seed quality (Abdul Rahman and Ellis, 2019). Due to such effects, seed longevity is considered a highly plastic trait (Leprince et al., 2017), with considerable variation among seed lots in longevity and hence vigour.

3 Post-harvest treatments to improve seed lot quality

Post-harvest treatments that improve the physiological quality of a seed lot are essentially either reducing variation in maturity within a sample or are promoting the progress of seed development-germination. Reducing seed-to-seed variation in maturity within a seed lot that has already been dried is through seed-sorting techniques, including seed grading, using equipment such as aspirators and screens, which sort based on size or weight. Optical colour sorters may remove immature individuals as well as off-types. More advanced seed sorters may be more precise in terms of assessing how well each individual seed matches the desired attributes for the seed lot, based on properties such as size, colour and chlorophyll fluorescence.

In terms of other post-harvest treatments to improve seed lot quality, as discussed above, it may be helpful to place the seeds under certain conditions (e.g. high humidity) immediately after harvest to allow seed development to continue, before drying the seeds for further processing. In terms of enhancing the vigour of dried seeds, seed priming is widely used, with the term 'priming' encompassing a wide range of treatments, including, for example, hydropriming, osmopriming, biopriming (Paparella et al., 2015), plasmapriming (Starič et al., 2020), and nanopriming (Song and He, 2021). These priming techniques improve the vigour of seed lots at planting. The mechanism(s) behind the different enhancement techniques is not fully understood, but in the case of, for example, hydro- and osmopriming, the treatment allows germination processes to start but not proceed so far that the seeds lose desiccation tolerance; the seeds are re-dried and then, when sown for germination, are able to germinate more quickly and uniformly as a seed lot because the germination processes have already commenced, thereby resulting in improved stand establishment. The priming treatment is normally applied within a short period of time before the seeds are distributed for sowing, because the treatment, in many cases, reduces the storage potential of the seeds (Fabrissin et al., 2021). The relationship between vigour and storability (Fig. 1), is therefore uncoupled for seed lots for which priming reduces longevity. The response may depend on the physiological age of the seeds when they are primed; there is some evidence that the longevity of seed lots with low vigour, whether it is due to immaturity or to ageing, is perhaps more likely to improve in response to a priming treatment, in terms of their subsequent longevity in storage (Powell et al., 2000; Butler et al., 2009). Thus, priming seeds to extend longevity and vigour in storage, rather than to improve vigour at planting, could be adopted, but it will probably require research to evaluate treatments at the individual seed lot level.

4 Maintaining viability and vigour during storage

It is well-recognised that seed viability and vigour are maintained for longer, the lower the moisture content of the seeds (to a critical level) and the temperature of storage (limits not well-defined). However, as shown in Fig. 1, vigour will be declining before the percentage viability is impacted; hence, to really preserve vigour, storage conditions would have to be strictly controlled, probably to a degree that is not cost efficient overall. Indeed, this is perhaps why vigour, and vigour testing, is of interest: because different seed lots will inevitably show differences in vigour despite identical handling and similarly high germination.

The relationships between seed longevity (ability to germinate after storage) and moisture content and temperature are described for a handful of species, based on the seed viability equations (Ellis and Roberts, 1980):

$$v = K_i - p/\sigma \tag{1}$$

in which

$$\sigma = 10^{K_E - C_W \log m - C_H t - C_Q t^2} \tag{2}$$

These equations together describe the viability, v, in probits after p days of storage at moisture content, m (% fresh weight), and temperature, t (°C). K_i is the initial viability of the seeds in probits and therefore reflects the storage potential of the seed lot at the start of storage (Fig. 1); K_E, C_W, C_H and C_Q are parameters that have to be solved for each species (but see below). Expressing viability in probits as in the first equation assumes that there is a normal distribution of seed deaths over time; σ is the standard deviation of this normal distribution. Transforming germination percentages to probit values and plotting these against the storage period in effect linearizes the seed survival curve (see columns and right-most axis in Fig. 1B), with the slope equivalent to $-1/\sigma$ (Hay et al., 2019). The moisture content as seeds are dried can be estimated based on Cromarty's equation (Cromarty et al., 1982), if the species' seed oil content (D_o; proportion of dry weight) has been determined or is in the database, and storage temperature (t) and relative humidity (RH) are known:

$$m_{\%\ \text{d.wt.}} = \frac{(1-D_o)\sqrt{-440 \times \ln\left(1-(RH/100)\right)}}{1.1+t/90} \tag{3}$$

The estimated moisture content, $m_{\%\ \text{d.wt.}}$, on a dry weight basis can be converted to a fresh weight basis, thus:

$$m_{\%\ \text{f.wt.}} = \frac{m_{\%\ \text{d.wt.}}}{1+\left(m_{\%\ \text{d.wt.}}/100\right)} \tag{4}$$

It is this moisture content, expressed as a percentage of fresh weight, which is used in the viability equation (Eq. 2).

If seeds are stored where there may be some fluctuation in seed moisture content and temperature, for example, if there are minor perturbations in temperature due to thermostat capability, effective values for these variables can be calculated, but it seems that the change in these variables themselves does not impact storage longevity (Hung et al., 2001; Xanthopoulos, 2020). Of course, an environment with moisture content and temperature variation may be an open environment where oxygen also becomes a factor that can influence longevity (see below). While the equations were developed based on germination after storage, they could be modified to express a vigour parameter, for example, rate (speed) of germination based on the modelling illustrated in Fig. 1, by for example, combining such a threshold model (Bradford et al., 1993) with the viability equations, as suggested by Hay et al. (2014). However, from the perspective of describing decline in vigour before a significant loss in viability, v and hence, K_i, would have to be high (>2-3 probits; Fig 1).

In applying the viability equations, it has been assumed that seed lots of a particular species will differ in longevity simply due to their initial viability, and not due to differences in 'baseline' longevity (i.e. K_E) or responses to changes in storage moisture content or temperature (Ellis and Roberts, 1981). That is, the parameters of Equation (2) are constant within a species. However, Lee et al. (2019), who carried out a genome-wide association study of seed longevity using seeds from 299 varieties of rice from the Indica group, found that there was significant variation in longevity when seeds were stored at the same moisture content and temperature that was not just due to variation in quality at the start of storage. This means that when using the viability equations to predict longevity periods, it is important to understand that the estimates provided come with considerable error. Nonetheless, they do provide an indication of the storability of seeds from particular species and of relative longevity across different species, accessions, or mutants (Fig. 3; and see, for example, Hay et al., 2003).

The viability equations were developed based on germination data for seeds stored under hermetic conditions; samples of seeds were equilibrated to specific moisture levels and then sealed inside air-tight packets, which were placed in ovens at, usually, elevated temperatures (typically up to 60°C). Elevated temperatures were used to ensure viability periods could be determined over reasonable experimental periods, especially for seeds at low moisture contents. Some seed storage experiments have used lower temperatures, but overall there is not much data for seeds stored below 0°C in particular (Dickie et al., 1990), except for seeds in genebank storage where there is generally, not a systematic experimental design and the data is not necessarily very reliable for modelling seed longevity (Hay et al., 2021). The confidence in the viability equations thus comes from the continuity in behaviour, which appears to hold over a wide temperature range (Dickie et al., 1990), but over a somewhat limited moisture content range. Seed longevity increases as moisture content is reduced until a certain critical moisture content, at which point, further drying has little or no further effect on longevity (Ellis et al., 1989, 1990, 1995). The critical value appears to be when seeds are in equilibrium with 10–15% relative humidity and, in moisture content terms, increases as storage temperature decreases, for example, from 4.0% moisture content (fresh weight basis) at 65.2°C to 5.4% at 30.2°C for seeds of *Trifolium pratense* L. (Ellis and Hong, 2006). Commercial seed lots are not normally dried to such low moisture contents, or stored hermetically, at least until they are packaged prior to sale to users.

Many of the ageing reactions that occur in seeds are driven by oxygen (Zhang et al., 2021). This, in addition to controlling seed moisture content,

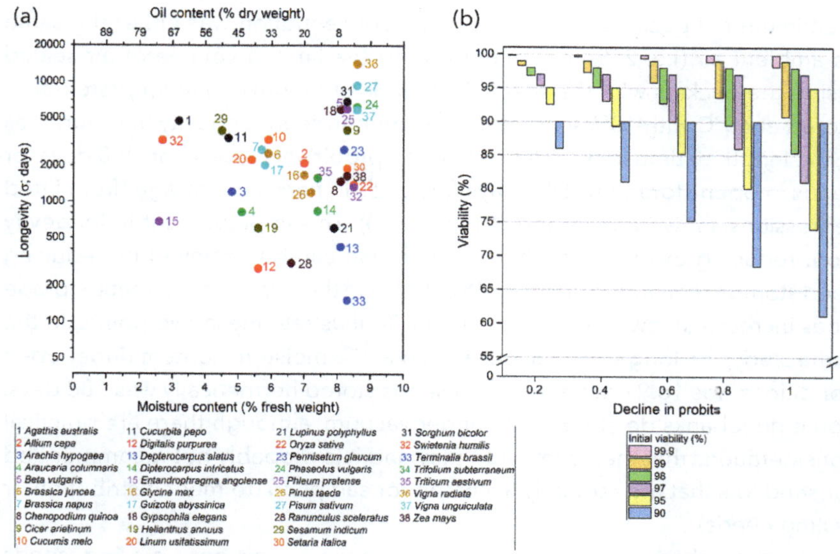

Figure 3 (a) Predictions of seed longevity for seeds of different species stored at 30% relative humidity and 20°C. Predictions are based on the seed viability equations (equation 1; Ellis and Roberts, 1980) and Cromarty's equation (to estimate seed moisture content based on storage conditions and seed oil content; Cromarty et al., 1982), as formulated in the Seed Information Database (SID; Royal Botanic Gardens Kew, 2021). The variation in seed moisture content is due to the variation in oil content of seeds from different species: oily seeds such as those of *Agathis australis* (64.6% oil content) have a lower moisture content (% fresh weight basis) at a given relative humidity and temperature than starchy seeds such as those of *Oryza sativa* (2.2%). N.B. In SID, the presented oil contents are as sourced, and not necessarily accurate determination of oil content on a dry weight basis as used in Cromarty's equation (eqn. 3). There are other species with the viability equation parameters solved, but for which an oil content was not available; a few species were also excluded because the oil contents available in the database differed markedly within the species. Longevity is expressed as the predicted time, σ, for viability to fall by one probit. In (b), the meaning of the probit decline is illustrated in % terms, for different initial viabilities. For example, if the initial seed lot viability is 99.9%, a decline of 1 probit during storage means that the final viability would be 98.2% (pink bar, right-most series); if the initial viability is 90%, a decline of 1 probit means that the final viability would be 61.1% (blue bar). Similarly, if the initial viability is 99%, a decline of 0.2 probits during storage means that the final viability would be 98.3% (orange bar, left-most series); if the initial viability is 95%, a decline of 0.2 probits means that the final viability would be 92.6% (yellow bar).

is another reason why it is recommended that seeds in long-term storage should be stored in air-tight containers, especially since the damaging effect of free availability of oxygen on storability increases at low seed moisture contents (Ellis and Hong, 2007). Ellis and Hong (2007) stored seeds either in an 'open environment' with free availability of oxygen, or in heat-sealed aluminium foil packets. Thus, the oxygen concentration was either the same as ambient air (i.e. 21%) or limited due to the limited volume of air sealed inside the packets with the seeds. For seeds of timothy (*Phleum pratense* L.) stored at 45°C with 10% seed moisture content, seed longevity (measured as σ, the length of time for viability to fall by 1 probit) increased from 9.0 days for seeds in open storage to 64.9 days for seeds in hermetic storage (from fitted regressions shown in Ellis and Hong (2007)). This improvement in longevity from reducing oxygen availability is less than can be achieved by reducing seed storage moisture content by 1%, although the benefit of hermetic storage does increase at lower moisture content. To illustrate these two points, in the same study, the longevity of seeds stored at 9% moisture content under open conditions was 16.9 days and that of seeds stored hermetically was 188 days. Some genebanks do store seeds under vacuum, although there are practical considerations that then come into play and it is probably not recommended for seed lots that are regularly accessed for sampling (to meet distribution or testing needs).

Further characterization of the response to oxygen, storing seeds with different partial pressures of oxygen would be possible, although in reality, it is perhaps unlikely that seeds would ever be stored at oxygen concentrations other than those described above, i.e. the same as the concentration of oxygen in air, limited oxygen availability due to air-tight packaging in which the volume of air is limited, or under vacuum (minimal oxygen). The proviso to this is experiments in which seeds are stored in tanks at high pressure and elevated oxygen ('EPPO storage' referring to the elevated partial pressure of oxygen), which may help the understanding of the mechanisms of seed ageing (Groot et al., 2012; Buijs et al., 2020; Hourston et al., 2020).

5 Monitoring and other ways to predict viability and vigour during storage

While in theory, the seed viability equations could be used to predict the viability of a seed lot during storage, for reasons mentioned above (Section 4), but also due to sampling error associated with the initial germination result (which is required as a parameter of the viability equations), a prediction derived in such a way is not likely to be very accurate when applied to real seed-storage scenarios, especially storage of large bulk lots in warehouses. Furthermore, the parameters of the equation that describe the effect of changes in moisture content and temperature on the longevity of seeds of a particular species, have only been solved for relatively few species, while it is known that seed longevity can vary widely across different species

(Walters et al., 2005; Probert et al., 2009; Colville and Pritchard, 2019). It is also important to recognize that the symptoms of seed ageing, when they appear depending on the physiological age of the seeds, will vary between species (Wijayasinghe, 2017). Thus, some testing for viability, and vigour if necessary, will be required for stored seeds. A germination test is the most common method for assessing viability and some indication of vigour could also be made by scoring the seeds for germination at regular intervals during the germination test, to get an understanding of speed and uniformity of germination (Fig. 1; Table 1). There has been increasing interest in automating that process over recent years, involving image capture and analysis, and then statistical modelling (Wagner et al., 2011; Colmer et al., 2020; Genze et al., 2020; Merieux et al., 2021). Such systems could also be used for the germination test that is conducted as part of an accelerated ageing or controlled deterioration vigour test (Matthews et al., 2012; Table 1). Viability, and in some cases vigour (notably soya bean), may also be determined using a topographical tetrazolium test (Table 1; ISTA, 2021).

Fu et al. (2015) discussed the potential of a range of other measures that correlate with viability and could therefore be used as an alternative to germination testing, in the context of monitoring the viability of seed lots in genebank storage. In the case of genebanks, non-destructive testing is particularly desirable, since it would minimize the consumption of seeds and ultimately reduce rates of seed regeneration required to ensure sufficient quantities of seeds in storage. Possibilities include assessment of changes in DNA and RNA (Fu et al., 2015; Fleming et al., 2019; Fu et al., 2020; Saighani et al., 2021); measurement of volatiles released into the headspace above a seed sample (Colville et al., 2012; Michalak et al., 2021); quantification of seed metabolic rate (Dalziell and Tomlinson, 2017); and spectral-based analysis, including NIR and hyperspectral imaging (Ma et al., 2020; Zhang et al., 2020). The latter are perhaps particularly advantageous as they could be truly non-destructive, whereas some of the other methodologies require imbibing seeds. The potential of such tests that detect signs of ageing that correlate with viability and vigour, will depend on the validity of the correlations, probably across diverse cultivars, and the degree of risk that is accepted in terms of using an indirect measure of viability and vigour. This risk might present a cost that is countered by having a non-destructive test and/or one that is relatively fast or otherwise resource-efficient. Indeed there is perhaps more benefit of such testing for commercial seed lots than in genebank management, since genebanks do not usually rejuvenate accessions until germination has fallen to 85% or less (Hay and Whitehouse, 2017), whereas seed companies are concerned by the loss in viability and vigour that occurs well before this stage (Fig. 1).

6 Conclusion

One of the challenges in seed longevity research is knowing how to compare the longevity of seeds of different species. Different laboratories use different combinations of moisture content and temperature to measure seed longevity; some store seeds in sealed containers, others in relatively 'open' systems (e.g. over a salt solution); equations used to model seed survival curves vary; and some researchers use p_{50}, meaning, sensu strictu, the time to 50% viability during storage, as the measure of longevity, while others use a half-life p_{50} (time for viability to halve during storage, which, based on the curve in Fig. 1B, can be seen to be expected to vary depending on initial viability). Some of this is discussed in Hay et al. (2019, 2022). Furthermore, laboratory/experimental storage conditions are likely to be very different from commercial or genebank seed storage conditions. For example, to understand whether seeds of species A are likely to be shorter-lived in genebank storage than seeds of species B, without any data from the actual storage environment, it would perhaps make sense to use the same storage system as seeds in the genebank, i.e. most probably, storage in aluminium foil packets. To get data in a reasonable time frame, the moisture content and temperature will need to be elevated, but if the seeds of the two species differ in seed oil content, it does not make sense to equilibrate them to the same moisture content before sealing them inside the aluminium foil packets for storage since the water activity of the seeds at the aging temperature will differ (Vertucci and Roos, 1991). Conversely, attempting to ensure that the seeds of the different species will have the same water activity in a hermetic ageing environment requires knowledge of the relationships between moisture content and relative humidity. Perhaps the only way to address this question is to gather more information for seeds stored in diverse environments with respect to moisture (including water activity), temperature, and oxygen, and rely on modelling to make more meaningful comparisons possible. This will require significant effort. Based on ageing experiments in open storage (at 60% relative humidity and 45°C), Probert et al. (2009) compared the longevity of seeds from 195 species and derived predictions of relative longevity based on climate at the seed source and whether or not the seeds were endospermic. Further efforts could also focus on understanding how seed longevity rankings (and actual values) vary for seeds of diverse species stored in such different environments.

7 Future trends in research

Historically, interest in seed vigour has focused on traditional agricultural and horticultural crops, due to the commercial significance of seed vigour in relation to crop production. Although in seed testing, vigour is often considered a seed lot property (hence the relations shown in Fig. 1), the genetic basis of vigour traits, such as cold or heat tolerance at germination, is being revealed for more and more species, as different crops are adapted for cultivation in new environments and/or using different technologies. It is possible that current vigour tests will have to be re-evaluated for new varieties with enhanced vigour traits. It will also be important to understand how these vigour traits impact storability. Interest in the development of germination and vigour tests for seeds of minor/orphan crops and for seeds from wild species, especially those of interest for restoration, is also increasing. Understanding the relative longevity of seeds of these species will also be required. In the case of species found to have seeds with very short longevity in conventional storage systems, alternative storage techniques may be developed. Work will continue to understand the molecular basis of seed longevity and with this may come new ways to store seeds to maintain viability and vigour, and to predict/monitor seed lot viability and vigour during storage.

8 Where to look for further information

Key information and research in this area can be found at the following organisations:

- Association of Official Seed Analysts/Society of Commercial Seed Technologists (AOSA / SCST; https://analyzeseeds.com/).
- International Society for Seed Science (ISSS; https://seedscisoc.org/) and its official journal, *Seed Science Research* (https://www.cambridge.org/core/journals/seed-science-research).
- International Seed Testing Association (ISTA; https://www.seedtest.org/en/home.html) and its official journal, *Seed Science and Technology* (https://www.ingentaconnect.com/content/ista/sst).
- Society for Ecological Restoration (SER; https://www.ser.org/).

9 Acknowledgements

Many thanks to Alan G. Taylor and Michael Loss, School of Integrative Plant Science, Cornell University, USA for help compiling Table 1.

Case study – rice seed development and drying

Modern varieties of rice are shatter-resistant; the seeds remain on the plant long after mass maturity. Various studies of rice seed development have shown that mass maturity occurs at 18–25 days after 50% anthesis/flowering (Ellis et al., 1993; Kameswara Rao and Jackson, 1996a,b,c). These studies have also shown how seed physiological quality, measured by storage potential, continues to increase after mass maturity. In producing seeds for the International Rice Genebank at the International Rice Research Institute (IRRI) in the Philippines, it was recommended that seeds intended for storage in the genebank should be harvested at 35 days after flowering (DAF) (Kameswara Rao and Jackson, 1996b,c). It is standard practice to then dry seeds intended for long-term conservation in a drying room running at low humidity and cooled; at IRRI, the genebank drying room was running at 15% relative humidity and 15°C. However, a preliminary study provided some evidence that these drying conditions were not optimal for the physiological quality of the seeds (Crisostomo et al., 2011). Further research then investigated this response in more detail.

As reported in Whitehouse et al. (2015), seeds of 20 accessions from different rice variety groups were produced following normal genebank practices. Immediately after harvest, the seeds were subdivided into lots that were subjected to different drying treatments: immediately placed in the genebank drying room or initially intermittently dried using a heated air batch dryer (reaching 45°C; 8 h per day with seeds sealed inside an air-tight container overnight (16 h)) for up to 6 days, before final equilibrium drying in the drying room. Although it was intended that the seeds for each variety would be harvested at 35 days after 50% anthesis (DAA), for logistical reasons and based on the experience of the field team, the seeds of the different varieties were actually harvested at any time between 24 and 48 DAF. However, in terms of the response to the different drying treatments, it became apparent that the chronological age (DAA) of the seeds at harvest was less important than the moisture content of the seeds at harvest. If the seeds were harvested at high moisture content, there was an advantage in terms of subsequent longevity (measured using standard seed storage experiments) from drying at the higher temperature (Figure 4). These and subsequent results (Whitehouse et al., 2018) illustrate the importance of the maturation drying stage of seed development for the physiological quality of the harvested seeds. In the case of rice, if the seeds are not able to proceed through this stage, because ambient conditions are not conducive to seed drying on the plant, post-harvest drying at a higher temperature, which perhaps simulates 'ideal' on plant drying conditions, is better for the physiological quality of the seeds than immediately placing the seeds to dry in a cool environment.

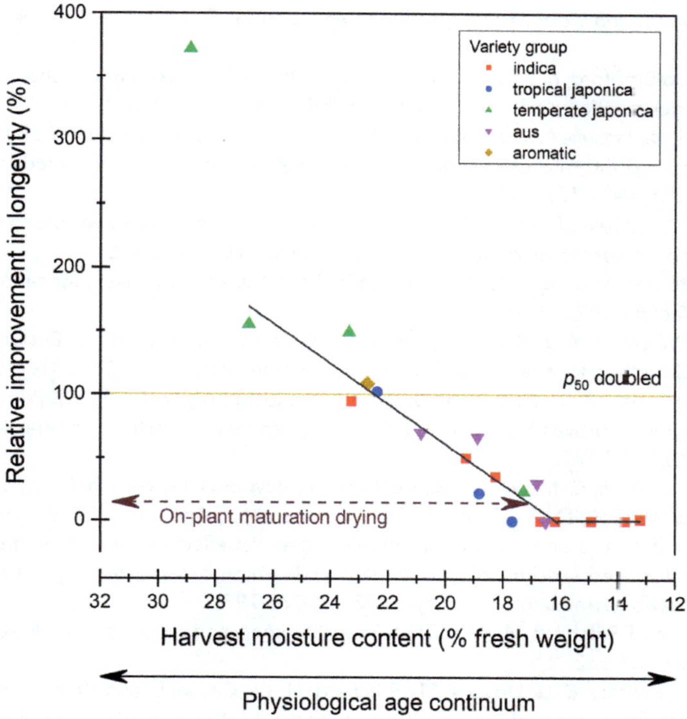

Figure 4 Freshly harvested rice seeds from different variety groups were either immediately dried at 15% RH and 15°C or initially dried at high temperature. Seed storage experiments were then carried out using seeds from each treatment and the time when seed viability was reduced to 50% (p_{50}) estimated. The relative improvement in longevity from high temperature drying is expressed as the percentage improvement in the p_{50} relative to that of seeds immediately dried at 15% RH and 15°C. Thus, a 100% relative improvement in longevity represents a doubling in p_{50}. Seeds that progressed through the maturation drying stage of seed development had progressed further through the physiological age continuum and therefore did not have such a significant response to drying at the higher temperature. Adapted from Whitehouse et al. (2015).

10 References

Abdul Rahman, S. M. and Ellis, R. H. (2019). Seed quality in rice is most sensitive to drought and high temperature in early seed development, *Seed Sci. Res.* 29(4), 238–249.

Abud, H. F., Cicero, S. M. and Gomes Junior, F. G. (2018). Radiographic images and relationship of the internal morphology and physiological potential of broccoli seeds, *Acta Sci. Agron.* 40(1), e34950.

AOSA (2009). *Seed Vigor Testing Handbook*, Association of Official Seed Analysts, Ithaca, NY.

Bello, P. and Bradford, K. J. (2021). Relationships of Brassica seed physical characteristics with germination performance and plant blindness, *Agriculture* 11(3), 220.

Bradford, K. J., Tarquis, A. M. and Duran, J. M. (1993). A population-based threshold model describing the relationship between germination rates and seed deterioration, *J. Exp. Bot.* 44(7), 1225–1234.

Bradford, K. J., Bello, P., Fu, J.-C. and Barros, M. (2013). Single-seed respiration: a new method to assess seed quality, *Seed Sci. Technol.* 41(3), 420–438.

Buckley, W. T. and Huang, J. (2011). An ethanol-based seed vigour assay for canola, *Seed Sci. Technol.* 39(2), 510–526.

Buijs, G., Willems, L. A. J., Kodde, J., Groot, S. P. C. and Bentsink, L. (2020). Evaluating the EPPO method for seed longevity analyses in Arabidopsis, *Plant Sci.* 301, 110644.

Butler, L. H., Hay, F. R., Ellis, R. H., Smith, R. D. and Murray, T. B. (2009). Priming and re-drying improves the survival of *Digitalis purpurea* L. seeds in storage, *Ann. Bot.* 103(8), 1261–1270.

Colmer, J., O'Neill, C. M., Wellsy, R., Bostrom, A., Reynolds, D., Websdale, D., Shiralagi, G., Lu, W., Lou, Q., Le Cornu, T., Ball, J., Renema, J., Andaluz, G. F., Benjamins, R., Penfield, S. and Zhou, J. (2020). SeedGerm: a cost-effective phenotyping platform for automated seed imaging and machine-learning based phenotypic analysis of crop seed germination, *New Phytol.* 228(2), 778–793.

Colville, L. and Pritchard, H. W. (2019). Seed life span and food security, *New Phytol.* 224(2), 557–562.

Colville, L., Bradley, E. L., Lloyd, A. S., Pritchard, H. W., Castle, L. and Kranner, I. (2012). Volatile fingerprints of seeds of four species indicate the involvement of alcoholic fermentation, lipid peroxidation, and Maillard reactions in seed deterioration during ageing and desiccation stress, *J. Exp. Bot.* 63(18), 6519–6530.

Crisostomo, S., Hay, F. R., Reaño, R. and Borromeo, T. (2011). Are the standard conditions for GeneBank drying optimal for rice seed quality?, *Seed Sci. Technol.* 39(3), 666–672.

Cromarty, A. S., Ellis, R. H. and Roberts, E. H. (1982). *The Design of Seed Storage Facilities for Genetic Conservation*, International Board for Plant Genetic Resources, Rome.

Cruz, J. D., Martins, C. C., Jeromini, T. S. and da Silva, G. Z. (2020). Production fields and physiological quality of *Panicum maximum* cv. mombasa seeds, *Biosci. J.* 36(6), 2050–2059.

Dalziell, E. L. and Tomlinson, S. (2017). Reduced metabolic rate indicates declining viability in seed collections: an experimental proof-of-concept, *Conserv. Physiol.* 5(1), cox058.

De Medeiros, A. D., Pinheiro, D. T., Xavier, W. A., da Silva, L. J. and dos Santos Dias, D. C. F. (2020). Quality classification of *Jatropha curcas* seeds using radiographic images and machine learning, *Ind. Crop. Prod.* 146, 112162.

Dickie, J. B., Ellis, R. H., Kraak, H. L., Ryder, K. and Tompsett, P. B. (1990). Temperature and seed storage longevity, *Ann. Bot.* 65(2), 197–204.

Ellis, R. H. (2019). Temporal patterns of seed quality development, decline, and timing of maximum quality during seed development and maturation, *Seed Sci. Res.* 29(2), 135–142.

Ellis, R. H. and Hong, T. D. (2006). Temperature sensitivity of the low-moisture-content limit to negative seed longevity-moisture content relationships in hermetic storage, *Ann. Bot.* 97(5), 785–791.

Ellis, R. H. and Hong, T. D. (2007). Seed longevity – moisture content relationships in hermetic and open storage, *Seed Sci. Technol.* 35(2), 423–431.

Ellis, R. H. and Roberts, E. H. (1980). Improved equations for the prediction of seed longevity, *Ann. Bot.* 45(1), 13–30.

Ellis, R. H. and Roberts, E. H. (1981). The quantification of ageing and survival in orthodox seeds, *Seed Sci. Technol.* 9, 373–409.

Ellis, R. H., Hong, T. D. and Roberts, E. H. (1989). A comparison of the low-moisture-content limit to the logarithmic relation between seed moisture and longevity in twelve species, *Ann. Bot.* 63(6), 601–611.

Ellis, R. H., Hong, T. D., Roberts, E. H. and Tao, K.-L. (1990). Low moisture content limits to relations between seed longevity and moisture, *Ann. Bot.* 65(5), 493–504.

Ellis, R. H., Hong, T. D. and Jackson, M. T. (1993). Seed production environment, time of harvest, and the potential longevity of seeds of three cultivars of rice (*Oryza sativa* L.), *Ann. Bot.* 72(6), 583–590.

Ellis, R. H., Hong, T. D. and Roberts, E. H. (1995). Survival and vigour of lettuce (*Lactuca sativa* L.) and sunflower (*Helianthus annuus* L.) seeds stored at low and very-low moisture contents, *Ann. Bot.* 76(5), 521–534.

Fabrissin, I., Sano, N., Seo, M. and North, H. M. (2021). Ageing beautifully: can the benefits of seed priming be separated from a reduced lifespan trade-off?, *J. Exp. Bot.* 72(7), 2312–2333.

Finch, H. J. S., Samuel, A. M. and Lane, G. P. F. (2014). *Lockhart and Wiseman's Crop Husbandry Including Grassland* (8th edn.), Woodhead Publishing Limited, Cambridge; Waltham; Kidlington.

Finch-Savage, W. E. and Bassel, G. W. (2016). Seed vigour and crop establishment: extending performance beyond adaptation, *J. Exp. Bot.* 67(3), 567–591.

Fleming, M. B., Hill, L. M. and Walters, C. (2019). The kinetics of ageing in dry-stored seeds: a comparison of viability loss and RNA degradation in unique legacy seed collections, *Ann. Bot.* 123(7), 1133–1146.

Fu, Y.-B., Ahmed, Z. and Diederichsen, A. (2015). Towards a better monitoring of seed ageing under *ex situ* seed conservation, *Cons. Physiol.* 3, 26.

Fu, Y.-B., Ahmed, Z., Yang, H., Horbach, C. and Peterson, G. W. (2020). Patterns of mitochondrial DNA fragmentation in bread wheat (*Triticum aestivum* L.) seeds under ex situ GeneBank storage and artificial aging, *Genet. Resour. Crop Evol.* 67(8), 2023–2036.

Galletti, P. A., Carvalho, M. E. A., Hirai, W. Y., Brancaglioni, V. A., Arthur, V. and da Silva, C. B. (2020). Integrating optical imaging tools for rapid and non-invasive characterization of seed quality: tomato (*Solanum lycopersicum* L.) and carrot (*Daucus carota* L.) as study cases, *Front. Plant Sci.* 11, 577851.

Genze, N., Bharti, R., Grieb, M., Schultheiss, S. J. and Grimm, D. G. (2020). Accurate machine learningbased germination detection, prediction and quality assessment of three grain crops, *Plant Methods* 16(1), 157.

Groot, S. P. C., Surki, A. A., de Vos, R. C. H. and Kodde, J. (2012). Seed storage at elevated partial pressure of oxygen, a fast method for analysing seed ageing under dry conditions, *Ann. Bot.* 110(6), 1149–1159.

Hay, F. R., Davies, R. M., Dickie, J. B., Merritt, D. J. and Wolkis, D. M. (2022). More on seed longevity phenotyping. *Seed Sci. Res.* https://doi.org/10.1017/S0960258522000034.

Hay, F. R. and Probert, R. J. (1995). The effect of different drying conditions and maturity on desiccation tolerance and seed longevity in *Digitalis purpurea* L., *Ann. Bot.* 76, 639–647.

Hay, F. R. and Smith, R. D. (2003). Seed maturity: when to collect seeds from wild plants. In: Smith, R. D., Dickie, J. B., Linington, S. H., Pritchard, H. W. and Probert, R. J. (Eds), *Seed Conservation: Turning Science into Practice*, The Royal Botanic Gardens, Kew, UK, pp. 97–133.

Hay, F. R. and Probert, R. J. (2011). Collecting and handling seeds in the field. In: Guarino, L., Ramanatha Rao, V. and Goldberg, E. (Eds), *Collecting Plant Genetic Diversity: Technical Guidelines – 2011 Update*, Bioversity International, Rome.

Hay, F. R. and Whitehouse, K. J. (2017). Rethinking the approach to viability monitoring in seed genebanks, *Conserv. Physiol.* 5(1), cox009.

Hay, F. R., Mead, A., Manger, K. and Wilson, F. J. (2003). One-step analysis of seed storage data and the longevity of Arabidopsis thaliana seeds, *J. Exp. Bot.* 54(384), 993–1011.

Hay, F. R., Mead, A. and Bloomberg, M. (2014). Modelling seed germination in response to continuous variables: use and limitations of probit analysis and alternative approaches, *Seed Sci. Res.* 24(3), 165–186.

Hay, F. R., Timple, S. and van Duijn, B. (2015). Can chlorophyll fluorescence be used to determine the optimal time to harvest rice seeds for long-term GeneBank storage?, *Seed Sci. Res.* 25(3), 321–334.

Hay, F. R., Valdez, R., Lee, J. S. and Sta Cruz, P. C. (2019). Seed longevity phenotyping: recommendations on research methodology, *J. Exp. Bot.* 70(2), 425–434.

Hay, F. R., Whitehouse, K. J., Ellis, R. H., Sackville Hamilton, N. R., Lusty, C., Ndjiondjop, M. N., Tia, D., Wenzl, P., Santos, L. G., Yazbek, M., Azevedo, V. C. R., Peerzada, O. H., Abberton, M., Oyatomi, O., de Guzman, F., Capilit, G., Muchugi, A. and Kinjanjui, Z. (2021). CGIAR GeneBank viability data reveal inconsistencies in seed collection management, *Glob. Food Secur.* 30, 100557.

Hourston, J. E., Perez, M., Gawthrop, F., Richards, M., Steinbrecher, T. and Leubner-Metzger, G. (2020). The effects of high oxygen partial pressure on vegetable *Allium* seeds with a short shelf-life, *Planta* 251(6), 105.

Hung, L. Q., Hong, T. D. and Ellis, R. H. (2001). Constant, fluctuating and effective temperature and seed longevity: a tomato (*Lycopersicon esculentum* Mill.) exemplar, *Ann. Bot.* 88(3), 465–470.

ISTA (2021). *International Rules for Seed Testing 2021*, International Seed Testing Association, Bassersdorf, Switzerland.

Joosen, R. V., Kodde, J., Willems, L. A., Ligterink, W., van der Plas, L. H. and Hilhorst, H. W. (2010). GERMINATOR: a software package for high-throughput scoring and curve fitting of Arabidopsis seed germination, *Plant J.* 62(1), 148–159.

Kameswara Rao, N. and Jackson, M. T. (1996a). Seed production environment and storage longevity of japonica rices (*Oryza sativa* L.), *Seed Sci. Res.* 6(1), 17–21.

Kameswara Rao, N. and Jackson, M. T. (1996b). Effect of sowing date and harvest time on longevity of rice seeds, *Seed Sci. Res.* 7, 13–20.

Kameswara Rao, N. and Jackson, M. T. (1996c). Seed longevity of rice cultivars and strategies for their conservation in genebanks, *Ann. Bot.* 77(3), 251–260.

Kameswara Rao, N., Dulloo, M. E. and Engels, J. M. M. (2017). A review of factors that influence the production of quality seed for long-term conservation in genebanks, *Genet. Resour. Crop Evol.* 64(5), 1061–1074.

Kucukhuseyin, E. B., Memıs, N., Gokdas, Z., Groot, S. P. C. and Demir, I. (2021). Ethanol release as an indicator of seed vigour in radish, pepper, watermelon, aubergine, leek and onion seed lots, *Seed Sci. Technol.* 49(1), 1–6.

Lamichaney, A. and Maity, A. (2021). Implications of rising atmospheric carbon dioxide concentration on seed quality, *Int. J. Biometeorol.* 65(6), 805–812.

Lamichaney, A., Kumar Swain, D., Biswal, P., Kumar, V., Pratap Singh, N. and Krishna Hazra, K. (2019). Elevated atmospheric carbon-dioxide affects seed vigour of rice (*Oryza sativa* L.), *Environ. Exp. Bot.* 157, 171–176.

Lamichaney, A., Tewari, K., Basu, P. S., Katiyar, P. K. and Singh, N. P. (2021). Effect of elevated carbon-dioxide on plant growth, physiology, yield and seed quality of chickpea (*Cicer arietinum* L.) in Indo-Gangetic plains, *Physiol. Mol. Biol. Plants* 27(2), 251–263.

Lee, J. S., Velasco-Punzalan, M., Pacleb, M., Valdez, R., Kretzschmar, T., McNally, K. L., Ismail, A. M., Cruz, P. C. S., Sackville Hamilton, N. R. and Hay, F. R. (2019). Variation in seed longevity among diverse Indica rice varieties, *Ann. Bot.* 124(3), 447–460.

Leprince, O., Pellizzaro, A., Berriri, S. and Buitink, J. (2017). Late seed maturation: drying without dying, *J. Exp. Bot.* 68(4), 827–841.

Lesk, C., Rowhani, P. and Ramankutty, N. (2016). Influence of extreme weather disasters on global crop production, *Nature* 529(7584), 84–87.

Li, R., Chen, L., Wu, Y., Zhang, R., Baskin, C. C., Baskin, J. M. and Hu, X. (2017). Effects of cultivar and maternal environment on seed quality in *Vicia sativa*, *Front. Plant Sci.* 8, 1411.

Ma, T., Tsuchikawa, S. and Inagaki, T. (2020). Rapid and non-destructive seed viability prediction using near-infrared hyperspectral imaging coupled with a deep learning approach, *Comput. Electron. Agr.* 177, 105683.

Matthews, S., Wagner, M.-H., Kerr, L., McLaren, G. and Powell, A. A. (2012). Automated determination of germination time courses by image capture and early counts of radicle emergence lead to a new vigour test for winter oilseed rape (*Brassica napus*), *Seed Sci. Technol.* 40(3), 413–424.

Merieux, N., Cordier, P., Wagner, M. H., Ducournau, S., Aligon, S., Job, D., Grappin, P. and Grappin, E. (2021). ScreenSeed as a novel high throughput seed germination phenotyping method, *Sci. Rep.* 11(1), 1404.

Michalak, M., Plitta-Michalak, B. P., Nadarajan, J. and Colville, L. (2021). Volatile signature indicates viability of dormant orthodox seeds, *Physiol. Plant.* 173(3), 788–804, https://doi.org/10.1111/ppl.13465.

Nasehzadeh, M. and Ellis, R. H. (2017). Wheat seed weight and quality differ temporally in sensitivity to warm or cool conditions during seed development and maturation, *Ann. Bot.* 120(3), 479–493.

Nguyen, C. D., Chen, J., Clark, D., Perez, H. and Huo, H. A. (2021). Effects of maternal environment on seed germination and seedling vigor of *Petunia* × hybrida under different abiotic stresses, *Plants (Basel)* 10(3), 581.

Oliveira, K. R., Sampaio, F. R., Siqueira, G. S., Galvão, Í. M., Bennett, S. J., Gratão, P. L. and Barbosa, R. M. (2021). Physiological quality of soybean seeds grown under different low altitude field environments and storage time, *Plant Soil Environ.* 67(2), 92–98.

Padilha, M. S., Coelho, C. M. M. and De Andrade, G. C. D. (2020). Seed reserve mobilization evaluation for selection of high- vigor common bean cultivars, *Rev. Caatinga* 33(4), 927–935.

Paparella, S., Araújo, S. S., Rossi, G., Wijayasinghe, M., Carbonera, D. and Balestrazzi, A. (2015). Seed priming: state of the art and new perspectives, *Plant Cell Rep.* 34(8), 1281–1293.

Powell, A. A., Yule, L. J., Jing, H. C., Groot, S. P., Bino, R. J. and Pritchard, H. W. (2000). The influence of aerated hydration seed treatment on seed longevity as assessed by the viability equations, *J. Exp. Bot.* 51(353), 2031–2043.

Probert, R., Adams, J., Coneybeer, J., Crawford, A. and Hay, F. (2007). Seed quality for conservation is critically affected by pre-storage factors, *Aust. J. Bot.* 55(3), 326–335.

Probert, R. J., Daws, M. I. and Hay, F. R. (2009). Ecological correlates of ex situ seed longevity: a comparative study on 195 species, *Ann. Bot.* 104(1), 57–69.

Ranal, M. A. and de Santana, D. G. (2006). How and why to measure the germination process, *Braz. J. Bot.* 29(1), 1–11.

Royal Botanic Gardens Kew (2021). *Seed Information Database (SID)*, Version 7.1. http://data.kew.org/sid/ (Accessed June 2021).

Saighani, K., Kondo, D., Sano, N., Murata, K., Yamada, T. and Kanekatsu, M. (2021). Correlation between seed longevity and RNA integrity in the embryos of rice seeds, *Plant Biotechnol. (Tokyo)* 38(2), 277–283, https://doi.org/10.5511/plantbiotechnology.21.0422a.

Singh, K. D., Duddu, H. S. N., Vail, S., Parkin, I. and Shirtliffe, S. J. (2021). UAV-based hyperspectral imaging technique to estimate canola (*Brassica napus* L.) seedpods maturity, *Can. J. Remote Sens.* 47(1), 33–47.

Sinniah, U. R., Ellis, R. H. and John, P. (1998a). Irrigation and seed quality development in rapid-cycling brassica: seed germination and longevity, *Ann. Bot.* 82(3), 309–314.

Sinniah, U. R., Ellis, R. H. and John, P. (1998b). Irrigation and seed quality development in rapid-cycling brassica: soluble carbohydrates and heat-stable proteins, *Ann. Bot.* 82(5), 647–655.

Solberg, S. Ø, Yndgaard, F., Andreasen, C., von Bothmer, R., Loskutov, I. G. and Asdal, Å (2020). Long-term storage and longevity of orthodox seeds: a systematic review, *Front. Plant Sci.* 11, 1007.

Song, K. and He, X. (2021). How to improve seed germination with green nanopriming, *Seed Sci. Technol.* 49(2), 81–92.

Starič, P., Vogel-Mikuš, K., Mozetič, M. and Junkar, I. (2020). Effects of nonthermal plasma on morphology, genetics and physiology of seeds: a review, *Plants (Basel)* 9(12), 1736.

Thenkabail, P. S., Smith, R. B. and De Pauw, E. (2000). Hyperspectral vegetation indices and their relationships with agricultural crop characteristics, *Remote Sens. Environ.* 71(2), 158–182.

Vertucci, C. W. and Roos, E. E. (1991). Correspondence. Seed moisture content, storage, viability and vigour, *Seed Sci. Res.* 1, 277–279.

Wagner, M.-H., Demilly, D., Ducournau, S., Dürr, C. and Léchappé, J. (2011). Computer vision for monitoring seed germination from dry state to young seedlings, *Seed Test. Int.* 142, 49–51.

Walters, C., Wheeler, L. M. and Grotenhuis, J. M. (2005). Longevity of seeds stored in a GeneBank: species characteristics, *Seed Sci. Res.* 15(1), 1–20.

Whitehouse, K. J., Hay, F. R. and Ellis, R. H. (2015). Increases in the longevity of desiccation-phase developing rice seeds: response to high-temperature drying depends on harvest moisture content, *Ann. Bot.* 116(2), 247–259.

Whitehouse, K. J., Hay, F. R. and Ellis, R. H. (2018). Improvement in rice seed storage longevity from high-temperature drying is a consistent positive function of harvest moisture content above a critical value, *Seed Sci. Res.* 28(4), 332–339.

Wijayasinghe, M. (2017). *Seed Longevity in Storage of Alpine Plant Species*, Dottorato di Ricerca in Scienze della Terra e dell'Ambiente, CV NASSTEC, Università degli Studi di Pavia, Ciclo XXX (2014–2017).

Xanthopoulos, G. (2020). Viability modelling of seeds and sensitivity analysis under fluctuating temperature and moisture content, *J. Stored Prod. Res.* 89 101708.

Xu, J., Meng, J. H. and Quackenbush, L. J. (2019). Use of remote sensing to predict the optimal harvest date of corn, *Field Crops Res.* 236, 1–13.

Yadav, G. and Ellis, R. H. (2016). Development of ability to germinate and of longevity in air-dry storage in wheat seed crops subjected to rain shelter or simulated supplementary rainfall, *Seed Sci. Res.* 26(4), 332–341.

Yadav, G. and Ellis, R. H. (2018). Longevity of 285 seed lots of wheat in hermetic storage compared with independent estimates from the seed viability equation, *Seed Sci. Technol.* 46(2), 341–347.

Yadav, S. K., Jalink, H., Groot, S. P. C., van der Schoor, R., Yadav, S., Dadlani, M. and Kodde, J. (2015). Quality improvement of aged cabbage (*Brassica oleracea* var. *capitata*) seeds using chlorophyll fluorescence sensor, *Sci. Hortic. Amsterdam* 189, 81–85.

Zhang, T. T., Fan, S. X., Xiang, Y. Y., Zhang, S. J., Wang, J. H. and Sun, Q. (2020). Non-destructive analysis of germination percentage, germination energy and simple vigour index on wheat seeds during storage by Vis/NIR and SWIR hyperspectral imaging, *Spectrochim. Acta A Mol. Biomol. Spectrosc.* 239, 118488.

Zhang, K., Zhang, Y., Sun, J., Meng, J. and Tao, J. (2021). Deterioration of orthodox seeds during ageing: influencing factors, physiological alterations and the role of reactive oxygen species2, *Plant Physiol. Biochem.* 158, 475–485.

Chapter 9

Enhancing seed defence mechanisms against pathogens

John Hampton, Bio-Protection Research Centre, Lincoln University, New Zealand

1 Introduction

For the world's major crops, production losses from pest and pathogen damage have been estimated to range from 27% to 42%, and without protection this would increase to between 48% and 83% (Glare et al., 2016). For the past 60 years, farmers have relied heavily on the use of pesticides for crop protection. However, new synthetic pesticide leads are decreasing (Glare et al., 2012) and there is increasing pressure on the continuing use of chemical control because of concerns about environmental and mammalian safety (Westman et al., 2019; Glare et al., 2016). More sustainable and environmentally acceptable alternative strategies are required (Guarnizo et al., 2020).

Priming, defined by Mauch-Mani et al. (2017), as enhancing plant responsiveness to biotic stressors (Table 1) is an alternative strategy. Plants have the ability to defend themselves from pests or pathogens because they have an innate immune system or defence mechanisms that respond to pest or pathogen attacks (Guarnizo et al., 2020). These natural plant resistance interactions and mechanisms can be promoted by microbial-plant interactions (Enebe and Babalola, 2019) and non-toxic internal or environmental biostimulants (Westman et al., 2019). Priming agents may be microorganisms or chemicals, and priming can be applied to seeds (Singh et al., 2020; Westman et al., 2019).

http://dx.doi.org/10.19103/AS.2022.0105.13

Table 1 Definitions

Priming (1)	Methods for physiologically enhancing seed performance, viz seed pre-sowing controlled hydration methods (hydropriming, osmopriming) designed to enhance seed germination synchrony (McDonald, 2000).
Priming (2)	Enhancing plant responsiveness to biotic stressors (Mauch-Mani et al., 2017).
Biopriming (1)	Including biocontrol organisms in the priming (1) process as a method of delivering then to the crop and/or controlling pathogen growth during the priming process (Rocha et al., 2019; Halmer, 2006; Warren and Bennett, 1999).
Biopriming (2)	The treatment of seed with plant beneficial micro-organisms prior to seed sowing (Afzal et al., 2020), usually in a seed coating.
Defence priming	Enhanced sensitivity and responsiveness to stress that results from a prior experience and leads to increased resistance and/or stress tolerance (Westman et al., 2019; Mauch-Mani et al., 2017; Conrath et al., 2015).
Priming stimulus	A signal that prepares a plant for enhanced responsiveness by causing subtle physiological molecular and epigenetic changes (Mauch-Mani et al., 2017).
Priming reaction	Microbial/plant interaction(s) (Conrath et al., 2015) which leads to induced systematic resistance (ISR) and/or systemic acquired resistance (SAR) and subsequent expression of genes encoding pathogenesis-related proteins and defensins (Esparza-Reynoso et al., 2020).
Horizontal transmission	Movement of microbial reproductive structures (e.g. spores) from plant to plant by wind, rain or vectors (Lugtenberg et al., 2016).
Vertical transmission	Movement of microbes between plant generations via seeds following colonisation of inflorescence primordia and developing ovules (Perez et al., 2020).

In seed technology, priming is a broad term which refers to methods for physiologically enhancing seed performance (Table 1). Biopriming originally meant the process of including biocontrol organisms in this process (Warren and Bennett, 1999; Table 1), but increasingly the term is used to refer to the treatment of seed with plant beneficial microorganisms prior to sowing (Table 1), usually in a seed coating. However, as chemicals can also be used for seed priming, this chapter uses the term 'defence priming' (Table 1). This chapter will focus on defence priming from a seed's perspective.

Plant defence can be divided into layers (Table 2), which include the presence of (i) a physical barrier that invaders must penetrate, (ii) a recognition layer which involves signals activated by the presence of the invader, and (iii) a defence action layer (Table 2). The nature of these signals varies depending on the microbial communities present in different plant tissues (Conrath et al., 2015). A priming stimulus (Table 1) prepares a plant to activate rapid and/or

Table 2 Plant defence system - a three-layer mechanism

Layer 1	A physical barrier that microbes must penetrate: characterised by strong molecular bonds, cross-linking of complex sugar and fat molecules; consists of wax, cuticle layers, cell walls and stomata (Malinovsky et al., 2014).
Layer 2	Molecular; provides basal defence and recognises self from non-self structures; allows plant to respond to microbe-associated molecular patterns (MAMPs) or pathogen associated molecular patterns (PAMPs); leads to rapid and enhanced defence resulting from recognition of pathogen effectors (virulence factors) by their effect on host targets (Cook et al., 2015).
Layer 3	An adaptive effector-triggered immunity (ETI) activated by recognition of effectors via R proteins (Thomma et al., 2011), generally leading to a hyposensitive response at the infection site leading to disease resistance (Cui et al., 2015); signalling modules (receptor-like kinases and proteins, activated protein kinase) trigger biochemical, biophysical, and molecular change in the plant including reactive oxygen species (ROS) generation, stomatal closure, and defence gene activation (Tang et al., 2017).

Adapted from Nogueira-Lopez et al. (2020).For further information, refer to the review by Nogueira-Lopez et al. (2020).

strong defence responses (Conrath et al., 2006), allowing a faster and more robust activation of these defence responses upon challenge (Conrath et al., 2015). This priming reaction (Table 1) leads to two types of induced resistance: induced systemic resistance (ISR) and systemic acquired resistance (SAR). ISR is induced by beneficial interactions between plants and micro-organisms (Conrath et al., 2015). It is not associated with major changes in the expression of defence genes (Guarnizo et al., 2020), but is governed by jasmonic acid (JA) and ethylene (ET) pathways (Hammerschmidt, 2014). SAR is induced by pathogen infection or by the application of a chemical substance that confers resistance to a broad spectrum of pathogens and is dependent on salicylic acid (SA) (Mishra et al., 2012). SAR induction requires the accumulation of endogenous SA signals, which mediate the activation of a wide set of pathogenesis-related genes (Guarnizo et al., 2020). Defence response mechanisms (Table 2) are beyond the scope of this chapter. Refer to Conrath et al. (2015) for a review of the molecular mechanisms of priming and to Westman et al. (2019) for a meta-analysis of defence priming, which uses *Arabidopsis* as a model system.

2 Microbial priming agents

2.1 Endophytes

Endophytes are microorganisms that can be detected inside healthy plant tissues and are asymptomatic (Card et al., 2016; Lugtenberg et al., 2016). Some are acquired by horizontal infection, while others are vertically transmitted via seed (Perez et al., 2020). Endophytes with a long co-evolutionary history with their host play a pivotal role in regulating plant interactions with other microorganisms such as pathogens (May and Nelson, 2014). While the seed microbiome is receiving increasing attention (Nelson, 2018; Rochefort et al., 2019; Torres-Cortés et al., 2019), the biodiversity of seed-borne endophytes is still largely unexplored, but both bacterial and fungal endophytes have been reported in seeds of numerous plant species (see Shahzad et al., 2018).

2.1.1 Vertically transmitted endophytes

The most widely studied endophyte/host relationship is that for the symbiotic asexual *Epichloë* endophytes of cool-season grasses (Caradus and Johnson, 2020; Fernando et al., 2020; Perez et al., 2020). These endophytes produce insect deterrent alkaloids (Fernando et al., 2020; Perez et al., 2020), but these alkaloids are not considered to exhibit anti-fungal activity (Fernando et al., 2020). However, these authors did report antifungal activity for crude extracts from liquid cultures of two *Epichloë* strains from perennial ryegrass and demonstrated that individual metabolites unique to each strain were present *in vitro* and *in planta*. While fungal inhibitors including indoles and sesquiterpines are produced by sexual *Epichloë* species (Yue et al., 2000), the metabolites responsible for the anti-fungal activity of asexual *Epichloë* strains are yet to be identified (Fernando et al., 2020).

The role of vertically transmitted *Epichloë* endophytes in providing plant defences against pathogens has been less intensively studied. Caradus and Johnson (2020), Perez et al. (2020), and Xie et al. (2018) have provided recent reviews of host/*Epichloë*/pathogen interactions. Perez et al. (2020) conducted a meta-analysis which covered 18 host grass species, 11 *Epichloë* endophytes, and 22 fungal pathogen species. They reported a negative effect on pathogen growth (as did Fernando et al., 2020) and spore germination from *in vitro* studies, while *in planta* endophytes reduced disease severity and incidence. These responses mainly occurred on fungal pathogens which attack root and/or leaf tissues (Perez et al., 2020), and only occasionally on pathogens causing disruption to reproductive organs (Caradus and Johnson, 2020). Perez et al. (2020), and Xie et al. (2018) have reviewed the mechanism by which *Epichloë* endophytes could reduce pathogen infection, but evidence for activation of SAR and ISR is limited. Wang et al. (2016) reported an increased SA content in wild-rye plants containing *Epichloë* when exposed to two fungal pathogens, but there is as yet no direct evidence that *Epichloë* fungi can also regulate JA content (Xie et al., 2018). However, in tall fescue, there was an increased expression of the *TFFE41* gene. TFFE41 protein is very similar to omega-3 fatty acid desaturases, proteins involved in the biosynthesis of omega-3 trienoic fatty acids, the precursors of JA (Johnson et al., 2003). Whether TFFE41 protein leads to an increase in JA synthesis is not yet known (Bastias et al., 2017).

There is stronger evidence for *Epichloë* endophytes positively impacting the ability of the host to combat biotic stress (Xie et al., 2018). The activity of antioxidant enzymes including superoxide dismutase, peroxidase, and catalase was increased in grass plants containing *Epichloë* (Ma et al., 2015), while malondialdehyde content was significantly reduced (Ma et al., 2015), indicating an increased ability of the infected host plant to scavenge the reactive oxygen species (ROS) involved in host cell damage/death. The presence of *Epichloë* has also been associated with increased fungal gene expression in the host. The antifungal protein Efe-AfpA is accumulated in the host plant apoplast and has been suggested as a possible effector protein required for the interaction of the fungus with its host (Wang et al., 2021), suggesting that it may have a function in host disease resistance (Tian et al., 2017). Nissinen et al. (2019) concluded that modulation of fungal plant communities by *Epichloë* species appeared to be directed via fungal–fungal interactions rather than via modulation of SA- and JA-regulated plant response pathways. However, upregulation of defence priming has been reported (Wang et al., 2016) and defence priming via this mechanism cannot as yet be ruled out (Nissinen et al., 2019).

The biodiversity of seed-borne bacterial endophytes is as yet largely unknown, but numerous plant species contain diverse communities of bacterial endophytes (see Shahzad et al., 2018), many of which can enhance plant tolerance to biotic stresses (Shahzad et al., 2017). For example, Diaz-Herrera et al. (2016) isolated *Bacillus*, *Paenibacillus*, and *Pantoea* spp. from wheat seeds; all three genera include strains used as biocontrol agents, and isolates of *Pantoea* and *Paenibacillus* restricted the growth of the pathogen *Fusarium graminearum in vitro* (Diaz-Herrera et al., 2016). Ruiza et al. (2011) found that seed-endophytic bacteria in rice seeds were able to protect seedlings against infection by a *Curvularia* sp. by directly inhibiting the growth of the pathogen. Similarly, Mukherjee et al. (2020) isolated bacterial endophytes from chickpea seeds and demonstrated that several were able to increase the tolerance of chickpea plants to the pathogen *Fusarium oxysporum* f. sp. *ciceris* when treated *in vitro*. While vertical transmission of bacterial endophytes has been demonstrated in various plant species (Bergna et al., 2018; Shahzad et al., 2018; Gagne-Bourgue et al., 2013), more are reported to promote plant growth (e.g. Bergna et al., 2018) rather than increase host resistance to pathogen attacks.

For those seed-borne bacterial endophytes which can mitigate plant biotic stress, the mechanisms remain unclear. The production of extracellular enzymes, phytohormones, and secondary metabolites has been reported (Shahzad et al., 2018). Shahzad et al. (2017) suggested that *Bacillus amyloliquefaciens* induced ISR in tomato plants, while Mukherjee et al. (2020) noted that endophytic bacteria could induce increases in SA and JA. Matsumoto et al. (2021) identified a seed-endophytic bacterium, *Sphingomonas melonis*, which is accumulated and vertically transmitted in disease-resistant rice phenotypes, which can confer resistance in disease-susceptible phenotypes to the pathogen *Burkholderia plantarii* by producing anthranilic acid. They demonstrated that anthranilic acid interferes with the RpoS gene of the seed-borne pathogen, which is a central regulator of general stress response, allowing cells to survive environmental challenges and prepare for subsequent stresses. They suggested that this resulted in an impairment of the upstream cascades required for virulence factor biosynthesis. Naturally occurring seed endophytes may be synergistic components of innate immunity in their host plants (Matsumoto et al., 2021). Further research is required to determine the role of seed-borne bacterial endophytes in inducing plant defence priming.

While not related to defence priming, White et al. (2018) have recently reviewed the role of seed-endophytic bacteria in 'endobiome interference'. They hypothesised that microbes of the endobiome of a particular plant are adapted to the internal conditions of that plant, and the conditions in the endospheres of plants may differ between plant species. Inoculating these bacteria into different plant species reduced seedling growth and increased seedling mortality. For example, the bacterial endophyte *Micrococcus* isolated from tomato seeds reduced dandelion seedling root length and killed 80% of the seedlings. White et al. (2018) concluded that endobiome interference could become a method for the control of invasive or weedy plant species, allowing a reduction in herbicide use in crops.

2.1.2 Horizontally transmitted endophytes

Endophytes have the ability to colonise host roots, stems, and leaves. Those not transmitted vertically are transmitted horizontally; endophyte reproductive structures, usually spores, are moved by wind, rain splash, or vectors from plant to plant. Fungal and bacterial endophytes contribute plant fitness benefits, enabling the adaption of the plant to biotic stress (Sahu et al., 2019; Card et al., 2016). Examples of fungal endophytes able to protect host plants from pathogenic fungi were provided by Lugtenberg et al. (2016). They include *Piriformospora indica*, which enhances tolerance to various root and foliar pathogens (see Johnson et al., 2014); *Epicoccum nigrum* known for its biocontrol activity against *Sclerotinia*, *Pythium*, and *Monilia* spp. (Fávaro et al., 2012); and a *Penicillium brevicompactum* strain which suppressed seed-borne pathogens of barley (Murphy et al., 2015). Some fungal entomopathogens, originally considered only as insect pathogens, are now known to endophytically colonise a wide range of plant species (Vega et al., 2008) and are also antagonistic to some plant pathogens (Jaber and Ownley, 2017). For example, *Beauveria bassiana*, widely used as a commercial biopesticide, has been reported as an endophyte of several plants and has been shown to suppress diseases caused by fungal, bacterial, virus, and nematode plant pathogens (Jaber and Ownley,

2017). Examples of bacterial endophytes with activity against plant pathogens include *B. amyloliquefaciens* for biocontrol of *Rhizoctonia solani* in sugar beet (Karimi et al., 2016), *Bacillus subtilis* for biocontrol of *Bipolaris maydis* in maize (Ding at al., 2017), and an *Azospirillum* sp. for biocontrol of *Pseudomonas syringae* pv. *tomato* and *Botrytis cinerea* in tomato (Fujita et al., 2017).

Like vertically transmitted endophytes, horizontally transmitted endophytes produce secondary metabolites that may inhibit pathogen growth (antimicrobial compounds; see Lugtenberg et al. (2016) or induce systemic resistance (Enebe and Babalola, 2019). Raad et al. (2019) reported transcriptional reprogramming of plant defence pathways, with changes in the expression of genes related to pathogenesis, phytoalexin, and JA and SA signalling pathways in *Arabidopsis thaliana* plants colonised by endophytic *B. bassiana*. However, as already noted for *Epichloë* endophytes, the control mechanisms reported are predominantly for antimicrobial activity rather than defence priming, although Jaber and Ownley (2017) suggested a combination of mechanisms, rather than any one single mechanism that may be employed by endophytic fungi.

Gomez-Vidal et al. (2009) reported that colonisation of date palm by the entomopathogenic fungi *B. bassiana* and *Lecanicillium* spp. resulted in the upregulation of proteins involved in plant defence and stress responses, thus inducing a primed state. Hong et al. (2016) found that *Paenibacillus polymyxa*, when present in *A. thaliana* plants, was able to induce the expression of pathogenesis-related genes responsible for SA and JA signalling and defence pathways, and suggested this was the mechanism for the control of *P. syringae*.

2.1.3 Application of microbes which become endophytic

The rhizosphere, the zone between plant roots and soil, contains a vast array of microorganisms. Some are free-living species in the rhizosphere, while others are able to colonise plant roots (Hakim et al., 2021; Schweiger et al., 2021). The latter association is often beneficial for plants because the resultant endophyte-plant interactions can provide control of plant pathogens (Hakim et al., 2021; Cruz-Magalhaes et al., 2019). Numerous symbiotic bacteria described by Hakim et al. (2021) as intracellular plant-growth-promoting rhizobacteria, can colonise plant roots or nodules within the plant roots, and most are endophytic (Kumar and Dubey, 2020). Non-rhizobial bacterial endophytes include the bio-protection species *Bacillus*, *Brevibacillus*, and *Paenibacillus* among others (Hakim et al., 2021). However, an isolate of *Paenibacillus* which reduced the incidence and severity of *Xanthomonas campestris* pv. *campestris* in cabbage following seed application was not detected within the host tissues (Ghazalibiglar et al., 2016). More recently Rivas-Franco et al. (2020a,b) have demonstrated that the entomopathogen *Metarhizium* becomes endophytic in maize roots following application as a seed treatment, while the presence of *Trichoderma* fungi in plant roots following growth medium or seed treatment application has long been known (Brotman et al., 2013). Once present in the roots, whether endophytes can move into the aboveground tissues is less certain and may depend on endophyte–host interactions. Among the entomopathogens with plant protection abilities, some species of *Beauveria* have the ability to inhabit the aboveground plant tissues (Moonjely et al., 2016). Until recently *Metarhizium* spp. were considered to only associate with host root tissues (Rivas-Franco et al., 2020a), but in a new study, Cervantes et al. (2021) recovered *Metarhizium* from meadow fescue leaf tissue following seed coating. Similarly, seed-applied *Trichoderma atroviride* has been recovered from perennial ryegrass leaves (Daryaei et al., 2015). Singh et al. (2020) reported that the rhizobacterium *Pseudomonas aeruginosa* strain MF-30, which exhibited anti-fungal properties against *R. solani*, was able to colonise both roots and foliar tissues of maize plants following seed bio-priming. Whether endophyte presence in above-ground tissues is important or whether root colonisation is all that is required to prime plant defences is still being investigated. For *Trichoderma* spp., root colonisation can induce SR in tissues far from the pathogen colonisation site (Coppola et al., 2019; Mayo-Prieto et al., 2019). *Trichoderma* species induce plant defence priming though multiple mechanisms, regulated in part by bioactive molecules that cross-talk with the host defence hormonal pathways (Esparza-Reynoso et al., 2020). Microbe-associated molecular patterns (MAMPs) produced by *Trichoderma* spp., including peptaibols, cellulases, xylanases, cerato-platanins, swollenin, and other secondary metabolites activate the SA and ET signalling pathways, elicit systemic defence, JA and SA biosynthesis, and activate plant defence mechanisms (see Esparza-Reynoso et al., 2020). Rivas-Franco et al. (2020a) reported that following seed treatment with *Metarhizium anisopliae*, maize plants challenged by *F. graminearum* had elevated SA and JA concentrations in roots and shoots, indicating a potentially primed state. Singh et al. (2020) reported ISR following biopriming of maize seeds with *P. aeruginosa* as a result of a significantly higher induction and accumulation of antioxidant and defence-related enzymes in the maize plant, particularly phenylalanine ammonia lyase (PAL). PAL is the key enzyme that triggers ISR via the phenylpropanoid pathway in plants (Singh et al., 2020).

3 Chemical priming agents

3.1 Volatile organic compounds

Volatile organic compounds (VOCs) are low molecular weight compounds comprising aliphatic and aromatic hydrocarbons, ketones, alcohols, esters, aldehydes, terpenes, and other classes (Alfiky and Weisskopf, 2021). These secondary metabolites, because of their volatility, can move both above (Brilli et al., 2019) and below (Thomas et al., 2020) ground. They are important for antibiosis and signalling for symbiotic interaction and have a role in maintaining plant health through the direct suppression of plant pathogens, the induction of plant resistance against pathogens, and the promotion of plant growth (Thomas et al., 2020).

3.1.1 Plant volatile organic compounds

The possible role of plant VOCs (pVOCs) in agriculture to improve sustainable plant defence strategies was reviewed by Brilli et al. (2019). pVOCs can prime plant defences against insect pests and pathogens (Maurya, 2020; Brilli et al., 2019). pVOCs such as z-3 hexenyl acetate, methyl salicylate, and monoterpenes have been reported to prime resistance to pathogens by actively participating in the mechanisms leading to SAR (Dempsey and Klessig, 2012). A number of other pVOCs have inhibited germination and growth of plant pathogens (Quintana-Rodriguez et al., 2018; Simas et al., 2017), but most of these reports have come from laboratory studies, and as yet, there are few field studies (Brilli et al., 2019).

Maurya (2020) found that seeds exposed to indole, a herbivore-induced pVOC, were defence primed against herbivore attack. Indole is known to act as a defence priming cue (Ye et al., 2018; Erb et al., 2015). Maurya (2020) suggested that the direct application of pVOCs to seeds could provide a more viable priming-mediated solution to pest (and pathogen?) management than the in-field foliar or soil applications of pVOCs (see Brilli et al. (2019) for a discussion of the limitations and drawbacks of field pVOC application).

Seeds can also release VOCs (Zhang et al., 1995) as a result of physiological deterioration during ageing. Colville et al. (2012) reported that alcoholic formation was the major source of these VOCs. Interestingly, some of these VOCs (e.g. 2-butanone, benzaldehyde, D-Limonene) are also produced by microbes (Nogueira-Lopez et al., 2020; Thomas et al., 2020). Exposure of plants to 2-butanone and benzaldehyde produced by *Bacillus* spp. induced resistance to a bacterial pathogen of cucumber and tobacco, respectively (Thomas et al., 2020). Whether non-deteriorating seeds produce VOCs, which may have a role in defence priming, is yet to be determined.

3.1.2 Microbial volatile organic compounds

Soil microorganisms, both beneficial and pathogenic (Li et al., 2018), produce a large number of secondary metabolites enabling them to compete with other neighbouring microorganisms (Thomas et al., 2020) Their structural diversity explains their broad spectrum of activities (Macheleidt et al., 2016). Microbial VOCs can promote plant growth (Cruz-Magalhaes et al., 2019; Sharifi and Ryu, 2018), but some have the potential to be used for suppression of plant pathogens. For example, strains of *Bacillus*, *Acinetobacter*, *Burkholderia*, *Pseudomonas*, and *Streptomyces* produce bacterial VOCs (bVOCs), and strains of *Penicillium*, *Nodulisporium*, and *Trichoderma* produce fungal VOCs (fVOCs) with activity against bacterial or fungal plant pathogens (see Thomas et al., 2020).

Both bVOCs and fVOCs can directly suppress plant pathogens (Tyagi et al., 2020; Xie et al., 2018; Rajer et al., 2017), but they can also induce plant resistance (Thomas et al., 2020). For example, the bVOCs 2-butanone, albuterole, and 1,3-propanediol produced by strains of *B. subtilis* induce plant resistance by a mechanism dependent on JA and SA signalling pathways (Song et al., 2015), and the expression of resistance (RRSI) and pathogen-related proteins (Prla, Prlb) which are involved in SA signalling (Tahir et al., 2017). Similarly, the fVOCs M-cresol produced by *Cladosporium* and methyl benzoate produced by *Ampleomyces* induce JA signalling pathways under pathogen challenge, with the former also inducing SA signalling pathways (Naznin et al., 2014), while expression of genes involved in SA signalling was induced by dimethyl disulphide (Tyagi et al., 2020).

Species of *Trichoderma* can produce a vast array of fVOCs (Nogueira-Lopez et al., 2020), which have roles in both defence and signalling (Zeilinger et al., 2016). For example, Cruz-Magalhaes et al. (2019) isolated 29 fVOCs from *T. atroviride* strain IMI206040, with two, 2-heptanone and 6-pentyl-2*H*-pyran-2-one (6PP) being proportionally more abundant. *Trichoderma* VOCs are known to be able to activate plant defence mechanisms (Malmierca et al., 2015; Vinale et al., 2008a,b). fVOCs can enhance plant growth (Fig. 1) and defence (Fig. 2) (Garnica-Vergara et al., 2016; Kottb et al., 2015), the latter through enhancing activation of defence enzymes (El-Hasan and Buchenauer, 2009) and increasing expression of defence-related genes (Kottb et al., 2015). However, while *Trichoderma* produced VOCs have been identified through chemical screens (e.g. Cruz-Magalhaes et al., 2019), their roles during plant–fungus interactions have as yet been little studied (Zeilinger and Atanasova, 2020).

There are as yet no published reports of treating seeds with microbial VOCs to prime plant defence systems. However, the plant growth promotion in carrot (Fig. 1) and the response in radish to *R. solani* challenge (Fig. 2), both followed the application of an fVOC to seeds via enclosure in a polymer coating (Hampton et al., unpublished data).

3.2 Other chemical priming

Many natural and synthetic compounds can prime plants (see reviews by Mauch-Mani et al., 2017; Conrath et al., 2006). These include β-aminobutyric acid (BABA), probinazole, benzothiadazole (BTH), and SA, all of which can

Figure 1 Plant growth promotion after carrot seed treatment with a fVOC (left) compared to the untreated control (right). Photograph by Dr Wadia Kandula, Bio-Protection Research Centre, Lincoln University, New Zealand.

Figure 2 Reduction of the impact on radish of the soil-borne fungal pathogen *Rhizoctonia solani* by a fVOC. fVOC seed treatment allowed the production of a harvestable product (left) in comparison with the untreated control, which has failed to do so (right). Photograph by Dr Wadia Kandula, Bio-Protection Research Centre, Lincoln University, New Zealand.

induce resistance in plants to protect against a broad range of pathogens (Balmer et al., 2015; Yi et al., 2015), and secondary metabolites which mediate SAR, for example, JA, methyl salicylate, and glycerol-3-phosphate (Mauch-Mani et al., 2017).

There are many reports of using chemicals to prime seeds, but most investigated the effects on subsequent plant growth and development rather than on biotic stress resistance (Westman et al., 2019). Worrall et al. (2012) primed tomato seeds with JA or BABA. They found that plants grown from JA-treated seed had increased resistance to *B. cinerea*, while BABA seed treatment provided defence against powdery mildew caused by *Oidium neolycopersici*, and for both treatments, the priming responses were still evident 8 weeks after seedling emergence. These responses were associated with enhanced defence gene expression during pathogen attacks. Song et al. (2017) primed the seeds of pepper and cucumber with BTH and reported increased defence gene (*PR1* - pepper; *PR2* - cucumber) activity in the presence of *P. syringae* pv. *lachrymans*. Priming seeds of pearl millet (*Pennisetum glaucum* L.) with vitamins (thiamine, riboflavin, niacin) reduced the symptoms of the oomycete *Sclerospora graminicola* on plant tissues (Pushpalatha et al., 2011), while Martinez-Aguilar et al. (2021) primed bean seeds with 2,6,dichloro-isonicotinic acid (INA) (see Section 4). Westman et al. (2019) concluded that vitamins could be attractive agents for priming plants of various species when applied to seeds.

4 Transgenerational defence priming

Defence priming represents a type of immunological memory that allows plants to remember stressful situations (Mauch-Mani et al., 2017), and this memory can be inherited (Ramirez-Carrasco et al., 2017; Conrath et al., 2015). Mauch-Mani et al. (2017), described this as a 'transgenerational priming state - a priming state that is visible in the progeny of stimulated plants and mediated by epigenetic mechanisms'.

One of the first reports of transgenerational priming was that of Slaughter et al. (2012) who demonstrated that progeny of *A. thaliana* plants that had been primed with the elicitor BABA showed enhanced resistance to *P. syringae* pv. *tomato* as well as enhanced expression of defence-related genes. Ramirez-Carrasco et al. (2017) primed bean (*Phaseolus vulgaris*) plants with BABA and recorded a faster and stronger transcript accumulation of the *PvPR1* plant defence gene, which was transmitted to the progeny. This primed resistance to *P. syringae* pv. *phaseolicola* attack lasted for two generations.

While the above authors both applied the primary activator by spraying plant tissues, Martinez-Aguilar et al. (2021) used seed treatment. Bean seeds were primed with the resistance inducer 2,6dichoro-isonicotinic acid (INA), a synthetic SA analogue; this treatment induced resistance to *P. syringae* pv. *phaseolicola*, and this primed state was transferred epigenetically to the progeny.

Trichoderma fungi applied as a seed treatment or incorporated into the growing medium can also induce transgenerational defence priming (Morán-Diez et al., 2021), providing the plant with long-lasting resistance against biotic and abiotic stresses by balancing phytohormone-dependent pathways (SA, JA, ET, abscisic acid, auxin, and gibberellins) and modulating the level of growth and defence regulatory proteins (Hermosa et al., 2013). *Trichoderma* may be involved in the plant's epigenetic regulation through histone modifications and replacements, DNA hypomethylation and RNA-directed DNA methylation (Morán-Diez et al., 2021). This is an area of research worthy of further exploration.

Transgenerational defence priming has been demonstrated in different species, suggesting it may be widely distributed in plants (Ramirez-Carrasco et al., 2017). Epigenetic changes involved, including DNA methylation, histone methylation/acetylation, chromatin remodelling, and transmission of small RNAs (Martinez-Aguilar et al., 2016), offer the opportunity to identify the mechanisms by which epigenetic variation modifies plant gene regulation, and use this information in breeding for stress tolerance (Martinez-Aguilar et al., 2021). For seed production, transgenerational defence priming originating from seed treatment could become a key component in integrated pest management (Ramirez-Carrasco et al., 2017) and also provide a mechanism for mitigating the abiotic stresses resulting from climate change, particularly the impact of heat stress, which can reduce seed vigour and germination (Rashid et al., 2020).

5 Beneficial microbe delivery

The use of microorganisms for defence priming requires an effective association between the microorganism and the seed, either through endophytic colonisation during seed development on the plant, incorporation during seed priming (biopriming), the application of a seed coating, or possibly by injection of the microorganisms into seed (Bailly et al., 2020; Rocha et al., 2019). Afzal et al. (2020) and Rocha et al. (2019) reviewed seed coating and delivery systems, while Deshmukh et al. (2020) reviewed seed biopriming. Reichenberger et al. (2017) have a patent for seed injection of microorganisms into large seeds such as maize and sunflower, but according to Bailly et al. (2020) this method is so far restricted to a laboratory scale.

Endophytic colonisation of the host plant can be through an existing obligate association (entire life cycle occurs within the host; Card et al., 2016), injection of the endophyte into host tissue (Johnson et al., 2013; Mitter, 2020), seed priming (Mahmood and Kataoka, 2019; O'Callaghan, 2016), seed coating

(Mitter, 2020; O'Callaghan, 2016; Rivas-Franco et al., 2020b), root dipping (Brownbridge et al., 2012), soil drenching (Bressan and Borges, 2004), soil incorporation (Jaronski and Jackson, 2008), and spraying flowers (Mitter et al., 2017). Both soil and seed application systems place the microorganism close to the plant roots, which can then be colonised superficially and endophytically (Rivas-Franco et al., 2020b). This root association is sufficient to induce defence priming (Rivas-Franco et al., 2019, 2020a) and is a practical application method for non-vertically transmitted endophytes.

Vertically transmitted endophytes escape competition with soil microorganisms, and are in intimate contact with the plant tissue following seed germination. Mitter et al. (2017) described introducing a bacterial endophyte into plants by spray application to flowers, which allowed introduction of the bacteria into the ovary and resulted in its inclusion in the seed harvested – an induced vertical inheritance. While this endophyte is a growth promotor (Mitter et al., 2017) rather than an inducer of defence responses, the technique offers a possible mechanism for the introduction of microbes with the latter property into plants. However, whether the resulting vertical transmission is maintained over more than one generation is as yet unknown.

6 Summary and future trends

Defence priming is now recognised as a strategy to combat biotic stress and represents an approach to enhance plant protection in today's agricultural systems (Mauch-Mani et al., 2017; Conrath et al., 2015). Westman et al. (2019) concluded that 'seed defence priming may protect plants from an early age, promises to be both labour and resource efficient, poses very little environmental risk, and is thus both commercially and ecologically promising'.

However, before promise can be turned into commercial reality, more research is required in the following areas of seed defence priming:

- An improved understanding of the seed microbiota and possible roles in defence priming.
- The role of endophytes in contributing to defence priming.
- Can fungal and/or bacterial VOCs consistently be successfully used for defence priming?
- The molecular mechanisms involved in priming.
- Whether defence priming provides defence against any stress (i.e. biotic and/or abiotic), or is it more specific.
- Is defence priming vertically transmitted and therefore intergenerational?
- Are there as yet undiscovered defence priming agents?
- The role of priming in integrated pest management programmes.
- The economics of defence priming.

Knowledge emerging from answering these research questions will contribute to sustainable plant protection arising from seed defence priming

7 Where to look for further information

The following reviews are recommended:

- *Trichoderma*-plant interactions; Morán-Diez et al. (2021).
- The seed microbiome; Nelson (2018).
- Plant defence priming; Conrath et al. (2015), Mauch-Mani et al. (2017), Westman et al. (2019).
- Microbial inoculation of seed; O'Callaghan (2016).

The following books are recommended:

- New and Future Developments in Microbial Biotechnology and Bioengineering, V. K. Gupta, S. Zeilinger, H. B. Singh, I Druzhinina (Eds.), Elsevier, The Netherlands, 2020.
- Seed Endophytes, S. Verma, J. White Jr (Eds.), Springer, USA, 2019.

8 References

Afzal, I., Javed, T., Amirkhani, M. and Taylor, A. G. (2020). Modern seed technology: seed coating delivery systems for enhancing seed and crop performance, *Agriculture* 10(11), 526. doi: 10.3390/agriculture10110526.

Alfiky, A. and Weisskopf, L. (2021). Deciphering *Trichoderma*-plant-pathogen interactions for better development of biocontrol applications, *Journal of Fungi* 7(1), 61. doi: 10.3390/jof7010061.

Bailly, C., Bousquet, A., Braun, V., Buitink, J., Desbois-Vimont, C., Durand Tardif, M., Fougereux, J. A., Gaillard, A., Aurelia, G., Grappin, P., Hily, Y., Lejard, F., Malausa, T., Muriel, M., Multeau, C., Orgeur, G., Piovan, R., Poisson, A-S., Prefizi, C., Rajjou, L., Reboud, X., and Szambien, M. (2020). Towards seed protection using biocontrol strategies, Position paper, *Consortium Biocôntrole*/GIS BV, Paris, France.

Balmer, A., Pastor, V., Gamir, J., Flors, V. and Mauch-Mani, B. (2015). The 'prime-ome': Towards a holistic approach to priming, *Trends in Plant Science* 20(7), 443–452. doi: 10.1016/j.tplants.2015.04.002.

Bastias, D. A., Mostinez-Ghersa, M. A., Ballare, C. L. and Gundel, P. E. (2017). *Epichloe* fungal endophytes and plant defences: Not just alkaloids, *Trends in Plant Science* 22(11), 939–948. doi: 10.1016/j.tplants.2017.08.005.

Bergna, A., Cernava, T., Randler, M., Grosch, R., Zachow, C. and Berg, G. (2018). Tomato seeds preferably transmit plant beneficial endophytes, *Phytobiomes Journal* 2(4), 183–193. doi: 10.1094/PBIOMES-06-18-0029-R.

Bressan, W. and Borges, M. T. (2004). Delivery methods for introducing endophytic bacteria into maize, *BioControl* 49(3), 315–322. doi: 10.1023/B:BICO.0000025372.51658.93.

Brilli, F., Loreto, F. and Baccelli, I. (2019). Exploiting plant volatile organic compounds (VOCs) in agriculture to improve sustainable defense strategies and productivity in crops, *Frontiers in Plant Science* 10, 264. doi: 10.3389/fpls.2019.00264.

Brotman, Y., Landau, U., Cuadros-Inostroza, Á, Tohge, T., Fernie, A. R., Chet, I., Viterbo, A. and Willmitzer, L. (2013). *Trichoderma*-plant root colonisation: Escaping early plant defense responses and activation of the antioxidant machinery for saline stress tolerance, *PLoS Pathogens* 9(3), e1003221. doi: 10.1371/journal.ppat.1003221.

Brownbridge, M., Reay, S. D., Nelson, T. L. and Glare, T. R. (2012). Persistence of *Beauveria bassiana* (Ascomycota: Hypocreales) as an endophyte following inoculation of radiata pine seed and seedlings, *Biological Control* 61(3), 194–200.

Caradus, J. R. and Johnson, L. J. (2020). *Epichloe* fungal endophytes - From a biological curiosity in wild grasses to an essential component of resilient high performing ryegrass and fescue pastures, *Journal of Fungi* 6(4), 322. doi: 10.3390/jof6040322.

Card, S. D., Johnson, L. J., Teasdate, S. and Caradus, J. (2016). Deciphering endophyte behaviour - The link between endophyte biology and efficacious biological control agents, *FEMS Microbiology Ecology* 92(8), 1–19. doi: 10.1093/femsec/fiw114.

Cervantes, P., Brookes, J., Hampton, J. G. and Glare, T. R. (2021). *The effect of Metarhizium novozealandicum on the growth and culturable endomycobiome of meadow fescue*, Summer Studentship Report, Bio-Protection Research Centre, Lincoln University, New Zealand.

Colville, L., Bradley, E. L., Lloyd, A. S., Pritchard, H. W., Castle, L. and Kranner, I. (2012). Volatile fingerprints of seeds of four species indicate the involvement of alcoholic fermentation, lipid peroxidation, and Maillard reactions in seed deterioration during ageing and desiccation stress, *Journal of Experimental Botany* 63(18), 6519–6530. doi: 10.1093/jxb/ers307.

Conrath, U., Beckers, G. J., Flors, V., García-Agustín, P., Jakab, G., Mauch, F., Newman, M. A., Pieterse, C. M., Poinssot, B., Pozo, M. J., Pugin, A., Schaffrath, U., Ton, J., Wendehenne, D., Zimmerli, L. and Mauch-Mani, B. (2006). Priming: Getting ready for battle, *Molecular Plant-Microbe Interactions* 19(10), 1062–1071. doi: 10.1094/MPMI-19-1062.

Conrath, U., Beckers, G. J. M., Langenbach, C. J. G. and Jaskiewicz, M. R. (2015). Priming for enhanced defence, *Annual Review of Phytopathology* 53, 97–119. doi: 10.1146/annurev-phyto-080614-120132.

Cook, D. E., Mesarich, C. H. and Thomma, B. P. H. J. (2015). Understanding plant immunity as a surveillance system to detect invasion, *Annual Review of Phytopathology* 53, 541–563. doi: 10.1146/annurev-phyto-080614-120114.

Coppola, M., Diretto, G., Diglio, M. C., Woo, S. L., Giuliano, G., Molisso, D., Pennacchio, F., Lorito, M. and Rao, R. (2019). Transcriptome and metabolome reprogramming in tomato plants by *Trichoderma harzianum* strain T2 primes and enhances defense responses against aphids, *Frontiers in Physiology* 10, 745. doi: 10.3389/fphys.2019.00745.

Cruz-Magalhaes, V., Nieto-Jacobo, M. F., van Zijll de Jong, E., Rostás, M., Padilla-Arizmendi, F., Kandula, D., Kandula, J., Hampton, J., Herrera-Estrella, A., Steyaert, J. M., Stewart, A., Loguercio, L. L. and Mendoza-Mendoza, A. (2019). The NADPH oxidases Nox1 and Nox2 differentially regulate volatile organic compounds, fungistatic activity, plant growth promotion and nutrient assimilation in *Trichoderma atroviride*, *Frontiers in Microbiology* 9, 3271. doi: 10.3389/fmicb.2018.03271.

Cui, H., Tsuda, K. and Parker, J. E. (2015). Effector-triggered immunity: from pathogen perception to robust defence. *Annual Review of Plant Biology 66*, 487-511. doi:10.1146/annurev-arplant.050213-040012.

Daryaei, A., Falloon, R. F., Jones, E. E., Kandula, D. R. W., Ghazalibiglar, H., Alizadeh, H., Glare, T. R. and Hampton, J. G. (2015). Endophyte colonisation of perennial ryegrass by *Trichoderma atroviride, New Zealand Plant Protection* 68, 444. doi: 10.30843/nzpp.2015.68.5850.

Dempsey, D. A. and Klessig, D. F. (2012). SOS - too many signals for systemic acquired resistance?, *Trends in Plant Science* 17(9), 538–545. doi: 10.1016/j.tplants.2012.05.011.

Deshmukh, A. J., Jaiman, R. S., Bambharolia, R. P. and Patil, V. A. (2020). Seed biopriming - A review, *International Journal of Economic Plants* 7(1), 038–043. doi: 10.23910/2/2020.0359.

Diaz-Herrera, S., Grassi, C., Zawoznik, M. and Groppa, M. D. (2016). Wheat seeds harbour bacterial endophytes with potential as plant growth promotors and biocontrol agents of *Fusarium graminearum*, *Microbiological Research* 186–187, 37–43.

Ding, T., Su, B., Chen, X., Xie, S., Gu, S., Wang, Q., Huang, D. and Jiang, H. (2017). An endophytic bacterial strain isolated from *Eucommia ulmoides* inhibits southern corn leaf blight, *Frontiers in Microbiology* 8, 903. doi: 10.3389/fmicb.2017.00903.

El-Hasan, A. and Buchenauer, H. (2009). Actions of 6-pentyl-alpha-pyrone in controlling seedling blight caused by *Fusarium moniliforme* and inducing defense response in maize, *Journal of Phytopathology* 157(11–12), 697–707; doi: 10.1111/j.1439-0434.2009.01565.x.

Enebe, M. C. and Babalola, O. O. (2019). The impact of microbes in the orchestration of plant's resistance to biotic stress: A disease management approach, *Applied Microbiology and Biotechnology* 103(1), 9–25. doi: 10.1007/s00253-018-9433-3.

Erb, M., Veyrat, N., Robert, C. A., Xu, H., Frey, M., Ton, J. and Turlings, T. C. (2015). Indole is an essential herbivore-induced volatile priming signal in maize, *Nature Communications* 6, 6273. doi: 10.1038/ncomms7273.

Esparza-Reynoso, S., Pelagio-Flores, R. and Lopez-Bucio, J. (2020). Mechanism of plant immunity triggered by *Trichoderma*. In: *New and Future Developments in Microbial Biotechnology and Bioengineering*, Gupta, V. K., Zeilinger, S., Singh, H. B. and Druzhinina, I. (Eds). Elsevier, the Netherlands, pp. 57–73.

Fávaro, L. C., Sebastianes, F. L. deS. and Araujo, W. L. (2012). *Epicoccum nigrum* P16, a sugarcane endophyte, produces antifungal compounds and induces root growth, *PLoS ONE* 7, 336826. doi: 10.1371/journal.pone.0036826.

Fernando, K., Reddy, P., Hettiarachchige, I. K., Spangenberg, G. C., Rochfort, S. J. and Guthridge, K. M. (2020). Novel antifungal activity of *Lolium*-associated *Epichloe* endophytes, *Microorganisms* 8(6), 955. doi: 10.3390/microorganisms8060955.

Fujita, M., Kusajimo, M., Okumura, Y., Nakajima, M., Minamisawa, K. and Nakashita, H. (2017). Effects of colonisation of a bacterial endophyte, *Azospirillum* sp. B510, on disease resistance in tomato, *Bioscience, Biotechnology, and Biochemistry* 81(8), 1657–1662. doi: 10.1080/09168451.2017.1329621.

Gagne-Bourgue, F., Aliferis, K. A., Sequin, P., Rani, M., Samson, R. and Jabaji, S. (2013). Isolation and characterisation of indigenous endophytic bacteria associated with leaves of switchgrass (*Panicum virgatum* L) cultivars, *Journal of Applied Microbiology* 114(3), 836–853. doi: 10.1111/jam.12088.

Garnica-Vergara, A., Barrera-Ortiz, S., Munzo-Parra, E., Raya-González, J., Méndez-Bravo, A., Macías-Rodríguez, L., Ruiz-Herrera, L. F. and López-Bucio, J (2016). The volatile

6-pentyl-2H-pyran-2-one from *Trichoderma atroviride* regulates *Arabidopsis thaliana* root morphogenesis via auxin signalling and EHTYLENE INSENSITIVE 2 functioning, *New Phytologist* 209(4), 1496–1512. doi: 10.1111/nph.13725.

Ghazalibiglar, H., Hampton, J. G., van Zijll de Jong, E. and Holyoake, A. (2016). Is induced systemic resistance the mechanism for control of black rot in *Brassica oleracea* by *Paenibacillus* sp.?, *Biological Control* 92, 195–201. doi: 10.1016/j.biocontrol.2015.10.014.

Glare, T. R., Caradus, J., Gelernter, W. D., Jackson, T., Keyhani, N., Köhl, J., Marrone, P., Morin, L. and Stewart, A. (2012). Have biopesticides come of age?, *Trends in Biotechnology* 30(5), 250–258. doi: 10.1016/j.tibtech.2012.01.003.

Glare, T. R., Gwynn, R. L. and Moran-Diez, M. E. (2016). Development of biopesticides and future opportunities. In: *Microbial-based Biopesticides: Methods in Molecular Biology* 1477, Glare, T. R. and Moran-Diez, M. E. (Eds). Humana Press, New York, pp. 211–221.

Gomez-Vidal, S., Salinas, J., Tena, M. and Lopez-Llorca, L. V. (2009). Proteomic analysis of date palm (*Phoenix dactyifera* L.) responses to endophytic colonisation by entomopathogenic fungi, *Electrophoresis* 30(17), 2996–3005. doi: 10.1002/elps.200900192.

Guarnizo, N., Oliveros, D., Murillo-Arango, W. and Bermūdez-Cardona, M. B. (2020). Oligosaccharides: Defence inducers, their recognition in plants, commercial uses and perspectives, *Molecules* 25(24), 5972. doi: 10.3390/molecules25245972.

Hakim, S., Naqqash, T., Nawaz, M. S., Laraib, I., Siddique, M. J., Zia, R., Mirza, M. S. and Imran, A. (2021). Rhizosphere engineering with plant-growth-promoting microorganisms for agriculture and ecological sustainability, *Frontiers in Sustainable Food Systems* 5, 617157. doi: 10.3389/fsufs.2021.617157.

Halmer, P. (2006). Priming. In: *The Encyclopaedia of Seeds: Science Technology and Uses*, Black, M., Bewley, J. D. and Halmer, P. (Eds), CAB International, Wallingford, UK, pp. 533–537.

Hammerschmidt, R. (2014). Introduction: Definitions and term history. In: *Induced Resistance for Plant Defence*, Walters, D., Newton, A. and Lyon, G. (Eds). John Wiley and Sons, Oxford, UK, pp. 1–10.

Hermosa, R., Rubio, M. B., Cardoza, R. E., Nicolás, C., Monte, E. and Gutiérrez, S. (2013). The contribution of *Trichoderma* to balancing the costs of plant growth and defense, *International Microbiology* 16(2), 69–80. doi: 10.2436/20.1501.01.181.

Hong, C. E., Kwon, S. Y. and Park, J. M. (2016). Biocontrol activity of *Paenibacillus polymyxa* AC-1 against *Pseudomonas syringae* and its interaction with *Arabidopsis thaliana*, *Microbiological Research* 185, 13–21. doi: 10.1016/j.micres.2016.01.004.

Jaber, L. R. and Ownley, B. H. (2017). Can we use entomopathogenic fungi as endophytes for dual biological control of insect pests and plant pathogens?, *Biological Control* 116, 36–45. doi: 10.1016/j.biocontrol.2017.01.018.

Jaronski, S. T. and Jackson, M. A. (2008). Efficacy of *Metarhizium anisopliae* microsclerotial granules, *Biocontrol Science and Technology* 18(8), 849–863. doi: 10.1080/09583150802381144.

Johnson, L. J., Johnson, R. D., Schardl, C. L. and Panaccione, D. G. (2003). Identification of differentially expressed genes in the mutualistic association of tall fescue with *Neotyphodium coenophialum*, *Physiological and Molecular Plant Pathology* 63(6), 305–317. doi: 10.1016/j.pmpp.2004.04.001.

Johnson, L. J., de Bouth, A. C. M., Briggs, L. R., Caradus, J. R., Finch, S. C., Fleetwood, D. J., Fletcher, L. R., Hume, D. E., Johnson, R. D., Popay, A. J., Tapper, B. A., Simpson, W. R., Voisey, C. R. and Card, S. D. (2013). The exploitation of *Epichloe* endophytes for agricultural benefit, *Fungal Diversity* 60(1), 171–188. doi: 10.1007/s13225-013-0239-4.

Johnson, J. M., Alex, T. and Oelmüller, R. (2014). *Piriformospora indica*: The versatile and multifunctional root endophyte fungus for enhanced yield and tolerance to biotic and abiotic stress in crop plants, *Journal of Tropical Agriculture* 52, 103–122. doi: 10.3389/fmicb.2016.00332.

Karimi, E., Safaie, N., Shams-Baksh, M. and Mahmoudi, B. (2016). *Bacillus amyloliquefaciens* SB14 from rhizosphere alleviates *Rhizoctonia* damping-off disease on sugar beet, *Microbiological Research* 192, 221–230. doi: 10.1016/j.micres.2016.06.011.

Kottb, M., Gigoloshvili, T., Grobkinsky, D. K. and Piechulla, B. (2015). *Trichoderma* volatiles effecting *Arabidopsis* from inhibition to protection against phytopathogenic fungi, *Frontiers in Microbiology* 6, 995. doi: 10.3389/fmicb.2015.00995.

Kumar, A. and Dubey, A. (2020). Rhizosphere microbiome: Engineering bacterial competitiveness for enhancing crop production, *Journal of Advanced Research* 24, 337–352. doi: 10.1016/j.jare.2020.04.014.

Li, N., Alfiky, A., Wang, W., Islam, M., Nourollahi, K., Liu, X. and Kang, S. (2018). Volatile compound-mediated recognition and inhibition between *Trichoderma* biocontrol agents and *Fusarium oxysporum*, *Frontiers in Microbiology* 9, 2614. doi: 10.3389/fmicb.2018.02614.

Lugtenberg, B. J. J., Caradus, J. R. and Johnson, L. J. (2016). Fungal endophytes for sustainable crop production, *FEMS Microbiology Ecology* 92(12), 1–16. doi: 10.1093/femsec/fiw194.

Ma, M. Z., Christensen, M. J. and Nan, Z. B. (2015). Effects of the endophytic *Epichloe festucae* var. *lolii* of perennial ryegrass (*Lolium perenne*) on indicators of oxidative stress from pathogenic fungi during seed germination and seedling growth, *European Journal of Plant Pathology* 141(3), 571–583. doi: 10.1007/s10658-014-0563-x.

Macheleidt, J., Mattern, D. J., Fischer, J., Netzker, T., Weber, J., Schroeckh, V., Valiante, V. and Brakhage, A. A. (2016). Regulation and role of fungal secondary metabolites, *Annual Review of Genetics* 50(50), 371–392. doi: 10.1146/annurev-genet-120215-035203.

Mahmood, A. and Kataoka, R. (2019). Application of endophytes through seed priming. In: *Priming and Pretreatment of Seeds and Seedlings*, Hasanuzzamoan, M. and Fotopoulos, V. (Eds). Springer, Singapore, pp. 509–521.

Malinovsky, F. G., Fangel, J. U. and Willats, W. G. T. (2014). The role of the cell wall in plant immunity, *Frontiers in Plant Science* 5, 178. doi: 10.3389/fpls.2014.00178.

Malmierca, M. G., McCormick, S. P., Cardoza, R. E., Alexander, N. J., Monte, E. and Gutiérrez, S. (2015). Production of trichodiene by *Trichoderma harzianum* alters the perception of this biocontrol strain by plants and antagonized fungi, *Environmental Microbiology* 17(8), 2628–2646. doi: 10.1111/1462-2920.12506.

Martinez-Aguilar, K., Hernández-Chávez, J. L. and Alvarez-Veneges, R. (2021). Priming of seeds with INA and its transgenerational effect in common bean (*Phaseolus vulgaris* L.) plants, *Plant Science* 305, 110834. doi: 10.1016/j.plantsci.2021.110834.

Martinez-Aguilar, K., Ramirez-Carrasco, G., Hernández-Chávez, J. L., Barraza, A. and Alvarez-Venegas, R. (2016). Use of BABA and INA as activators of a primed state in the common bean (*Phaseolus vulgaris* L.), *Frontiers in Plant Science* 7, 653. doi: 10.3389/fpls.2016.00653.

Matsumoto, H., Fan, X., Wang, Y., Kusstatscher, P., Duan, J., Wu, S., Chen, S., Qiao, K., Wang, Y., Ma, B., Zhu, G., Hashidoko, Y., Berg, G., Cernava, T. and Wang, M. (2021). Bacterial seed endophyte shapes disease resistance in rice, *Nature Plants* 7(1), 60–72. doi: 10.1038/s41477-020-00826-5.

Mauch-Mani, B., Baccelli, I., Luna, E. and Flores, V. (2017). Defence priming: An adaptive part of induced resistance, *Annual Review of Plant Biology* 68, 485–512. doi: 10.1146/annurev-arplant-042916-041132.

Maurya, A. K. (2020). 'Role of plant volatile organic compounds (VOCs) in seed priming and plant herbivore interactions'. Unpublished PhD thesis, University of Louisville, Louisville, KY.

May, G. and Nelson, P. (2014). Defence mutualisms: Do microbial interactions within hosts drive the evolution of defensive traits?, *Functional Ecology* 28(2), 356–363.

Mayo-Prieto, S., Marra, R., Vinale, F., Rodriguez-Gonzales, A., Woo, S.L., Lorito, M., Gutierrez, S. and Casquero, P.A. (2019). Effect of *Trichoderma velutinum* and *Rhizoctonia solani* on the metabolome of bean plant (*Phaseolus vulgaris L.*), International Journal of Molecular Sciences 20(3), 549. doi:10.3390/ijms20030549.

McDonald, M. B. (2000). Seed priming. In: *Seed Technology and Its Biological Basis*, Black, M. and Bewley, J. D. (Eds). Sheffield Academic Press, Sheffield, UK, pp. 287–325.

Mishra, A. K., Sharma, K. and Misra, R. S. (2012). Elicitor recognition, signal transduction and induced resistance in plants, *Journal of Plant Interactions* 7(2), 95–120.

Mitter, B. (2020). Patents by inventor Birgit Mitter. https://patents.justia.com/inventor/birgit-mitter.

Mitter, B., Pfaffenbichler, N., Flavell, R., Compant, S., Antonielli, L., Petric, A., Berninger, T., Naveed, M., Sheibani-Tezerji, R., von Maltzahn, G. and Sessitsch, A. (2017). A new approach to modify plant microbiomes and traits by introducing beneficial bacteria at flowering into progeny seeds, *Frontiers in Microbiology* 8, 11. doi: 10.3389/fmicb.2017.00011.

Moonjely, S., Barelli, L. and Bidochka, M. J. (2016). Insect pathogenic fungi as endophytes, *Advances in Genetics* 94, 107–135. doi: 10.1016/bs.adgen.2015.12.004.

Morán-Diez, M. E., Martínez de Alba, Á. E., Rubio, M. B., Hermosa, R. and Monte, E. (2021). *Trichoderma* and the plant heritable priming responses, *Journal of Fungi* 7(4), 318. doi: 10.3390/jof7040318.

Mukherjee, A., Singh, B. K. and Verma, J. P. (2020). Harnessing chickpea (*Cicer arietinum* L.) seed endophytes for enhancing plant growth attributes and bio-controlling against *Fusarium* sp., *Microbiological Research* 237, 126469. doi: 10.1016/j.micres.2020.126469.

Murphy, B. R., Doohan, F. M. and Hodkinson, T. R. (2015). Persistent fungal root endophytes isolated from a wild barley species supress seed-borne infections in a barley cultivar, *BioControl* 60(2), 281–292.

Naznin, H. A., Kiyohara, D., Kimura, M., Miyazawa, M., Shimizu, M. and Hyakumachi, M. (2014). Systemic resistance induced by volatile organic compounds emitted by plant growth-promoting fungi in *Arabidopsis thaliana*, *PLoS ONE* 9(1), e86882. doi: 10.1371/journal.pone.0086882.

Nelson, E. B. (2018). The seed microbiome: Origins, interactions and impacts, *Plant and Soil* 422(1–2), 7–34. doi: 10.1007/s11104-017-3289-7.

Nissinen, R., Helander, M., Kumar, M. and Saikkonen, K. (2019). Heritable *Epichloe* symbiosis shapes fungal but not bacterial communities of plant leaves, *Scientific Reports* 9(1), 5253. doi: 10.1038/s41598-019-41603-5.

Nogueira-Lopez, G., Lawry, R., Echaide-Aquino, J. F. and Mendoza-Mendoza, A. (2020). Chemical communication between *Trichoderma* and plants. In: *New and Future Developments in Microbial Biotechnology and Bioengineering*, Gupta, V. K., Zeilinger, S., Singh, H. B. and Druzhinina, I. (Eds). Elsevier, the Netherlands, pp. 110–156.

O'Callaghan, M. (2016). Microbial inoculation of seed for improved crop performance: Issues and opportunities, *Applied Microbiology and Biotechnology* 100(13), 5729–5746. doi: 10.1007/s00253-016-7590-9.

Perez, L. I., Gundel, P. E., Zabalgogeazcoa, I. and Omacini, M. (2020). An ecological framework for understanding the role of *Epichloe* endophytes on plant defences against fungal diseases, *Fungal Biology Reviews* 34(3), 115–125. doi: 10.1016/j.fbr.2020.06.001.

Pushpalatha, H. G., Sudisha, J., Geetha, N. P., Amruthesh, K. N. and Shetty, H. (2011). Thiamine seed treatment enhances LOX expression, promotes growth and induces downy mildew disease resistance in pearl millet, *Biologia Plantarum* 55(3), 522–527. doi: 10.1007/s10535-011-0118-3.

Quintana-Rodriguez, E., Riveras-Macias, L. E., Adame-Alvarez, R. M., Torres, J. M. and Heil, M. (2018). Shared weapons in fungus-fungus and fungus-plant interactions? Volatile organic compounds of plant and fungal origin exert direct antifungal activity in vitro, *Fungal Ecology* 33, 115–121. doi: 10.1016/j.funeco.2018.02.005.

Raad, M., Glare, T. R., Brochero, H. L., Müller, C. and Rostás, M. (2019). Transcriptional reprogramming of *Arabidopsis thaliana* defence pathways by the entomopathogen *Beauveria bassiana* correlates with resistance against a fungal pathogen but not against insects, *Frontiers in Microbiology* 10, 615. doi: 10.3389/fmicb.2019.00615.

Rajer, F. U., Wu, H., Xie, Y., Xie, S., Raza, W., Tahir, H. A. S. and Gao, X. (2017). Volatile organic compounds produced by a soil-isolate, *Bacillus subtilis* FA26 induce adverse ultra-structural changes to the cells of *Clavibacter michiganensis* ssp. *sepedonicus*, the causal agent of bacterial ring rot of potato, *Microbiology* 163(4), 523–530. doi: 10.1099/mic.0.000451.

Ramirez-Carrasco, G., Martinez-Aguilar, K. and Alvarez-Veneges, R. (2017). Transgenerational defense priming for crop protection against plant pathogens: A hypothesis, *Frontiers in Plant Science* 8, 696. doi: 10.3389/fpls.2017.00696.

Rashid, M., Hampton, J. G., Shaw, M. L., Rolston, M. P., Khan, K. M. and Saville, D. J. (2020). Oxidative damage in forage rape (*Brassica napus* L.) seeds following heat stress during seed development, *Journal of Agronomy and Crop Science* 206(1), 101–117. doi: 10.1111/jac.12372.

Reichenberger, G., Pfaffenbichler, N., Mitter, B., Reising, J. and Brader, G. (2017). *WO2017162758A1 Patent. Seed Injection*.

Rivas-Franco, F., Hampton, J. G., Morán-Diez, M. E., Narciso, J., Rostás, M., Wessman, P., Jackson, T. A. and Glare, T. R. (2019). Effect of coating maize seed with entomopathogenic fungi on plant growth and resistance against *Fusarium graminearum* and *Costelytra giveni*, *Biocontrol Science and Technology* 29(9), 877–900, doi: 10.1080/09583157.2019.1611736.

Rivas-Franco, F., Hampton, J. G., Narciso, J., Rostás, M., Wessman, P., Saville, D. J., Jackson, T. A. and Glare, T. R. (2020a). Effects of a maize root pest and fungal pathogen on entomopathogenic fungal rhizosphere colonisation, endophytism and induction of plant hormones, *Biological Control* 150, 104347. doi: 10.1016/j.biocontrol.2020.104347.

Rivas-Franco, F., Hampton, J. G., Altier, N. A., Swaminathan, J., Rostás, M., Wessman, P., Saville, D. J., Jackson, T. A., Jackson, M. A. and Glare, T. R. (2020b). Production of microsclerotia from entomopathogenic fungi and use in maize seed coating as delivery for biocontrol against *Fusarium graminearum*, *Frontiers in Sustainable Food Systems* 4, 60628. doi: 10.3389/fsufs.2020.606828.

Rocha, I., Ma, Y., Souza-Alfonso, P., Vosatka, M., Freitas, H. and Oliveira, R. S. (2019). Seed coating: A tool for delivering beneficial microbes to agricultural crops, *Frontiers in Plant Science* 10, 1357. doi: 10.3389/fpls.2019.01357.

Rochefort, A., Briand, M., Marais, C., Wagner, M., Laperche, A., Vallée, P., Barret, M. and Sarniguet, A. (2019). Influence of environment and host plant genotype on the structure and diversity of the *Brassica napus* seed microbiota, *Phytobiomes Journal* 3(4), 326–336. doi: 10.1094/PBIOMES-06-19-0031-R.

Ruiza, D., Agaras, B., Werrab, P. D., Wall, L. G. and Valverde, C. (2011). Characterization and screening of plant probiotic traits of bacteria isolated from rice seeds cultivated in Argentina, *Journal of Microbiology* 49(6), 902–912. doi: 10.1007/s12275-011-1073-6.

Sahu, P. K., Singh, S., Gupta, A., Singh, U. B., Brahmaprakash, G. P. and Saxena, A. K. (2019). Antagonistic potential of bacterial endophytes and induction of systemic resistance against collar rot pathogen *Sclerotium rolfsii* in tomato, *Biological Control* 137, 104014. doi: 10.1016/j.biocontrol.2019.104014.

Schweiger, R., Padilla-Arizmendi, F., Nogueira-Lopez, G., Rostás, M., Lawry, R., Brown, C., Hampton, J., Steyaert, J. M., Müller, C. and Mendoza-Mendoza, A. (2021). Insights into metabolic changes caused by the *Trichoderma virens*-maize root interaction, *Molecular Plant-Microbe Interactions* 34(5), 524–537. doi: 10.1094/MPMI-04-20-0081-R.

Shahzad, R., Khan, A. L., Bilal, S., Asaf, S. and Lee, I. J. (2017). Plant growth-promoting endophytic bacteria versus pathogenic infections: An example of *Bacillus amyloliquefaciens* RWL-1 and *Fusarium oxysporum* f.sp *lycopersici* in tomato, *PeerJ* 5, e3107. doi: 10.7717/peerj.3107.

Shahzad, R., Khan, A. L., Bilal, S., Asaf, S. and Lee, I. J. (2018). What is there in seeds? Vertically transmitted endophytic resources for sustainable improvement in plant growth, *Frontiers in Plant Science* 9, 24. doi: 10.3389/FPLS.2018.00024.

Sharifi, R. and Ryu, C. M. (2018). Revisiting bacterial volatile-mediated plant growth promotion: Lessons from the past and objectives for the future, *Annals of Botany* 122(3), 349–358. doi: 10.1093/aob/mcy108.

Simas, D. L. R., de Amorim, S. H. B. M., Gourlart, F. R. V., Alviano, C. S., Alviano, D. S. and da Silva, A. J. R. (2017). Citrus species essential oils and their components can inhibit or stimulate fungal growth in fruit, *Industrial Crops and Products* 98, 108–115. doi: 10.1016/j.indcrop.2017.01.026.

Singh, S., Singh, U. B., Malviya, D., Paul, S., Sahu, P. K., Trivedi, M., Paul, D. and Saxena, A. K. (2020). Seed biopriming with microbial inoculant triggers local and systemic defense responses against *Rhizoctonia solani* causing banded leaf and sheath blight in maize (*Zea mays* L.), *International Journal of Environmental Research and Public Health* 17(4), 1396. doi: 10.3390/ijerph17041396.

Slaughter, A., Daniel, X., Flors, V., Luna, E., Hohn, B. and Mauch-Mani, B. (2012). Descendants of primed *Arabidopsis* plants exhibit resistance to biotic stress, *Plant Physiology* 158(2), 835–843. doi: 10.1104/pp.111.191593.

Song, G. C., Choi, H. K., Kim, Y. S., Choi, J. S. and Ryu, C. M. (2017). Seed defense biopriming with bacterial cyclodipeptides triggers immunity in cucumber and pepper, *Scientific Reports* 7(1), 14209. doi: 10.1038/s41598-017-14155-9.

Song, G. C., Choi, H. K. and Ryu, C. M. (2015). Gaseous 3-pentanol primes plant immunity against a bacterial speck pathogen, *Pseudomonas syringae* pv *tomato* via salicylic acid and jasmonic acid-dependent signalling pathways in *Arabidopsis*, *Frontiers in Plant Science* 6, 821. doi: 10.3389/fpls.2015.00821.

Tahir, H. A. S., Gu, Q., Wu, H., Raza, W., Safdar, A., Huang, Z., Rajer, F. U. and Gao, X. (2017). Effect of volatile compounds produced by *Ralstonia solanacearum* on plant growth promoting and systemic resistance inducing potential of *Baciullus* volatiles, *BMC Plant Biology* 17(1), 133. doi: 10.1186/s12870-017-1083-6.

Tang, D., Wang, G. and Zhou, J. M. (2017). Receptor kinases in plant-pathogen interactions: More than pattern recognition, *The Plant Cell* 29(4), 618–637. doi: 10.1105/tpc.16.00891.

Thomas, G., Withall, D. and Birkett, M. (2020). Harnessing microbial volatiles to replace pesticides and fertilisers, *Microbial Biotechnology* 13(5), 1366–1376. doi: 10.1111/1751-7915.13645.

Thomma, B. P. H. J., Nürnberger, T. and Joosten, M. H. A. J. (2011). Of PAMPs and effectors: The blurred PTI-ETI dichotomy, *The Plant Cell* 23(1), 4–15. doi: 10.1105/tpc.110.082602.

Tian, Z., Wang, R., Ambrose, K. V., Clarke, B. B. and Belanger, F. C. (2017). The *Epichloe festucae* antifungal protein has activity against the plant pathogen *Sclerotinia homeocarpa*, the causal agent of dollar spot disease, *Scientific Reports* 7(1), 5643. doi: 10.1038/s41598-017-06068-4.

Torres-Cortés, G., Garcia, B. J., Compant, S., Rezki, S., Jones, P., Préveaux, A., Briand, M., Roulet, A., Bouchez, O., Jacobson, D. and Barret, M. (2019). Differences in resource use lead to coexistence of seed-transmitted microbial populations, *Scientific Reports* 9(1), 6648. doi: 10.1038/s41598-019-42865-9.

Tyagi, S., Lee, K. J., Shukla, P. and Chae, J. C. (2020). Dimethyl disulphide exerts antifungal activity against *Sclerotinia minor* by damaging its membrane and induces systemic resistance in host plants, *Scientific Reports* 10(1), 6547. doi: 10.1038/s41598-020-63382-0.

Vega, F. E., Posada, F., Catherine Aime, M., Pava-Ripoll, M., Infante, F. and Rehner, S. A. (2008). Entomopathogenic fungal endophytes, *Biological Control* 46(1), 72–82. doi: 10.1016/j.biocontrol.2008.01.008.

Vinale, F., Sivasithamparam, K., Ghisalberti, E. L., Marra, R., Barbetti, M. J., Li, H., Woo, S. L. and Lorito, M. (2008a). A novel role for *Trichoderma* secondary metabolites in the interactions with plants, *Physiological and Molecular Plant Pathology* 72(1–3), 80–86. doi: 10.1016/j.pmpp.2008.05.005.

Vinale, F., Sivasithamparam, K., Ghisalberti, E. L., Marra, R., Woo, S. L. and Lorito, M. (2008b). *Trichoderma* -plant-pathogen interactions, *Soil Biology and Biochemistry* 40(1), 1–10. doi: 10.1016/j.soilbio.2007.07.002.

Wang, R., Luo, S., Clarke, B. B. and Belanger, F. C. (2021). The *Epichloë festucae* antifungal protein *Efe*-AfpA is also a possible effector protein required for the interaction of the fungus with its host grass *Festuca rubra* subsp. *rubra*, *Microorganisms* 9(1), 140. doi: 10.3390/microorganisms9010140.

Wang, X. Y., Qin, J. H., Chen, W., Zhou, Y., Ren, A. and Gao, Y. (2016). Pathogen resistant advantage of endophyte-infected over endophyte-free Leymus chinensis is strengthened by pre-drought treatment, *European Journal of Plant Pathology* 144(3), 477–486.

Warren, J. E. and Bennett, M. A. (1999). Bio-osmopriming tomato (*Lycopersicon esculentum* Mill) seeds for improved stand establishment, *Seed Science and Technology* 27, 489–499. doi: 10.1079/9780851994048.0477.

Westman, S. M., Kloth, K. J., Hanson, J., Ohlsson, A. B. and Albrectsen, B. R. (2019). Defence priming in *Arabidopsis* – A meta-analysis, *Scientific Reports* 9(1), 13309. doi: 10.1038/s41598-019-49811-9.

White, J. F., Kingsley, K. L., Butterworth, S., Brindisi, L., Gatei, J. W., Elmore, M. T., Verma, S. K, Yao, X. and Kowalski, K. P. (2018). Seed-vectored microbes: Their roles in improving seedling fitness and competitor plant suppression. In: *Seed Endophytes: Biology and Biotechnology*, Verma, S. and White, J. F. Jr. (Eds). Springer Cham, Heidelberg.

Worrall, D., Holroyd, G. H., Moore, J. P., Glowacz, M., Croft, P., Taylor, J. E., Paul, N. D. and Roberts, M. R. (2012). Treating seeds with activators of plant defense generates long-lasting priming of resistance to pests and pathogens, *New Phytologist* 193(3), 770–778. doi: 10.1111/j.1469-8137.2011.03987.x.

Xie, S., Zang, H., Wu, H., Uddin Rajer, F. and Gao, X. (2018). Antibacterial effects of volatiles produced by *Bacillus* strain D13 against *Xanthomonas oryzae* pv. oryzae, *Molecular Plant Pathology* 19(1), 49–58. doi: 10.1111/mpp.12494.

Ye, M. and Veyrat, N., Xu, H., Hu, L., Turlings, T. C. J. and Erb, M. (2018). An herbivore-induced plant volatile reduces parasitoid attraction by changing the smell of caterpillars, *Science Advances* 4(5), eaar4767. doi: 10.1126/sciadv.aar4767.

Yi, S. Y., Min, S. R. and Kwon, S. Y. (2015). NPRI is instrumental in priming for the enhanced flg22-induced MPK3 and MPK6 activation, *Plant Pathology Journal* 31(2), 192–194. doi: 10.5423/PPJ.NT.10.2014.0112.

Yue, Q., Miller, C. J., White, J. F. and Richardson, M. D. (2000). Isolation and characterisation of fungal inhibitors from *Epichloe festucae*, *Journal of Agricultural and Food Chemistry* 48(10), 4687–4692. doi: 10.1021/jf990685q.

Zeilinger, S. and Atanasova, L. (2020). Sensing and regulation to mycoparasitism – Relevant processes in *Trichoderma*. In: *New and Future Developments in Microbial Biotechnology and Bioengineering*, Gupta, V. K., Zeilinger, S., Singh, H. B. and Druzhinina, I. (Eds). Elsevier, the Netherlands, pp. 39–55.

Zeilinger, S., Gruber, S., Bansal, R. and Mukherjee, P. K. (2016). Secondary metabolism in *Trichoderma* - chemistry meets genomics, *Fungal Biology Reviews* 30(2), 74–90. doi: 10.1016/j.fbr.2016.05.001.

Zhang, M., Yajima, H., Umezawa, Y., Nakagawa, Y. and Esashi, Y. (1995). GC-MS identification of volatile compounds evolved by dry seeds in relation to storage conditions, *Seed Science and Technology* 23, 59–68.

Chapter 10

Advances in seed priming techniques

Henry Bruggink, Incotec Europe, The Netherlands

1 Introduction

Seed priming is a technique involving seed hydration and drying with the aim of accelerating and synchronising germination and, if present, releasing dormancy. This can lead to a more uniform crop stand and possibly increased yield, both under normal conditions and under conditions of abiotic and biotic stress. Priming of commercial seed lots is used in seed technology to improve seed vigour, but the technique is also used in other areas such as seed banks or restoration of forests (Brancalion et al., 2010; Peraza-Villarreal et al., 2018).

The physiological process of priming is described in Fig. 1, which shows water uptake during germination and priming (Lutts et al., 2016). When water uptake is unlimited, seeds will first absorb water (Phase I), after which there is a lag period (Phase II). Having already started in Phase I and continuing in Phase II, cellular and biochemical processes prepare the seed for germination (Weitbrecht et al., 2011; Steinbrecher and Leubner-Metzger, 2017). Phase III begins when the root tip protrudes through the seed coat and the seedling starts to grow.

In priming, water uptake is limited to a level at which Phase III cannot take place. This means that Phase II can be extended to a period in which all, or at

http://dx.doi.org/10.19103/AS.2022.0105.15

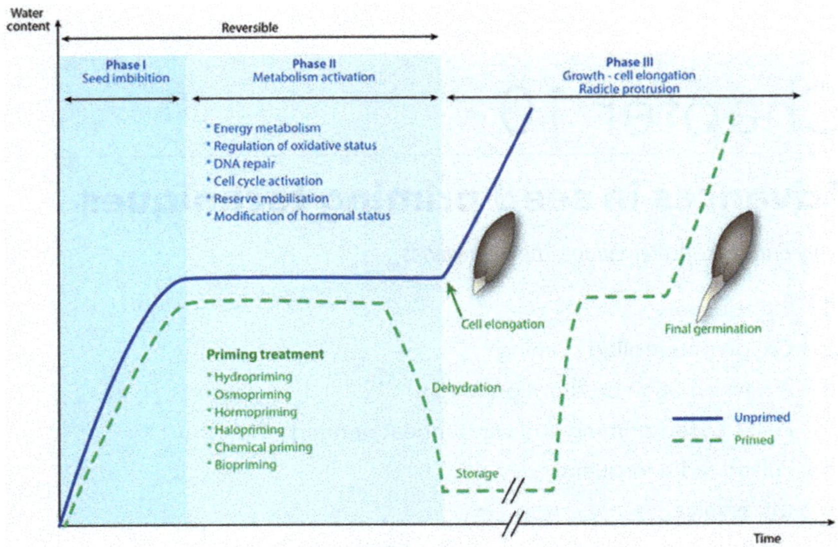

Figure 1 Germination phases and changes in seed water content in unprimed seeds and during seed priming. From Lutts et al. (2016).

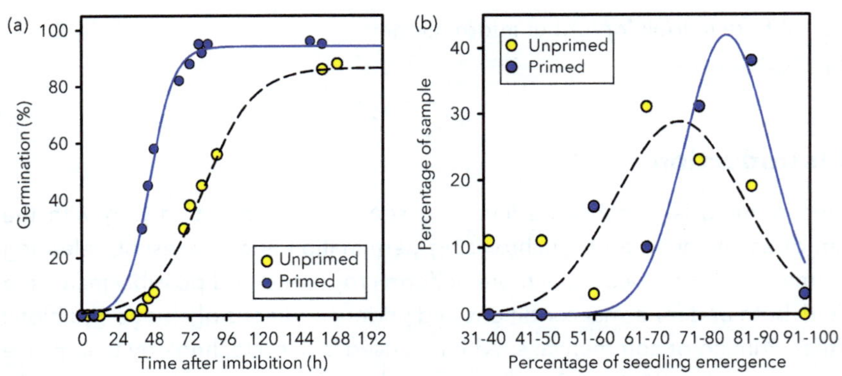

Figure 2 Priming improves speed, percentages and homogeneity of germination (a) and reduces variation in seedling emergence (b). (a) Germination of unprimed and primed seed, imaginary data. (b) The frequency of percentage seedling emergence in decile categories of osmotically primed and unprimed carrot seeds sown in 37 seedbed environments. Data are originally from Finch-Savage and Bassel (2016).

least most of the seeds, are ready to germinate at more or less the same time. At the end of the priming process, the seeds are dried back to a safe storage moisture content. Drying is important since it allows the seed company to store the seed and sell it at the appropriate time. When the primed seeds are then re-imbibed, they can germinate much faster and more uniformly (Fig. 1 (right part in dashed green) and Fig. 2).

The status of elevated moisture level needed for a priming to be effective can be obtained by three completely different techniques. The first technique is osmotic priming, which was developed by Heydecker et al. in 1973 and comprises osmotically controlling the water supply to seeds using an aqueous solution of a water-soluble polymer (e.g. polyethylene glycol, polyalcohol or polysaccharide) or an aqueous salt solution.

The second technique developed by Rowse (1994, 2006) comprises supplying seeds with water by spraying water in a rotating drum. The rate at which the water is sprayed during drum priming is such that the surface of the seeds remains sufficiently dry, while the seed moisture content is raised to the desired level. Comparable to this is the patent of Boukens and de Boer (2002). In this patent, seeds are first incubated in an environment which is saturated with water (i.e. by spraying, immersion or on moist filter paper), to a water potential greater than −10 MPa (and preferably between −8 MPa and −1.2 MPa). The final amount of water needed to reach the target priming moisture is absorbed from a gas phase. By controlling the relative humidity of the air, the seed is hydrated to the required level of 98% RH or higher.

The third technique involves a solid medium (e.g. a shale material like Agro-Lig) from which the water is taken up. By mixing seed, solid material and water in a certain ratio, the moisture status of the seed can be set. Several patents have been filed for this solid matrix priming technique, starting with Taylor in 1990. In 2002, Tsujimoto et al. obtained a patent in which superabsorbent polymers are used to control water uptake. Even though the mechanism of water uptake is claimed to be different from osmotic priming or solid matrix priming, this is considered to be a technical variation rather than a completely new technique.

Besides moisture level, the other important parameters to control are temperature and duration of priming. It is usual to prime at a temperature at which the germination proceeds more or less optimally. However, for species that germinate optimally at a relatively high temperature, priming at this temperature can allow microorganisms to replicate much faster than at a more moderate temperature, which is not desirable for seed health. For such species, a lower temperature may be more practical, although germination will proceed more slowly and consequently, the duration of priming will be extended.

For a seed population that does not have dormancy, priming should be long enough to synchronise the slowest seeds with the fastest ones in the population. However, priming should not take too long since the fastest seeds may be at the brink of radicle emergence and experience damage during drying at the end of the process. The effect of such 'overpriming', which can also be caused by too high a moisture level, is a decrease in germination. This can occur directly after priming (Capron et al., 2000) or after storage (Bruggink et al., 1999; Gupta et al., 2008).

When seeds are dormant, priming may take longer because, in addition to breaking dormancy, synchronisation still has to take place. However, there are exceptions. Lettuce seed imbibes (i.e. takes up water) and germinates rapidly at moderate temperatures but becomes dormant when seed is imbibed at higher temperatures (thermodormancy). For this species, priming is not performed to improve the speed or synchronicity of germination, but to increase the upper temperature limit, i.e. the temperature beyond which germination rapidly drops. This means that a relatively short priming may be sufficient for lettuce to break thermodormancy whilst, at the same time, rapid and uniform germination is reached.

Priming thus has three main components:

- water content;
- temperature; and
- duration.

Additional factors that might influence the success of priming are:

- the availability of oxygen;
- the presence or absence of light; and
- the addition of different compounds to the priming solution.

Oxygen is needed to support the biochemical processes that take place in the seed. Light is a stimulant that can decrease photodormancy. Compounds like gibberellic acid (GA_3) or potassium nitrate (KNO3) (Andreoli and Kahn, 1998; Karimi and Varyani, 2016), ethylene (Corbineau et al., 2014) or karrikins (Antala et al., 2020) can help to decrease dormancy or they can be added just to boost the germination of the seed.

Priming is primarily used to increase germination speed and to synchronise germination. However, it also has other applications. One of these is to prepare seeds or plants for future biotic and abiotic stress conditions, like insect damage, drought or cold stress. This works not only during or shortly after germination but for a much longer period of vegetative growth. In Thomas and Puthur (2017), these two different priming aspects are taken together. Seed priming is defined as the induction of a particular physiological state in plants by treating the seeds prior to germination with natural and/or synthetic compounds. However, although the impact of priming on increased speed and synchronicity may affect subsequent plant growth and final yield of a crop, it is not necessarily the result of creating the kind of physiological changes that make a plant more resistant to stress.

This chapter focuses on recent developments in priming from the point of view of the seed industry, based on patents or applications. Patents filed for specific crops or using specific components are beyond the scope of this chapter. What complicates the evaluation of new techniques is that service providers and seed companies are very secretive about the way they process seed. This means that patent evaluation in most cases can only be done by consulting the scientific literature.

For more general reading about research on priming we refer the reader to some general reviews (Paparella et al., 2015; Lutts et al., 2016; Waqas et al., 2019).

2 Oxygen-controlled priming

In 2008, Van Duijn and Van Asbrouck published a patent application: Oxygen-controlled priming of seed. This so-called Oxyprime® method is, in contrast to other priming methods, not based on control of seed metabolism by controlling seed water content. Seed metabolism is controlled by exact setting of the oxygen level below the Critical Oxygen Pressure (COP) values and restricting the total oxygen amount to prevent overpriming.

Oxyprime® basically comprises a one-time determination of the COP value of a seed species, followed by keeping the seeds for a given time in pure water with a carefully controlled oxygen level below the COP value. The primed seeds are then dried and ready for storage and sale. Seeds also may be imbibed in water containing a carefully determined amount of oxygen. Priming automatically stops and is completed when the oxygen supply is exhausted. Variations in these protocols provide for a multitude of fine-tuning options.

An example of this priming treatment is given for sugar beet seed (*Beta vulgaris*; Fig. 3). Sugar beet seeds with a relatively slow germination and low

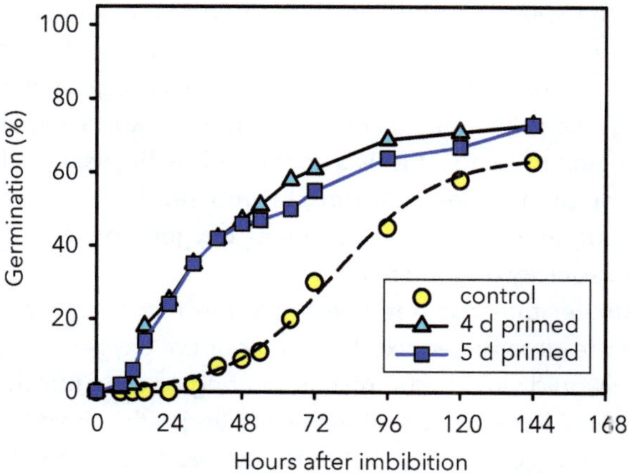

Figure 3 Germination of non-primed (control) and oxygen-controlled primed seeds of sugar beet. Redrawn from Van Duijn and Van Asbrouck (2020).

germination percentage (Fig. 3, control) were used. The COP value of the seeds during germination was about 4%, as determined by oxygen consumption rate germination experiments in a Q2 machine which measures the oxygen consumption of individual seeds. Seeds are placed in a microtiter plate on top of a wetted paper disc or agarose. The plate is then closed by a foil with oxygen-sensitive dots, with a dot inside each cell. By scanning the dots, the oxygen level in each cell can be measured non-invasively.

In the priming process, 344 seeds were imbibed on wet filter paper under a continuous flow of air with high humidity (to prevent seeds from drying) containing 1% oxygen. This treatment was continued for different time periods at 20°C in the dark. After 4 or 5 days of incubation, half of the seeds were collected from the boxes. None of them had germinated. All seeds were surface dried with filter paper and subsequently packed in dry filter paper for further drying for 4–5 days. After that, the seeds were tested for germination in 96 well plates (one seed per well) on agarose surface (15 ml of 0.5 % agarose per plate) at 20°C. In addition to primed seeds non-primed control seeds (192) were also germinated. Figure 3 shows the germination curves of the three groups of seeds. It can be clearly seen that both the 4 and the 5 days primed seeds show faster and more complete germination as compared to non-primed control seeds.

However, although there is an increase in germination speed and in total germination, the effect on synchronous germination appears to be somewhat limited. In research performed at Incotec, several species (amongst others tomato (*Solanum lycopersicum*), lettuce (*Lactuca sativa*), oilseed rape (*Brassica napus*) and pepper (*Capsicum annuum*)) were all primed at an oxygen content of 2.0–2.5%, in a drum priming-like setup. In most cases, the Oxyprime® resulted in an increase in germination speed and sometimes in total germination, albeit less than with standard Incotec priming techniques. Priming at a low oxygen level reduced the speed of physiological processes, resulting in a reduction in progression of germination. Extending the duration of priming did not compensate for this reduced activity and, in some cases, germination occurred during priming. The priming effect of the Oxyprime® was always less than that of in-house Incotec priming. This was confirmed by Bujalski et al. (1993) who found that, after priming leek (*Allium porrum*) seeds with different oxygen levels, only seeds treated with air or higher oxygen concentrations showed enhanced synchronicity of germination.

One parameter that was not investigated in this study was the effect of priming with a pre-determined fixed total amount of oxygen. It is possible that such a method would yield better results, nearing or even matching the result of 'normal' priming. It may be possible that starting with a fixed total amount of oxygen will result in germination of the fastest seeds, or, when less oxygen is used, lack of synchronisation of the slowest seeds. More research is needed to establish the boundaries of the Oxyprime® treatment and to determine if it is a valuable priming method for certain species.

3 A new variation on drum priming

More recently, a new variation on drum priming has been patented by Shen (2018; 2017 and 2020). In the first patent, seeds are fully imbibed and then dried back to the target priming moisture content. In a second patent, imbibition is not complete but proceeds to a higher moisture level than used for the priming, after which the seed is dried back to the target priming moisture level. Both are in contrast to the original drum priming patent, where such an overshoot is avoided. Accidental imbibition to a supra-optimal moisture may have occurred occasionally but this would have been repaired quickly and was never officially described.

Besides moisture level, there are other differences compared with standard drum priming. First, it is stated that the drying process hardens the seed coat which either prevents or inhibits germination. Second, priming takes only 1–4 days which is, for at least some of the species in the patent, shorter than usual.

The effect of this priming method on seed and seedling vigour traits of several species is shown in Table 1. It is clear that germination was improved, especially the speed of germination, expressed as Mean Germination Time. The question is, however, if standard full-length priming would have given a better improvement. For bell pepper (*Capsicum annuum*), a priming duration of slightly more than 3 days is insufficient to synchronise the slowest seeds with fast-germinating seeds, suggesting that longer priming might have given an even better result.

The method can only be used with water, but other compounds may also be added. In this sense, the technique is not very different from the three basic priming technologies, especially drum priming. The statement that the drying step hardens the seed coat is also not supported by experimental evidence. Drying must be very limited otherwise no or a limited priming effect would be observed. Furthermore, moisture always seeks the state of equilibrium, making it very unlikely that hardening, if it occurs, would last long enough to prevent germination. Germination does not occur either because the moisture level is just too low or because the duration of priming is too short.

It should be noted that these patents are primarily to protect priming equipment developed by the companies concerned rather than the particular technique. It does seem, therefore, that the patents were filed mainly to protect this priming machine, allowing priming to be done automatically, rather than the priming method.

Table 1 Priming effect on germination and seedling emergence of representative species. Priming consisted of the following treatment: seeds were immersed in water for up to 250 min. to reach 30% (wheat, scots pine) or 50% moisture (fresh weight basis, bell pepper, Kentucky blue grass) then held at their respective water content for the indicated period of time. MGT = mean germination time.

			Germination				Soil emergence			
Category	Species	Treatment	%	% increase	MGT (h)	% decrease	First count (days)	% at first count time (control)	Seedling length (mm)	Seedling fresh weight (g)
Vegetable	Bell pepper	Control	86		164		15	27	20	0.4
		Primed 75 h	94	8	105	36	12	87	55	1.1
Crop	Wheat	Control	86		88		4	25	50	1.4
		Primed 24 h	86	0	65	19	4	73	77	1.5
Forest	Scots pine	Control	95		108		10	5	45	0.9
		Primed 60 h	99	4	90	17	9	43	56	1.1
Grass	Kentucky blue grass	Control	87		237		11	6	13	0.4
		Primed 84 h	93	5	138	42	7	67	53	0.6

Data are from Shen (2018).

4 Pellet-based priming and vapour heat priming

In 2015, Peter obtained a patent for priming as part of seed pelleting. The process starts with a pelleting step, after which the pellets are dried to a pre-determined moisture content of 15–55% and stored for a period of 30 minutes to 24 hours at 20°C to 36°C, to initiate germination or priming. The pelleting substance comprises at least one oxygen donor and a catalytically active initiator. Peroxide is an example of an oxygen donor. Initiators can be chosen from metal salts, chelates or trace element fertilisers. At the end of the priming period, the pellets are dried to a storage moisture content (5–10%).

The method resembles the previously described patents by Shen in that, during the pelleting process, the seed moisture content becomes higher than the priming moisture. However, the period in which the wet seed is stored after drying to a pre-determined moisture content is shorter. Unfortunately, no examples are provided in the patent. The maximum time during which the beneficial effects of priming take place is 24h which appears already quite short for a crop like sugar beet but is certainly not long enough for many other species. That makes the method more a stimulation of the germination than a true synchronisation.

4.1 Vapour heat priming

A relatively new technique is vapour heat priming or ThermoSeed®, a patent for which was filed in 2015 and granted in 2020 (Breukink et al.). ThermoSeed® is originally a treatment to kill or decrease the level of microorganisms by exposure to hot, humidified air (Alness et al., 2004). Seed is exposed to a treatment atmosphere with a relative humidity of 50% or more and a temperature of 50–100°C, for a period of 1 second to 30 minutes. The method not only has a disinfection effect but also a positive effect on germination, albeit not as much as standard priming. An example for oats (*Avena sativa*) is given in Fig. 4, where it can be seen that the effect of the priming is quite variable and, in a few cases, even negative. This can probably be ascribed to variation between seed lots.

Although the patent does not give an explanation, the priming effect may be related to a partial breakdown of the structures surrounding the embryo. The seeds are not dormant, but these structures can have a delaying effect on germination. Vapour heat priming decreases the resistance of these structures, allowing for faster germination. Similarly, the method also works on seeds that have a hard seed coat that does not allow for uptake of water (Finch-Savage and Leubner-Metzger, 2006), like some members of the Fabaceae family. In Fig. 5 the patented method results in a decrease in hard seeds and an increase in total germination of bean seeds (*Phaseolus vulgaris*). Here, the effect of the priming concerns a partial release of physical dormancy.

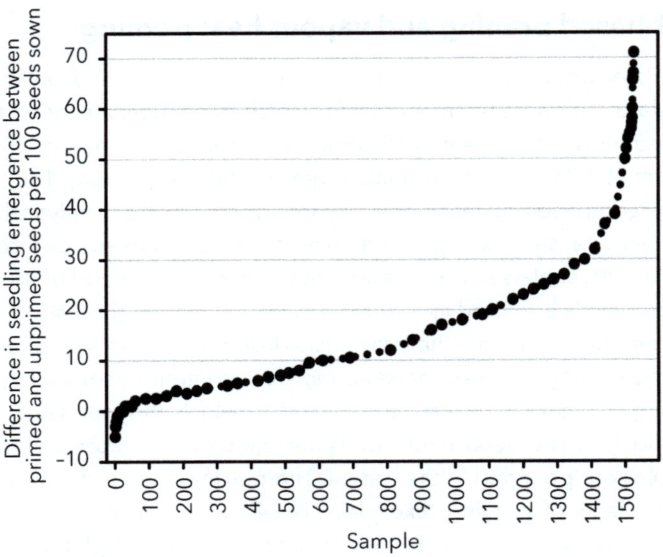

Figure 4 The effect of vapour heat on the number of emerged oat seedlings. Data are expressed as the difference between primed and unprimed lots per 100 seeds sown in potting soil under cold stress. 1527 samples were tested and are sorted on the X-axis according to the outcome of the germination test. Data redrawn from Breukink et al. (2020).

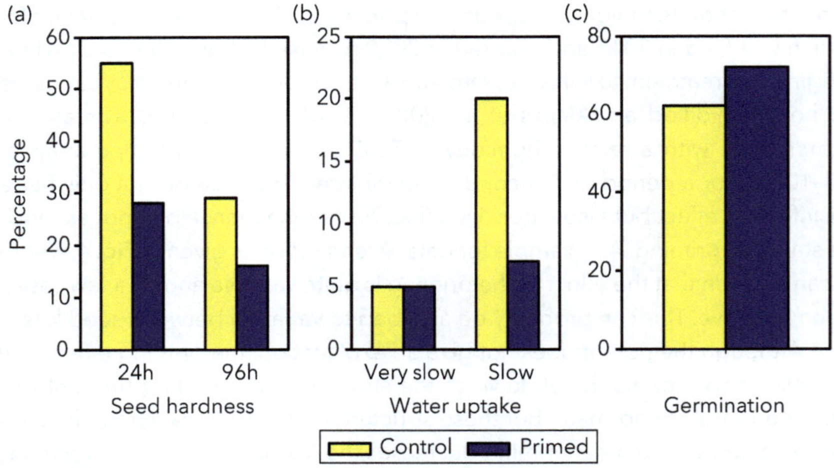

Figure 5 The effect of vapour heat priming on the seed hardness (a), water uptake (b) and total germination (c) of bean seeds. Seed hardness was assessed as the proportion of seeds with water uptake less than 5% of fresh weight after 24h. Very slow water uptake was assessed as the proportion of seeds with water uptake after 24h < 5 % and no germination after 96 h. Slow water uptake is the same but with germination after 96h. Data are redrawn from Breukink et al. (2020).

5 Pulsed radio frequency

Several studies describe the use of physical methods to enhance germination. Magneto-priming and irradiation with microwaves or ionising radiation are the most promising pre-sowing treatments (Araújo et al., 2016). One patented technology in this area consists of applying a pulsed radio frequency (PRF) electrical field to seeds (Sluijter and Teixeira, 2014, 2019). The technology is claimed to improve germination rate, germination capacity, and homogeneity of germination, as well as improved plant vigour. The germination of aged seed lots was also improved. The difference from other techniques using electrical fields is that most use high to very high voltages, which can be dangerous for the operator and may lead to heating of biological material (Jonas, 1955; Lagunas-Solar et al., 2002). This technique generates limited heat by using low voltage and pulsing the radiation with a high frequency.

It is important that seeds are partly hydrated. Several hydration methods are described, from spraying the seeds to priming seeds in an aqueous solution, with a preferred water content of 30–40% (dry weight). In the experiments described in the patent, however, it does not appear that the seeds are really primed, but just soaked for a limited amount of time. In that sense, the method cannot be compared directly with conventional priming, where the imbibition period is usually much longer. Furthermore, after PRF treatment, seeds should not be stored longer than a few days, although after 7 days of storage the PRF effects were not diminished. This makes the treatment complicated, since normally primed seeds can be stored for a much longer time.

An example of application of this method is to improve the seedling establishment of fava (*Vicia faba*) bean seeds after ageing. As shown in Fig. 6, PRF2 (30V/cm for 20 min.) increased the proportion of normal seedlings compared to control seeds. Shorter (PRF1) or longer treatments (PRF3) with an electric field of the same strength were damaging as higher proportions of dead seeds were observed, showing that the beneficial effects of PRF depend on duration in combination with field strength. However, the fact that 10 min. of PRF (PRF1) resulted in a high level of dead seeds was not in line with the result of the longer treatment of 20 min. where fewer dead seeds were found. Either the seeds were not dead, or sampling was done incorrectly.

PRF was shown to have an additive role in breaking dormancy of apple 'Ligol' seed (Grzesik et al., 2017). Apple (*Malus domesticus*) seeds were exposed to PRF for 1 h., at a constant voltage of 25 V, pulse repetition rates 4 Hz (pulses per second) with a pulse duration of 20 ms. PRF applied before or during stratification, either alone or together with selected phytohormones and chemical compounds, increased the percentage and speed of seed germination of apple 'Ligol' as well as the dynamics of seedlings growth (Grzesik et al., 2017).

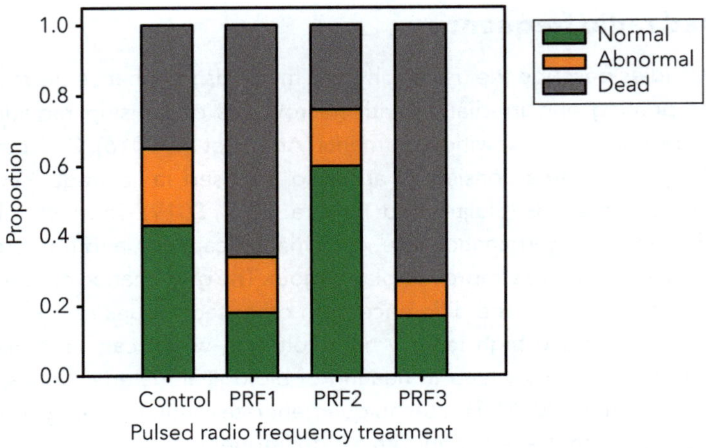

Figure 6 The effect of three different pulsed radio frequency (PRF) treatments on the seedling establishment of aged seeds of *Vicia faba*. PRF1: 23 V/cm for 20 min., PRF2 30 V/cm for 20 min., PRF3 33 V/cm for 20 min. Aged seeds were obtained by incubating them at 70% relative humidity for one week at room temperature followed by three days at 45°C. Green = normal seedings, orange = abnormal seedlings, grey = dead seeds. Data represent the averages of 4 replicates of 100 seeds. Redrawn from Sluijter and Teixeira (2019).

Another study applied PRF to seeds of beetroot (*Beta vulgaris*), lettuce (*Lactuca sativa*) and garden dill (*Anethum graveolens*) (Janas et al., 2019). The seeds were moistened up to a 35% moisture content in air-tight glass bottles for 4 h. in minimal amounts of distilled water at 20°C. Thereafter the seeds were subjected to PRF for 30 min., 60 min. or 120 min., at a voltage of 5 V or 25 V, a pulse repetition rate of 2 Hz or 4 Hz and with a pulse duration of 10 ms or 20 ms. PRF was shown to have a protective effect on seeds by a reduction in the incidence of seed-borne fungi and improvement in seed health and germination capacity. Since germination speed was not affected, it could be asked if the treatment really promotes germination, or whether improved germination is the result of decreasing the level of microorganisms on the seed.

6 Ultrasonics

There are several patents or applications in the area of sonication. In 1999 a patent was obtained by Shors et al., in which maize seed (*Zea mays*) germinated faster after sonication in a solution bubbled with helium or argon. Ultrasonic energy at 15–30 kHz frequency generates cavitational forces through collapse of micro-bubbles at the surface of the seed. In one example, individual kernels were sonicated after a 10-minute soak in 5 ml argon-saturated water at 15°C. After soaking or sonication, the kernels were blotted dry and placed on water-saturated filter paper in a petri dish, which was placed in an incubator at 25°C in the dark. Although only small numbers of seeds of maize were tested, the effect on the speed of germination was significant with a *P*-value less than 0.01 (Fig. 7). Shors et al. (1999) proposed various explanations for increased water uptake, including changing the pericarp structure, physical or biochemical changes in the seed (e.g. the production of free radicals), or an effect of the electrical and light energy released by the cavitation effect (sonoluminescence).

The invention of Schultheiss et al. (2009) describes a much more generic method, in which pressure pulses or acoustic shock waves are released on any type of plant material in a solution, including seeds. The hypothesis is that cells within the treated tissue are activated, thereby releasing growth factor proteins or other chemicals promoting growth and thus accelerating germination or plant growth. A similar approach is that of Redding (2013, 2016), which uses a sonication and imbedding process to improve seed uptake of water and/or other beneficial substances like gibberellic acid. The two patent applications describe treatment of wet and dry seed, respectively. Both applications were denied, in part because of perceived similarity with the existing Schultheiss patent. Schultheiss et al. (2009) refer to initiating physiological processes and avoiding damage by not creating cavitation bubbles. In contrast, Redding (2013, 2016) ascribes positive effects to physical changes to the seed coat caused by cavitation, that allow the seed to take up water, and possibly dissolved substances, at an enhanced rate.

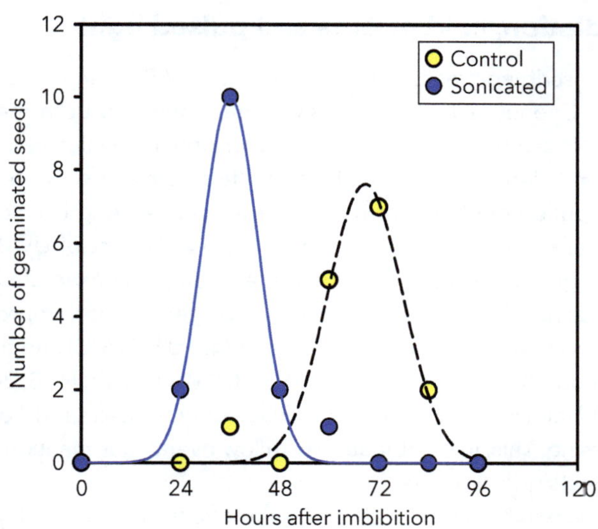

Figure 7 The effect of sonication on the germination of maize kernels. Redrawn from Shors et al. (1999).

Numerous publications describe the effects of ultrasonic treatments, as reviewed by Nazari and Eteghadipour (2017). They show that, in many cases, there is an increase in germination percentage (e.g. Nazari et al., 2015; Liu et al., 2016). Several publications also found an increase in the speed of germination (e.g. Goussous et al., 2010; Miano et al., 2016). Nazari and Eteghadipour (2017) explain how ultrasound works through the formation of micro-bubbles in the water that surrounds the seeds. This cavitation causes mechanical pressure on the seeds, leading to increased cell wall fluidity and the formation of micro-pores and micro-cracks which increase the ability to absorb water and oxygen. This is in direct contrast with Schultheiss et al. (2009) who suggest their method avoids the formation of cavitation bubbles.

The effect of ultrasonication is obvious in dormant seed lots. Seeds that would not germinate, or germinate much slower, because water uptake is hindered, can be stimulated to faster and greater germination by a more rapid water uptake, as in *Medicago scutellata* (Nazari et al., 2015) and common lambsquarter (*Chenopodium album*; Babael-Ghaghelestany et al., 2020). However, although ultrasonic treatment can increase and speed up the germination process, it is doubtful whether it can be compared with true priming. True priming is not only about speeding up germination but also about synchronisation of germination. With the proper ultrasonic treatment, germination can become faster, but the treatment will have the same effect on all the seeds of a seed lot and is too short to obtain a true synchronisation effect.

7 UV-B radiation, cold plasma and pulsed light

The patent describing treatment of seeds with UV-B (Wargent, 2016, 2020; Wargent and Cooney, 2017) is extensive with respect to both methods and examples. UV-B can be applied during or after priming, which can be done in water or polyethylene glycol. The claimed effects are not only an increase in germination, but other effects such as increased radicle length, cold tolerance, drought tolerance and flavonoid content. The main effect of UV-B treatment is to promote growth and stress tolerance, and only one example is given on seed germination. This example (Table 2) is of germination improvement of maize (*Zea mays*), which was imbibed in water for 20 h. (16°C) and then treated with different durations of UV-B light (3.19×10^{-5} W cm^{-2} s^{-1}, for 8, 24 and 32 h. of treatment). Unfortunately, germination data for untreated seed, i.e. unsoaked seed, are missing. This, in combination with low overall germination, makes the proposed effects questionable.

A lot of research has been done on the effect of UV-B light, particularly on dry seeds, because depletion of the stratospheric ozone layer leads to an increased level of UV-B light on the earth's surface. In a review by

Table 2 The effect of UV-B treatment on the germination of maize seeds (var. NZ yellow F1 hybrid). Seeds were immersed in water at 16°C in the dark for 20 h and transferred to 500 µmol m^{-2} s^{-1} of continuous red/blue light. UV-treated seeds received 3.19 x 10^{-5} W cm^{-2} s^{-1} of UV-B light for the indicated period of time.

	% germination		% increase compared to control
UV-B duration (h)	Untreated	UV-B treated	
8	45	73	27
24	27	64	37
32	45	55	9

Data are from Wargent (2020).

Thomas and Puthur (2017), only 1 of the 13 experiments reviewed showed an increase in germination. The effects of UV-B radiation on common chickweed (*Stellaria media*), dandelion (*Taraxacum officinale*), downy brome (*Bromus tectorum*), green foxtail (*Setaria viridis*), redstem filaree (*Erodium cicutarium*), and spotted catsear (*Hypochaeris radicata*) were studied by Dai and Upadhyaya (2002). UV-B radiation (4, 7, or 11 kJ m^{-2} d^{-1}) did not influence seed germination of any species except downy brome. In an experiment by Foroughbakhch Pournavab et al. (2019) on four different species (soybean (*Glycine max*), wheat (*Triticum aestivum*), sunflower (*Helianthus annuus*) and pine (*Pinus maximartinezii*), all treatments (varying doses) had a negative effect on the percentage of normal seedlings. Farokh et al. (2010) found an increase in the speed of germination of safflower (*Carthamus tinctorius*) seeds, but with markedly retarded subsequent growth of seedlings. The increase in germination speed is probably due to UV-B photons (280–320 nm) being more energetic than visible light photons (> 400 nm) and thus having a stronger effect on the surface of plant cells (Kovács and Keresztes, 2002), what causes the breakdown of the seed coat. Overall, it seems that the effects of UV-B on germination are limited at best and that most effects are on subsequent seedling development.

7.1 Cold plasma

Cold atmospheric plasma is the result of heating a gas using dielectric barrier discharges reactors, where the electron gas reaches a temperature much hotter while the rest of gas (ions and neutral atoms) stays at room temperature. As well as electrons, cold plasma produces reactive oxygen species, reactive nitrogen species, UV radiation and local electric fields. It can interact with water, thereby changing its chemical composition, resulting in production of hydrogen peroxide, nitrite, nitrate or ammonia or consumption of bicarbonate ions (Judée et al., 2018).

The first reported case of plasma application to seeds was in a patent by Krapivina et al. (1994). Cold atmospheric pressure plasma generated from a mixture of inorganic gases (atmospheric air, oxygen and nitrogen) was applied to soybean (*Glycine max*) seeds for 5 s to 300 s. Germination and growth were enhanced after treatment at specific power and treatment times (Figure 8). Longer treatment times or higher power resulted in a decrease of germination, indicating that an optimum treatment exists.

The beneficial effect of cold plasma treatment is an increase in the wettability of lentil (*Lens culinaris*), bean (*Phaseolus vulgaris*) and wheat (*Triticum sp. C9*), which resulted in faster imbibition and increased speed of germination (Bormashenko et al., 2012). Stolárik et al. (2015) found that in pea (*Pisum sativum*) the seed coat changed due to cold plasma (as revealed by SEM analysis), resulting in increased permeability and higher germination. In addition, growth characteristics and endogenous hormones changed, which correlated with increased growth of the pea seedlings. Finally, in rice (*Oryza sativa*), cold plasma treatment helped plants to recover cell turgidity under salinity stress (Sheteiwy et al., 2019). However, longer exposure times resulted in genotoxic effects in barley (*Hordeum vulgare*), indicating that appropriate exposure conditions need to be determined, to preserve germination and avoid lipid peroxidation and DNA damage (Pet'ková et al., 2021). An overview of these and other effects has been provided by Waskow et al. (2021).

A variation on cold atmospheric plasma is the use of gas plasma-activated water (GPAW) for priming seeds, as described in the patent application of Grainge

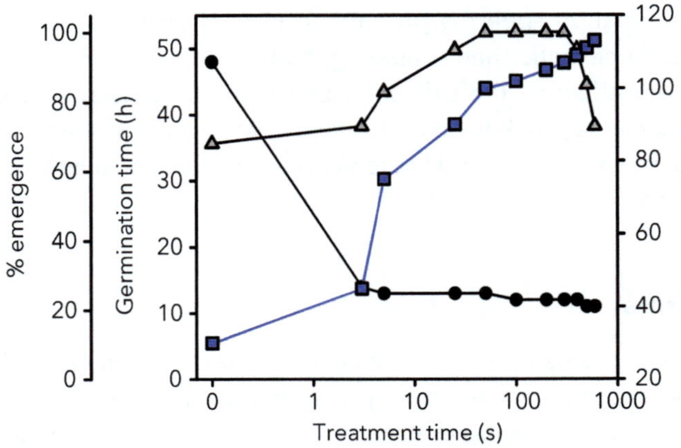

Figure 8 The effect of cold plasma treatment using oxygen (pressure 1.5 Torr, glow discharge was ignited by high-frequency voltage at 6.25 MHz to the electrodes using a power of 0.5 W/cm³ for the indicated period) on water uptake, germination time and seedling emergence of soybean seeds. Data are from Krapivina et al. (1994).

et al. (2021). Seed is soaked in water that is treated with gas plasma without reaching a moisture level at which germination can occur. This is done either by restricting the volume of gas plasma-activated water, or by restricting the duration of the imbibition period. The inventors think that exposure to GPAW promotes embryo growth, endosperm weakening and reduction of the resistance of the seed coat or fruit wall. This resembles findings with cold atmospheric plasma. However, in addition to these effects, a more tolerant response to ageing and a preserved seed vigour has also been reported. The method is therefore called EPOWER (Enhanced Priming Outcome With Enhanced Retainment of shelf life). Table 3 shows germination results of control, hydroprimed and EPOWER primed species, without and with artificial ageing. For all species, EPOWER (gas plasma-activated water) treated seeds have a better ageing tolerance (determined in an accelerated ageing test) than hydroprimed seeds. However, for all species except leek (*Allium porrum*), the germination of the control is substandard compared to industrial norms. For lettuce (*Lactuca sativa*), the germination of the unaged control is even lower (> 30%) than the germination of the aged control, indicating that the seed, at the time of treatment, was probably too fresh. After priming, a germination percentage of 77.4% does not meet the lettuce industry standard in Western Europe which is 96%.

Table 3 The effect of gas plasma-activated water priming and hydropriming on germination of aged and unaged seeds. For the ageing treatment, seeds were incubated for 7 days, except for lettuce (4 days). GPAW = gas plasma-activated water; RH = relative humidity.

		Germination (%)		
Species	Ageing conditions (Temp. (°C)/RH (%))	Non-primed	GPAW primed	Hydroprimed
Cabbage	Unaged control	72 ± 1	77 ± 5	83 ± 1
	Aged: 40/60	46 ± 4	76 ± 3	56 ± 3
Wild radish	Unaged control	44 ± 8	81 ± 12	76 ± 7
	Aged: 40/60	30 ± 7	82 ± 11	60 ± 9
Lettuce	Unaged control	50 ± 8	77 ± 3	71 ± 9
	Aged: 50/70	80 ± 5	37 ± 5	12 ± 1
Swede	Unaged control	11 ± 4	75 ± 2	36 ± 4
	Aged: 42/70	11 ± 6	66 ± 2	42 ± 4
Carrot	Unaged control	22 ± 5	97	84 ± 2
	Aged: 42/70	14 ± 3	84 ± 3	75 ± 3
Beetroot	Unaged control	0	78 ± 3	75 ± 3
	Aged; 42/70	0	47 ± 6	27 ± 4
Leek	Unaged control	95 ± 2	97 ± 0	99 ± 2
	Aged: 42/80	0	46 ± 7	31 ± 6

Data are from Grainge et al. (2021).

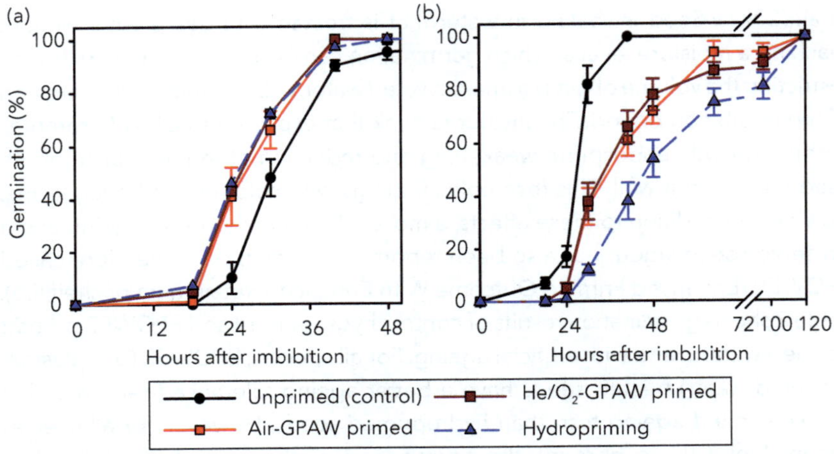

Figure 9 The effect of priming using two cold plasma treatments (Air-GPAW and He/O_2-GPAW priming) and hydropriming on the germination of lettuce seeds. (A) unaged seeds (B) after priming, seeds were aged by incubating them for 3 days at 70% RH, 50°C before recording germination. GPAW = Gas plasma-activated water. Data are from Grainge et al. (2021).

The data in Table 3 suggest that the initial quality of seed lots was already low. However, the priming effects are clear, and EPOWER priming is similar to hydropriming in performance with better retainment of ageing tolerance. A second experiment found that lettuce does reach a high germination and less severe ageing effects from EPOWER priming compared to hydroprimed seed (Fig. 9).

Cold plasma treatment is an environmentally-friendly method because it does not produce contaminated residues. However, much of the outcome of the plasma-seed priming appears to depend on both the plasma setup and seed physical characteristics. What has been studied so far is a combination of plasma device geometries and treatment methods that are able to alter the seed coat. However, treatments lack standardisation. While academic results suggest that there is potential in this technology as a proof-of-concept, we have little information on unpublished negative results, giving the false impression that this priming method is simple to establish and scale up. Further research is needed to explore the pros and cons of cold plasma treatments and to determine whether adding a cold plasma treatment would bring commercial benefits in seed priming operations.

7.2 Pulsed light

The EP patent by Hittin (2003) describes a method to expose seeds to light that is a simulation of the solar visible spectrum and in which the light is pulsed at a frequency of about 10 pulses per minute to 150 pulses per minute. The period of exposure ranges from 30 s to 900 s. Although the purpose of the invention is to accelerate germination and to promote the growth of the crop, the patents do not show how the treatment affects these characteristics and no results are given. Although almost 60 field trials have been conducted, none of this has been reported in the scientific literature.

8 Assessing new priming techniques

If the purpose of priming is to obtain both faster and more uniform germination, then most of the new techniques that have been patented over the past years cannot be considered true priming. Only a few technologies last long enough to obtain synchronisation. The Oxyprime® technique (Van Duijn and van Asbrouck, 2013, 2020) is long enough for synchronisation. The lack of oxygen appears to limit optimal physiological development in the same way as priming based on water limitation. The variation in oxygen levels may allow for better synchronisation before all the oxygen is depleted. The gas plasma-activated water technique (Grainge et al., 2021) can also prime seeds long enough to obtain good synchronisation. When water limitation is used, priming resembles drum priming and can be long enough to obtain both faster germination as well as improved synchronisation. The technique of Shen (2018, 2017 and 2020) is currently too short for many species but, with the appropriate duration, could obtain the same results as the three basic techniques (although it most resembles drum priming).

The duration of priming in pellet (Peter, 2015) is not long enough to obtain synchronous germination. It will speed up germination but even when the maximum duration of 24h. is used, there will be very limited synchronisation. With the vapour heat priming technique (Breukink et al., 2020), an improvement in germination is possible. However, data on synchronisation are lacking and seem hard to obtain since the treatment is far too short for physiological synchronisation. What seems to happen here is that the barrier of the seed coat or other surrounding structures is decreased, resulting in increased and slightly faster germination. For hard seeds, this technique has shown its value, but no synchronisation is obtained.

The PRF, ultrasonic and UV-B treatments can affect the speed of germination but are also too short to achieve synchronisation of germination. Of these three techniques, ultrasonication (Shors et al., 1999; Schultheiss et al., 2009) does seem to have the most effect on both speed of germination and final germination. However, UV-B (Wargent, 2020) seems to affect post-germination processes rather than germination itself. An issue with all three techniques is which settings should be used. In a regular priming the key variables are moisture level, temperature and duration. With the other techniques, more parameters are added that determine the intensity of the treatment e.g. frequency and power. Only very specific settings yield the intended effect.

Combining traditional priming with one of these techniques may lead to even faster, more synchronous germination and possibly beneficial effects on the growth characteristics of the resulting plant.

In terms of sustainability, several priming methods can be done using just water. Of the three basic techniques, drum priming (Rowse, 1994, 2006) is the most sustainable. Both osmotic priming and solid matrix priming produce significant amounts of waste. In drum priming, all kinds of substances may be added, making priming less sustainable. Of the most promising new techniques discussed in this chapter, OxyPrime®, the new variation on drum priming and gas plasma-activated water priming are potentially sustainable, since no additives are needed. Where in the Oxyprime patent (Van Duijn and van Asbrouck, 2013, 2020) no additives are mentioned, the Shen patent (2018, 2017 and 2020) explicitly mentions the possible addition of gibberellins, 6-benzylaminopurine, plant nutrients and/or salts; such additions would make the priming less sustainable. No environmentally damaging additives are needed in gas plasma-activated water priming (Grainge et al., 2021).

With the priming in pellet patent (Peter, 2015), sustainability is more difficult to achieve. The process depends on the materials used for producing the pellet and the pellet requires an oxygen donor and initiator. Vapour heat priming (Breukink et al., 2020) is very sustainable since only water is used to obtain the priming effect. The PRF patent (Sluijter and Teixeira, 2014, 2019) is sustainable in principle although, in some cases, the solution may be enriched with potassium nitrate, e.g. 0.2% KNO_3 is used in Zinnia priming. The ultrasonication patents of Shors et al. (1999) and Schultheiss et al. (2009) describe different methods, either creating cavitation or not. In the method of Shors et al. (1999), a neon or argon gas is added to the treatment liquid, whereas Schultheiss et al. (2009) use a non-defined solution. Both methods are therefore potentially sustainable, depending on what is added to the solution. The UV-B treatments as described in the patent of Wargent (2020), take place in water or in PEG (which does have sustainability issues). The technique described in the patent of Hittin (2003) is also potentially sustainable.

Reducing the level of microorganisms present on a seed is an aspect of sustainability, when it leads to reduced or no use of plant protection products. In this sense vapour heat priming or ThermoSeed® is sustainable. The main effect of PRF is arguably disinfection rather than priming. Ultrasound has also been used for disinfection of seeds (Rotter, 2018). Gas plasma-activated water priming can result in a certain degree of disinfection, as is stated in the gas plasma patent of Krapivina et al. (1994).

9 Future trends

It is hard to imagine any really new priming techniques that affect both germination and synchronisation. Even the recent patent application of

Grainge et al. (2021) is largely based on either hydropriming or drum priming. The only difference is the addition of compounds, created by the gas plasma, to the priming solution. In general, true priming requires a method to control the imbibition level of a seed lot, either by osmotics, or by matrix, or by the absolute amount of water that is added. Only then is it possible to achieve the duration that enables synchronisation of the seed lot. All other patents or applications aim mainly at changing the structures that envelope the embryo, or at changing the physiology of the embryo which may affect germination but more often affect subsequent seedling growth and plant development. Having said this, there may be added value in newer priming techniques. Any technique in which the resistance of the barrier of the seed coat, fruit wall or endosperm is decreased can result in a faster and more synchronous germination when it is combined with one of the three regular priming techniques. With similar effects on germination, but with the claim of better storability, gas plasma-activated water priming may add a new dimension to the priming, although this needs to be explored in more detail. Finally, new techniques can improve the performance of the embryo or the seedling, increasing the value of regular priming.

In commercial applications by either a seed company or a service provider, the seed lots of a single species can express a huge variation in germination and other characteristics. Such variations can be attributed to production-related issues as well as genetic background. Seed companies have the advantage that they know the history of the seed they want to prime, whereas service providers do not. In order to be able to prime all seed lots with one priming protocol, a choice has to be made for a protocol that gives an acceptable result for all seed lots. Alternatively, for each seed lot of a species the optimal priming conditions can be determined beforehand using a so-called test priming. Such test priming requires extra seed and time, both for priming and testing. A fast test method for untreated seed to predict the optimal priming protocol is desirable and should ideally be non-destructive. Although several methods have been published, these are applicable for a limited number of species only, like the analysis of the 11-S globulin in sugar beet seed (Job et al., 1997). More research is needed to be able to predict the parameter settings of an optimal priming protocol.

10 Where to look for further information

The following articles provide a good overview of germination, dormancy and priming:

- Araújo, S. S., Paparella, S., Dondi, D., Bentivoglio, A., Carbonera, D. and Balestrazzi, A. (2016). Physical Methods for Seed Invigoration: Advantages and Challenges in Seed Technology. Frontiers in Plant Science 7, 646. https://doi.org/10.3389/fpls.2016.00646.

- Finch-Savage, W. E. and Leubner-Metzger, G. (2006). Seed dormancy and the control of germination. New Phytologist 171, 501-523. https://doi.org/10.1111/j.1469-8137.2006.01787.x.
- Finch-Savage, W. E. and Bassel, G. W. (2016). Seed vigour and crop establishment: extending performance beyond adaptation. Journal of Experimental Botany, 67, 567–591. https://doi.org/10.1093/jxb/erv490.
- Gupta, A., Dadlani, M., Kumar, A., Roy, M., Naseem, M., Choudhary, V. K. and Miti, R. K. (2008). Seed Priming: The Aftermath. International Journal of Agriculture, Environment and Biotechnology 1, 199-209.
- Lutts, S., Benincasa, P., Wojtyla, L., Kubala, S. S., Pace, R., Lechowska, K., Quinet, M. and Garnczarska, M. (2016). Seed Priming: New Comprehensive Approaches for an Old Empirical Technique. In: New Challenges in Seed Biology – Basic and Translational research Driving Seed Technology. Eds. Araújo, S. and Balestrazzi, A. https://doi.org/10.5772/61583.
- Paparella, S., Araújo, S. S., Rossi,G., Wijayasinghe, M., Carbonera, D. and Balestrazzi, A. (2015). Seed priming: state of the art and new perspectives. Plant Cell Reports 34, 1281 - 1293. https://doi.org/10.1007/s00299-015-1784-y.
- Waqas, M., Korres, N. E., Khan, M. D., Nizami, A.-S., Deeba, F., Ali, I. and Hussain, H. (2019). Advances in the Concept and Methods of Seed Priming. In: Priming and Pretreatment of Seeds and Seedlings. Eds. Hasanuzzaman, M. and Fotopoulos, V. https://doi.org/10.1007/978-981-13-8625-1_2.
- Weitbrecht, K., Müller, K. and Leubner-Metzger, G. (2011). First of the mark: early seed germination. Journal of Experimental Botany 62, 3289-3309. https://doi.org/10.1093/jxb/err030.

Information on patent applications and patents can be obtained from: www.espacenet.com.

11 References

Alness, K., Andersson, S. and Bergman, S. (2004). US 6 350 409 B1: Heat disinfection of seeds. *EP* 0(904), 114B1.

Andreoli, C. and Kahn, A. A. (1998). Matriconditioning integrated with gibberellic acid to hasten seed germination and improve stand establishment of pepper and tomato. *Pesquisa Agropecuária Brasileira* 34(10), 1953–19582

Antala, M., Sytar, O., Rastogi, A. and Brestic, M. (2020). Potential of Karrikins as novel plant growth regulators in agriculture. *Plants* 9(1), 43.

Araújo, Sde S., Paparella, S., Dondi, D., Bentivoglio, A., Carbonera, D. and Balestrazzi, A. (2016). Physical methods for seed invigoration: Advantages and challenges in seed technology. *Frontiers in Plant Science* 7, 646. https://doi.org/10.3389/fpls.2016.00646.

Babael-Ghaghelestany, A., Alebrahim, M. T., MacGregor, D. R., Khatami, S. A. and Farzaneh, R. H. N. (2020). Evaluation of ultrasound technology to break seed dormancy of common lambsquarters (*Chenopodium album*). *Food Science and Nutrition* 8(6), 2662–2669. https://doi.org/10.1002/fsn3.1547.

Bormashenko, E., Grynyova, R., Bormashenko, Y. and Drori, E. (2012). Cold radiofrequency plasma treatment modifies wettability and germination speed of plant seeds. *Scientific Reports* 2, 741. https://doi.org/10.1038/srep00741.

Boukens, M. A. N. and de Boer, J. (2002). B1: method and apparatus for priming seed. *EP* 1(043), 926.

Brancalion, P. H. S., Tay, D., Novembre, A. DdL. C., Rodrigues, R. R. and Marcos-Filho, J. (2010). Priming of pioneer tree *Guazuma ulmifolia* (Malvaceae) seeds evaluated by an automated computer image analysis. *Scientia Agricola* 67(3), 274–279. https://doi.org/10.1590/S0103-90162010000300004.

Breukink, J. W., Bruggink, H., Forsberg, G. R., Mizuguchi, T., Arvidsson, S. O.. and Andersson, S. G. (2020). US 10 542 662 B2: Seed priming. *EP* 3(107), 366B1.

Bruggink, G. T., Ooms, J. J. J. and van der Toorn, P. (1999). Induction of longevity in primed seeds. *Seed Science Research* 9(1), 49–53. https://doi.org/10.1017/S0960258599000057.

Bujalski, W., Nienow, A. W., Maude, R. B. and Gray, D. (1993). Priming responses of leek (*Allium porrum* L.) seeds to different dissolved oxygen levels in the osmoticum. *Annals of Applied Biology* 122(3), 569–577. https://doi.org/10.1111/j.1744-7348.1993.tb04058.x.

Capron, I., Corbineau, F., Dacher, F., Job, C., Côme, D. and Job, D. (2000). Sugarbeet seed priming: Effects of priming conditions on germination, solubilization of 11-S globulin and accumulation of LEA proteins. *Seed Science Research* 10(3), 243–254. https://doi.org/10.1017/S0960258500000271.

Corbineau, F., Xia, Q., Bailly, C. and El-Maarouf-Bouteau, H. (2014). Ethylene, a key factor in the regulation of seed dormancy. *Frontiers in Plant Science* 5, Article 539. https://doi.org/10.3389/fpls.2014.00539.

Dai, Q. and Upadhyaya, M. K. (2002). Seed germination and seedling growth response of selected weedy species to ultraviolet-B radiation. *Weed Science* 50(5), 611–615. http://www.jstor.org/stable/4046697.

Farokh, P., Mahmoodzad, H. and Satari, T. N. (2010). Response of seed germination of safflower to UV-B radiation. *Research Journal of Environmental Sciences* 4(1), 70–74. https://doi.org/10.3923/rjes.2010.70.74.

Finch-Savage, W. E. and Bassel, G. W. (2016). Seed vigour and crop establishment: Extending performance beyond adaptation. *Journal of Experimental Botany* 67(3), 567–591. https://doi.org/10.1093/jxb/erv490.

Finch-Savage, W. E. and Leubner-Metzger, G. (2006). Seed dormancy and the control of germination. *New Phytologist* 171(3), 501–523. https://doi.org/10.1111/j.1469-8137.2006.01787.x.

Foroughbakhch Pournavab, R., Bacópulos Mejía, E., Benavides Mendoza, A., Salas Cruz, L. and Ngangyo Heya, M. (2019). Ultraviolet radiation effect on seed germination and seedling growth of common species from Northeastern Mexico. *Agronomy* 9(6), 269. https://doi.org/10.3390/agronomy9060269.

Goussous, S. J., Samarah, N. H., Alqudah, A. M. and Othman, M. O. (2010). Enhancing seed germination of four crop species using an ultrasonic technique. *Experimental Agriculture* 46(2), 231-242. https://doi.org/10.1017/S0014479709991062.

Grainge, G., Steinbrecher, T., Nakabayashi, K. and Leubner, G. (2021). *WO 2021/094755 A1: Gas Plasma Activated Water Seed Treatment.*

Grzesik, M., Górnik, K., Janas, R., Lewandowki, M., Romanowska-Duda, Z. and van Duijn, B. V. (2017). High efficiency stratification of apple cultivar Ligol seed dormancy by phytohormones, heat shock and pulsed radio frequency. *Journal of Plant Physiology* 219, 81-90. https://doi.org/10.1016/j.jplph.2017.09.007.

Gupta, A., Dadlani, M., Kumar, A., Roy, M., Naseem, M., Choudhary, V. K.. and Miti, R. K. (2008). Seed priming: The aftermath. *International Journal of Agriculture, Environment and Biotechnology* 1, 199–209.

Heydecker, W., Higgins, J. and Gulliver, R. L. (1973). Accelerated germination by osmotic seed treatment. *Nature* 246(5427), 42–44. https://doi.org/10.1038/246042a0.

Hittin, Z. H. (2003). Process and apparatus for promoting the germination of plant seeds and the production of agricultural crops. *EP* 1(162), 871B1, US 6 615 538 B2.

Janas, R., Górnik, K., Grzesik, M., Romanowska-Duda, Z. and van Duijn, B. (2019). Effectiveness of pulsed radio frequency in seed quality improvement of vegetable plant species. *International Agrophysics* 33, 463–471. https://doi.org/10.31545/intagr/108953.

Job, C., Kersulec, A., Ravasio, L., Chareyre, S., Pepin, R. and Job, D. (1997). The solubilization of the basic subunit of sugarbeet seed 11-S globulin during priming and early germination. *Seed Science Research* 7(3), 225–244. https://doi.org/10.1017/S0960258500003585.

Jonas, H. (1955). *US 2 721 713 A: Method of Treating Seeds by High Frequency Fields.*

Judée, F., Simon, S., Bailly, C. and Dufour, T. (2018). Plasma-activation of tap water using DBD for agronomy applications: Identification and quantification of long lifetime chemical species and production/consumption mechanisms. *Water Research* 133, 47–59. https://doi.org/10.1016/j.watres.2017.12.035.

Karimi, M. and Varyani, M. (2016). Role of priming technique in germination parameters of calendula (*Calendula officinalis* l.) seeds. *Journal of Agricultural Sciences, Belgrade* 61(3), 215–226. https://doi.org/10.2298/JAS1603215K.

Kovács, E. and Keresztes, A. (2002). Effect of gamma and UV-B/C radiation on plant cells. *Micron* 33(2), 199–210. https://doi.org/10.1016/s0968-4328(01)00012-9.

Krapivina, S. A., Filippov, A. K., Levitskaya, A. B. and Bakhvalov, A. (1994). *US 5 281 315: Gas Plasma Treatment of Plant Seeds.*

Lagunas-Solar, M., Zeng, N. X. and Essert, T. K. (2002). *WO 02/39786 A2. Method and System for Inhibiting Pathogenic and Spoilage Activity in Products and the Products Treated Thereby.*

Liu, J., Wang, Q., Karagić, Đ, Liu, X., Cui, J., Gui, J., Gu, M. and Gao, W. (2016). Effects of ultrasonication on increased germination and improved seedling growth of aged grass seeds of tall fescue and Russian wildrye. *Scientific Reports* 6, 22403. https://doi.org/10.1038/srep22403.

Lutts, S., Benincasa, P., Wojtyla, L., Kubala, S. S, Pace, R., Lechowska, K., Quinet, M.. and Garnczarska, M. (2016). Seed priming: New comprehensive approaches for an old empirical technique. In: Araújo, S. and Balestrazzi, A. (Eds). *New Challenges in Seed Biology - Basic and Translational Research Driving Seed Technology.* https://doi.org/10.5772/61583.

Miano, C. A., da Costa Pereira, J., Castanha, N., da Matta Júnior, M. D. and Augusto, P. E. D. (2016). Enhancing mung bean hydration using the ultrasound technology: Description of mechanisms and impact on its germination and main components. *Scientific Reports Nature* 6, 388996. https://doi.org/10.1038/srep38996.

Nazari, M. and Eteghadipour, M. (2017). Impacts of ultrasonic waves on seeds: A mini-review. *Agricultural Research and Technology: Open Access Journal* 6(3), 555688. https://doi.org/10.19080/ARTOAJ.2017.06.555688.

Nazari, M., Sharififar, A. and Asghari, H. R. (2015). Medicago scutellata seed dormancy breaking by ultrasonic waves. *Plant Breeding and Seed Science* 69(1), 15–24. https://doi.org/10.1515/plass-2015-0002.

Paparella, S., Araújo, S. S., Rossi,G., Wijayasinghe, M., Carbonera, D. and Balestrazzi, A. (2015). Seed priming: State of the art and new perspectives. *Plant Cell Reports* 34(8), 1281–1293. https://doi.org/10.1007/s00299-015-1784-y.

Peraza-Villarreal, H., Sánchez-Coronado, M. E., Lindig-Cisneros, R., Tinoco-Ojanguren, C., Velázquez-Rosas, N., Cámara-Cabrales, L. and Orozco-Segovia, A. (2018). Seed priming effects on germination and seedling establishment of useful tropical trees for ecological restoration. *Tropical Conservation Science* 11, 1–15. https://doi.org/10.1177/1940082918817886.

Peter, B. (2015). B1: Method for pre-germinating seed. *EP* 2(002), 702.

Pet'ková, M., Švubova, R., Kyzek, S., Medvecká, V., Slováková, Ľ, Ševčovičová, A. and Gálova, E. (2021). The effects of cold atmospheric pressure plasma on germination parameters, enzyme activities and induction of DNA damage in barley. *International Journal of Molecular Sciences* 22(6), 2833. https://doi.org/10.3390/ijms22062833.

Redding, B. K., Jr. (2013). *WO 2013/184169 A9: Ultrasonically Enhanced Seed Germination System*.

Redding, B. K. (2016). *WO 2016/191 585 A1, (2018), US 2018/0160629 A1: Ultrasonic Treatment of Seeds*.

Rotter, S. (2018). *US 10 080 370 B2: Ultrasound Cleaning Method With Suspended Nanoparticles*.

Rowse, H. R. (1994). B1: Methods of priming seed. *EP* 0(254), 569.

Rowse, H. R. (2006). B2: Priming seeds. *EP* 0(608), 915.

Schultheiss, R., Schaden, W., Warlick, J. and Goeschl, R. (2009). *US 7 600 343 B2: Method of Stimulating Plant Growth*.

Shen, T. (2017). B1: Improved method for seed priming. *EP* 3(174), 378.

Shen, T. (2018). B1, Improved method for seed priming. *EP* 2(991), 468.

Shen, T. (2020). *US 10 645 861 B2: Method for Seed Priming*.

Sheteiwy, M. S., An, J., Yin, M., Jia, X., Guan, Y., He, F. and Hu, J. (2019). Cold plasma treatment and exogenous salicylic acid priming enhances salinity tolerance of *Oryza sativa* seedlings. *Protoplasma* 256(1), 79-99. https://doi.org/10.1007/s00709-018-1279-0.

Shors, J. D., Soll, D. R., Daniels, K. J. and Gibson, D. P. (1999). *US 5 950 362: Method for Enhancing Germination*.

Sluijter, M. E. and Teixeira, A. J. L. (2014). *US 8 703 757 B2, (2019). EP* 2(111), 095 B1: Seed Treatment With PRF.

Steinbrecher, T. and Leubner-Metzger, G. (2017). The biomechanics of seed germination. *Journal of Experimental Botany* 68(4), 765–783. https://doi.org/10.1093/jxb/erw428.

Stolárik, T., Henselová, M., Martinka, M., Novák, O., Zahoranová, A. and Černák, M. (2015). Effect of low temperature plasma on the structure of seeds, growth and metabolism of endogenous phytohormones in pea (*Pisum sativum* L.). *Plasma Chemistry and Plasma Processing* 35(4), 659–676. https://doi.org/10.1007/s11090-015-9627-8.

Taylor, A. G. (1990). *US 4 912 874: Solid Matrix Priming of Seed.*

Thomas, D. T. T. and Puthur, J. T. (2017). UV radiation priming: A means of amplifying the inherent potential for abiotic stress tolerance in crop plants. *Environmental and Experimental Botany* 138, 57–66. https://doi.org/10.1016/j.envexpbot.2017.03.003.

Tsujimoto, T., Sato, H. and Matsushita, S. (2002). B1: Method of improving seed germination. *EP* 0(686), 340.

Van Duijn, A. and Van Asbrouck, J. (2020). US 8 381 438 B2: Oxygen-controlled priming of seed. *EP* 2(154), 942 B1, (2013).

Waqas, M., Korres, N. E., Khan, M. D., Nizami, A.-S., Deeba, F., Ali, I. and Hussain, H. (2019). Advances in the concept and methods of seed priming. In: Hasanuzzaman, M. and Fotopoulos, V. (Eds). *Priming and Pretreatment of Seeds and Seedlings*. Singapore: Springer. https://doi.org/10.1007/978-981-13-8625-1_2.

Wargent, J. J. (2016). *WO 2016/043605 A1: Method of Seed Treatment and Resulting Products.*

Wargent, J. J. (2020). *US 10 750 691 B2: Methods of Seed Treatment and Resulting Products.*

Wargent, J. J. and Cooney, L. J. (2017). *WO 2018/037281 A1: System, Device and Methods of Seed Treatment.*

Waskow, A., Howling, A. and Furno, I. (2021). Mechanisms of plasma-seed treatments as a potential seed processing technology. *Frontiers in Physics* 9, 617345. https://doi.org/10.3389/fphy.2021.617345.

Weitbrecht, K., Müller, K. and Leubner-Metzger, G. (2011). First of the mark: Early seed germination. *Journal of Experimental Botany* 62(10), 3289–3309. https://doi.org/10.1093/jxb/err030.

Index

http://dx.doi.org/10.19103/AS.2022.0105.06

Ingram Content Group UK Ltd.
Milton Keynes UK
UKHW050021290623
424106UK00005B/9

9 781786 769176